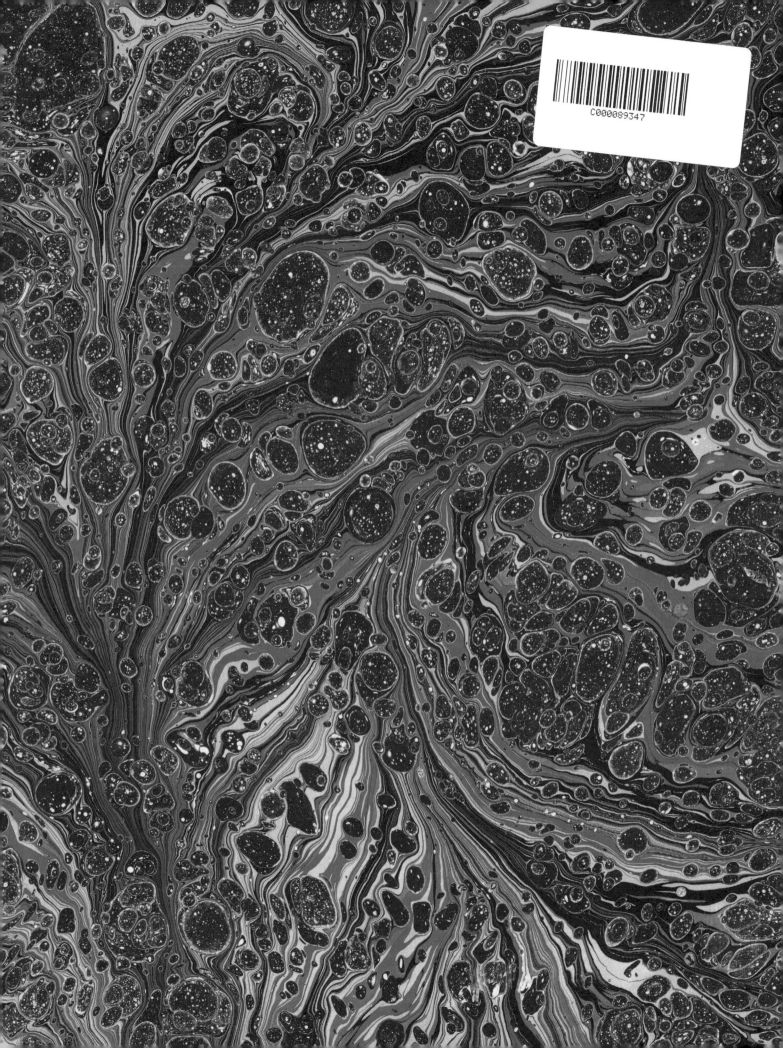

Tupolev Tu-4

Tupolev Tu-4
The First Soviet Strategic Bomber

Yefim Gordon
Dmitriy Komissarov
Vladimir Rigmant

Schiffer Publishing Ltd ®

4880 Lower Valley Road • Atglen, PA 19310

Book design by Polygon Press Ltd., Moscow.
Type set in Times New Roman/Square 721 BT

ISBN: 978-0-7643-4797-9
Printed in China

Published by Schiffer Publishing, Ltd.
4880 Lower Valley Road
Atglen, PA 19310
Phone: (610) 593-1777; Fax: (610) 593-2002
E-mail: Info@schifferbooks.com

For our complete selection of fine books on this and related subjects, please visit our website at www.schifferbooks.com. You may also write for a free catalog.

This book may be purchased from the publisher. Please try your bookstore first.

We are always looking for people to write books on new and related subjects. If you have an idea for a book, please contact us at proposals@schifferbooks.com.

Schiffer Publishing's titles are available at special discounts for bulk purchases for sales promotions or premiums. Special editions, including personalized covers, corporate imprints, and excerpts can be created in large quantities for special needs. For more information, contact the publisher.

Contents

Acknowledgments

The authors wish to express their gratitude to the Tupolev Public Limited Co., which supplied documentary materials on the Tu-4 and Tu-70/Tu-75, as well as to Mikhail Gribovskiy and to Nigel Eastaway, who provided access to the materials of the Russian Aviation Research Trust (RART). The book is illustrated by photos from Yefim Gordon, Maxim Bryanskiy, Neil Lewis, Peter Davison, and Marcus Fülber, as well as from the archives of Tupolev PLC, the M. M. Gromov Flight Research Institute (LII), RART, the personal archives of Yefim Gordon, Sergey and Dmitriy Komissarov, and from Internet sources.

Color artwork is by Aleksandr Gavrilov. Line drawings are by Tupolev PLC and the late Ivnamin G. Sultanov.

Another sortie is due to start soon for the crew of Tu-4 '28 Blue'.

Introduction

The epic story of how the Tu-4 bomber came into existence is one of the thrilling pages in the history of aircraft technology development in the Soviet Union. Events relating to this program that took place in 1943-49 were, to a great extent, instrumental in shaping the Soviet aircraft industry in the period following the end of WWII and in promoting its subsequent leap to the forefront of the world's jet-powered aviation; they also laid the foundation of the strategic parity between the USA and the USSR which secured the survival of mankind in the era of global nuclear confrontation.

Almost certainly, before 1943, neither the Soviet military nor the political leaders of the USSR had any information on the work that was being conducted in the USA at that time with a view to creating the latest "superbomber" – the Boeing B-29 Superfortress – which became the mainstay of American strategic air power a few years later. Soviet intelligence agencies were engrossed in penetrating the secrets of the American nuclear program and studying the work that was being conducted in the USA in the fields of electronics, jet aviation, and other breakthrough technologies in the field of military hardware; thus, the B-29 escaped their attention and, most importantly, the attention of the leaders in Moscow. The situation changed in 1943, after Eddie Rickenbacker's visit to the USSR. This visit can be regarded as the starting point for the story of how the B-29 became the Tu-4.

Captain Edward V. Rickenbacker, a former U.S. Army Air Corps fighter pilot who had fought with distinction in WWI, became President Franklin D. Roosevelt's advisor on aviation matters during WWII. Thus, he was sent on a mission to the Soviet Union for the purpose of sizing up the situation in Soviet aviation and seeing how the Russians utilized the American aviation materiel supplied under the Lend-Lease Agreement. This visit put the foreign relations department of the Soviet Air Force (VVS – *Voyenno-vozdooshnyye seely*) in a rather ticklish situation. On the one hand, given his status as Presidential advisor, Rickenbacker was entitled to an appropriate retinue, at the level of generals at least. On the other hand, after WWI he had not been promoted and was still at the rank of Captain; therefore, according to diplomatic canons, he could not be escorted by an officer of a rank superior to his own. As a result, the person selected to accompany him was a young officer, Capt. A. I. Smoliarov, who had recently graduated from the Soviet Air Force Academy. He spoke fairly good English, was well versed in aviation technology, and had a reasonably good knowledge of electronics in the bargain. As subsequent events showed, the foreign relations department of the Air Force could

hardly have made a better choice; no general, even the highest-ranking one, could have furnished the Soviets with such detailed initial information on the B-29 as this modest young Captain of the Air Force. In this particular case, the famous maxim coined by the Soviet leader Iosif V. Stalin, "personnel is the key factor", proved correct, thanks to Smoliarov's excellent work and Rickenbacker's talkativeness.

During his trip to the USSR, Rickenbacker visited a number of Soviet aircraft factories and Air Force units, including those operating American-supplied aircraft. A specially-equipped Douglas C-47 Skytrain was used for travelling from one airfield to another. The conditions on board were fairly austere for that kind of travel, but the lack of creature comforts was more than recompensed by the hosts' cordiality and an unlimited supply of liquor. During the flights, Smoliarov and Rickenbacker whiled away the time by discussing all sorts of subjects; quite naturally, aviation matters were predominant among them. During one such conversation, the American guest mentioned a new American "superbomber" – the Boeing B-29 Superfortress – and informed Smoliarov about its high performance. As he was supposed to, Smoliarov filed a detailed report on this visit to his next-in-command after Rickenbacker departed to the U.S. In this report, he cited the information on the new American machine, committing to paper in the minutest detail everything Rickenbacker had said about the B-29 during his visit.

This part of the report evoked interest on the part of Smoliarov's direct superiors. The name of a general who was Smoliarov's boss was substituted for the original signature on the report (curiously, the text of the report, which was in the first person, remained unchanged and the pronoun "I" came to refer to the general) and the report was sent to the higher echelon. Its materials, now suitably classified, began to make the rounds of important persons' studies in the People's Commissariat of Aircraft Industry (NKAP – *Narodnyy komissariaht aviatsionnoy promyshlennosti*); in the various organisations and institutes under NKAP; and in the commands of the Air Force, the Red Army, and the intelligence services. The initial source of the information was soon forgotten, and the American B-29 bomber became the object of a systematic search for information through all available official and unofficial channels. For the Soviet military and political leaders, the B-29 became a reality and an eventual threat at a later stage, as well as a standard the attainment of which the Soviet aircraft industry was to be geared.

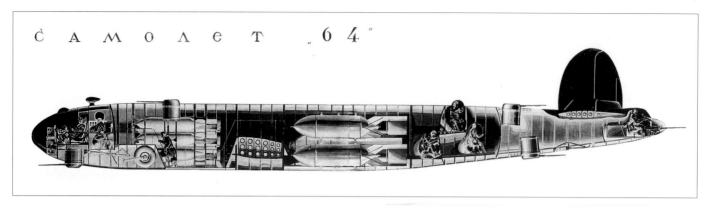

С А М О Л Е Т „6 4"

Above: A drawing from the advanced development project (ADP) documents showing the internal layout of the "aircraft 64" (Tu-10) strategic bomber – the early parabolic-nose version. Note the five powered cannon barbettes and the way the nosewheel well encroaches on the forward bomb bay, the bombs being placed on either side of it.

Left: This artist's impression shows the initial version of the "64" with elliptical vertical tails and a B-29 style parabolic nose.

Left: A later project version of the "64" with taller tapered vertical tails, a revised forward fuselage featuring raised pilot seats with individual canopies, and revised engine nacelles with dorsal air scoops.

For starters, attempts were made – unsuccessfully – to obtain the B-29 legally from the USA within the framework of the Lend-Lease Agreement, under which various war materiel was delivered to the USSR. On 19 July 1943, Maj-Gen. V. I. Belyayev, the chief of the Soviet military mission in the USA, sent an enquiry to the appropriate U.S. authorities concerning the possibility of supplying the Lockheed P-38 Lightning and Republic P-47 Thunderbolt fighters, the Consolidated B-24 Liberator bomber, and the B-29. Out of this list, the Soviets received only the P-47, and then only in small numbers. The Americans repeatedly turned down Soviet requests for four-engined heavy bombers, being of the opinion that the Red Army Air Force had no real need for them to carry out its mainly tactical operations at the Soviet-German front.

The Soviet Union made another try during its preparations for entering the war against Japan in the spring of 1945. Being aware that the Anglo-American allies were interested in the USSR entering the war in the Far East at the earliest possible date, on 28 May

1945, Soviet military representatives requested the delivery of as many as 120 B-29s with a view to using them in the forthcoming operations against Japan; yet again, not a single Superfortress was supplied. Right from the start of the war, the Soviet Union repeatedly asked the USA for Boeing B-17 Flying Fortress and B-24 bombers, and each time the USA turned down the requests – primarily because the Allies had placed their bets on a strategic air offensive in Europe in their war against Germany and needed the bombers themselves in ever-increasing numbers. Besides, the Allies did not want to create future problems for themselves, being convinced that WWII was bound to be followed by a Third World War – and not being sure which side of the front the traditionally anti-Western Russia would be on. In the West, the brief flirtation of the Soviet Communists with the German National Socialists in 1939-41 was still fresh in everyone's memory.

Here, it should be noted that it is the USSR that has the distinction of being the first to create a strategic bomber force. By the

This three-view from the ADP documents depicts the initial version of the "64" bomber. The aircraft was 28,575m (93ft 9in) long and stood 8,725m (28ft 7¹⁄₂ in) tall when parked, with a wing span of 39.0m (127ft 11⁷⁄₁₆ in) and a tailplane span of 10.0 m (32ft 9⁴³⁄₆₄ in). The wheel track and the distance between the two inboard engines' propeller shafts was 7.6m (24ft 11⁷⁄₃₂ in), while the distance between the inner and outer engines' propeller shafts was 4.4m (14ft 5⁷⁄₃₂ in). Note the single-wheel nose gear unit and the dihedral stabilizers.

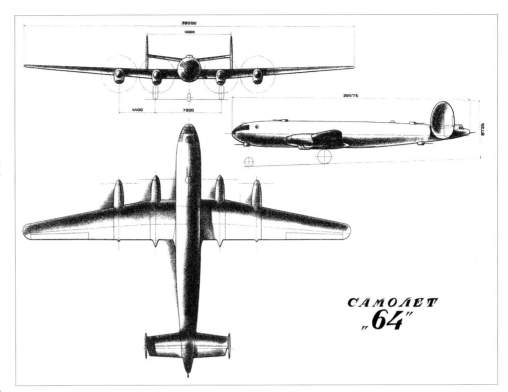

A three-view of the "64" from the later ADP version featuring individual "bug-eye" canopies for the pilots, a twin-wheel nose gear and zero-dihedral stabilizers. The dimensions had changed as well, with a length of 29,975m (98ft 4⁷⁄₆₄ in), a wing span of 42.0m (137ft 9³⁵⁄₆₄ in), a tailplane span of 10.6m (34ft 9²¹⁄₆₄ in), and a height on ground of 7,215m (23ft 8³⁄₆₄ in). The wheel track is 8.35m (27 ft 4⁴⁷⁄₆₄ in), the distance between the two inboard engines' propeller shafts 8.0m (26ft 2³¹⁄₃₂ in), and the distance between the inner and outer engines' propeller shafts 4.75m (15ft 7 in). The drawing is marked *Sekretno* (Secret).

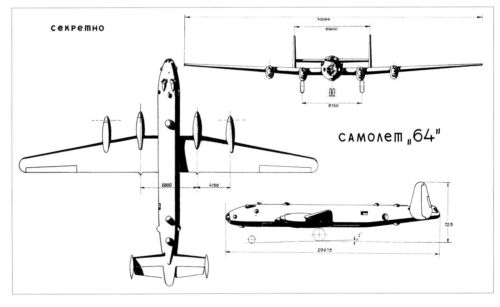

early 1930s, the Soviet military top echelon had worked out a sufficiently clear concept concerning strategic use of Long-Range Aviation (that is, a strategic bomber force). The main emphasis in these plans was made on the use and expeditious development of multi-engined heavy bombers. Thanks to the incredible exertion of the national economy and mobilization of the young Soviet aircraft industry's main resources, the Soviet Union succeeded in launching series production of the TB-3 (ANT-6) four-engined bomber developed by the design bureau headed by Andrey Nikolayevich Tupolev. More than 800 such bombers were built in the 1930s, equipping the world's first air force units capable of accomplishing major strategic tasks on their own. However, the TB-3's design was based on mid-1920s technology and was already obsolescent by the

time the bomber reached production status and achieved initial operational capability. It was no longer on par with the then-current state of the art in aviation technology and, most importantly, with the rapidly developing anti-aircraft defence systems available to the leading aviation powers.

In light of the new realities, in the early 1930s, the Soviet aircraft design bureau started work on a modern high-speed/high-altitude long-range bomber broadly equivalent to the Boeing B-17. The prototype of the B-17's Soviet counterpart, the TB-7 (ANT-42), developed under the leadership of Tupolev's aide, Vladimir M. Petlyakov, took to the air at the end of 1936. However, a whole range of technical and organizational problems, coupled with the general disarray in Soviet society and industry in the wake of mass

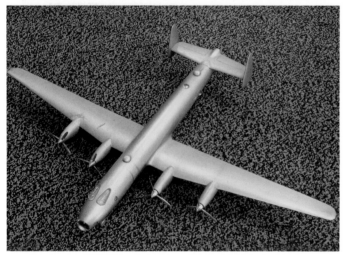

Three views of a desktop model of the late twin-tail version "64" showing the curious-looking "bug-eye" canopies for the pilots and the separate glazing of the navigator's station.

purges of the late 1930s and further compounded by complete uncertainty and incessant "ideological vacillations" in the Stalinist leadership's approach to the role of strategic aircraft in a future war, delayed the production entry of the TB-7; limited production did not begin until the eve of WWII. When Nazi Germany invaded the USSR on 22 June 1941, starting the Great Patriotic War, massed strategic operations were no longer on the agenda – the front required thousands of relatively simple and cheap machines. Hence, the TB-7 (renamed Pe-8 in 1942 to reflect Petlyakov's lead-

ing role in its development) was built on a very small scale – only 93 TB-7 aircraft were produced in the USSR between 1936 and 1945; for various reasons, only 60 to 70% of these saw active service. This does not compare with the thousands of American B-17s, B-24s, and British Avro Lancasters, and even the hundreds of German Focke-Wulf Fw 200 Condors and Heinkel He 177 Greifs. Yet, in its overall performance, the TB-7 was no worse than the B-17 (at any rate, at the time when the two prototypes made their appearance in the mid-1930s), and given better circumstances, could have become the mainstay of the Red Army Air Force's Long-Range Aviation (ADD – *Aviahtsiya **dahl'nevo deystviya***).

At the beginning of the Great Patriotic War, the Soviet heavy bomber units lost a good deal of their aircraft, in part during the first days, which were marked by disarray and panic, and in part when wastefully used for tactical strikes against the German armor and motorized infantry units that were mounting a swift offensive into the inner regions of the country. The few strategic air operations undertaken by the ADD during the initial period of the war pursued primarily limited political objectives and were not always efficiently planned and executed, resulting in considerable losses both of materiel and personnel. As a consequence of all this, by the beginning of 1942, the strength of Soviet long-range bomber units was cut by two-thirds. In the spring of 1942, taking into account the experience of the war and the opinions of Gen. Aleksandr Ye. Golovanov and many officers of this branch, Stalin grouped all the Long-Range Aviation units into a separate structure and subordinated it directly to himself. From the spring of 1942 onwards, the Soviet Supreme Command was in possession of a powerful instrument for the conduct of strategic operations: the ADD, headed by a dynamic newly-appointed Commander-in-Chief, Aleksandr Golovanov. In December 1944, the ADD lost its status as an independent branch of the Soviet Armed Forces, becoming the 18th Air Army and reporting to the Air Force command, but was reinstated in April 1946 as the DA (***Dahl'nyaya** aviahtsiya*, which again means Long-Range Aviation).

By the time the war ended, the units of the Long-Range Aviation, as a tool of the Supreme Command, represented a sufficiently effective and mobile force. However, in its composition and equipment, this force primarily corresponded to the character of the fighting on the Soviet-German front, where the outcome of a battle and the question of victory or defeat was determined mainly in the tactical or minor strategic depth of the enemy's defense. Accordingly, the machines equipping the units reflected to a certain extent the tasks posed for it. By the time the war ended, the lion's share of Long-Range Aviation's 2,000 aircraft was made up of twin-engined bombers, such as the Il'yushin DB-3 and IL-4 (DB-3F); the North American B-25 Mitchell supplied under the Lend-Lease Agreement; a small number of Yermolayev Yer-2 bombers; and a vast number of Lisunov Li-2 *Cab* transports (a derivative of the Douglas DC-3 built in the USSR under licence; in this context, the bomber-configured Li-2VP). They were supplemented by Pe-8s, plus a handful of damaged B-17s and B-24s that were abandoned by the USAAF in Eastern Europe and subsequently appropriated by the advancing Red Army in 1944 and restored to flying condition. This bomber armada could tackle its tasks with a fair degree of success in the Great Patriotic War, but it was obviously unsuit-

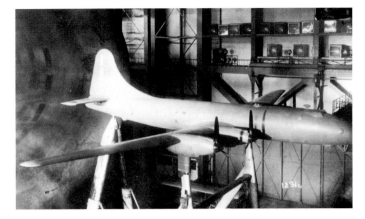

Top and above: This model in TsAGI's T-101 wind tunnel shows another version of the "64" (with a single fin and rudder). The shape of the vertical tail is remarkably similar to that of the B-29.

Top and above: A model of the 64 (the version with tapered twin vertical tails) in the same wind tunnel.

able for the new postwar situation involving an eventual confrontation with the extremely powerful U.S. war machine.

It should be noted that already in the course of WWII, the Soviet leaders had turned their attention to the problem of creating a prospective multi-engined bomber for the ADD/DA. There were several reasons for that. The Allies' shuttle raids against Germany involving strategic bombers showed the Soviet leaders what a potent force emerged when hundreds of thousands of heavy combat aircraft were brought together within strong units. Second, as a reaction to intelligence information on the first successes of the American nuclear program supplied by Soviet undercover agents from the USA, work was sped up on the creation of the "Red super-bomb". The future "wonder weapon" required an adequate delivery vehicle, the Pe-8 being obviously unsuited for this role. Design work on a new heavy bomber had to be initiated. What should it be like? Very appropriately, more or less credible information on the B-29 became available at that moment. It was the B-29 that became a beacon to which Soviet designers would turn their attention when creating a new bomber.

From the autumn of 1943 onwards, three Soviet design bureaus were tasked with creating new four-engined high-speed high-altitude long-range bombers by NKAP. A design bureau headed by Iosif F. Nezval' was tasked with a thorough modernization of the production Pe-8; in the course of 1944, it prepared a project envisaging a considerable improvement of the bomber's performance.

Also in 1944, Nezval' initiated project work on a long-range high-speed/high-altitude bomber similar in its design features and performance to the Boeing B-29. During the same period, Vladimir M. Myasishchev's design bureau worked on two projects – the DVB-202 and DVB-302 (*dahl'niy vysotnyy bombardirovshchik* – long-range high-altitude bomber) – which, in their approach to technical problems, were likewise very close to the B-29 and even bore a strong external similarity to it. At the same time, OKB-240 (*opytno-konstrooktorskoye byuro* – experimental design bureau), headed by Sergey V. Il'yushin, was working on a project of a similar four-engined bomber designated IL-14 (the first aircraft to have this designation; not to be confused with the postwar twin-engined airliner). All these projects did not progress further than the preliminary design stage and were shelved when the work on the Tu-4 started. It was Andrey N. Tupolev's design bureau (by then known as OKB-156) that had progressed furthest in creating a Soviet equivalent to the B-29.

In September 1943, OKB-156 was instructed to evolve an advanced development project (ADP) and build a mock-up of a high-altitude heavy bomber powered by the 2,200-hp Shvetsov M-71TK-M 14-cylinder radial engines. The aircraft was to feature pressurized cabins and defensive armament comprising cannons, rather than machine-guns. The aircraft was to possess the following basic performance:

Shown here in the early 1970s, Andrey N. Tupolev (left) and his former aide Vladimir M. Myasishchev were the chief competitors in the first Soviet strategic bomber program.

- maximum speed at 10,000m (32,810ft), 500km/h (311mph);
- range at the speed of 400km/h (249mph) with a full bomb load, 5,000km (3,108 miles);
- same speed, with a bomb load of 7,000-8,000kg (15,435-17,640lbs), 6,000km (3,729 miles);
- bomb bay capacity, 10,000kg (22,050lbs).

In designing a heavy bomber similar in class and performance to the B-29, the Tupolev OKB was to make allowance for the established indigenous production techniques, materials, and equipment that were available in the USSR at that time. The new machine,

An artist's impression of the Myasishchev DVB-202 (VM-22) bomber. The aircraft looked almost like a high-wing version of the B-29, especially with ASh-72 radials, as shown here. The wing position accounted for the tall main gear units.

This artist's impression shows a version of the DVB-202 powered by liquid-cooled Vee-12 engines from Mikulin OKB.

A provisional three-view of the DVB-202 with ASh-72 engines.

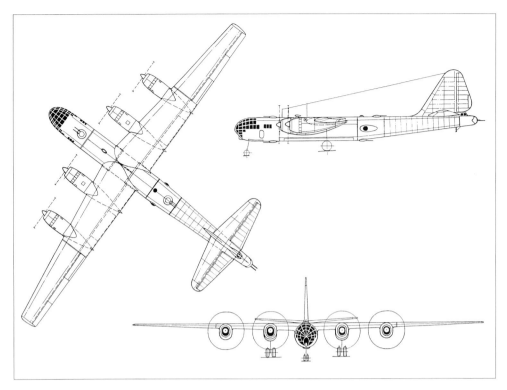

which received the manufacturer's designation "aircraft 64" and the official designation Tu-10 (the first aircraft thus designated), was projected in three versions: a bomber, an airliner (which had a separate in-house designation, "aircraft 66"), and a military transport derivative of the latter. An assessment of the specific operational requirement undertaken within the framework of the design studies showed that the "64", while possessing (as stipulated) dimensions similar to those of the ANT-42, would have to be twice as heavy. This posed a number of difficult problems for the designers and structural strength specialists, since the loads in structural members of the "aircraft 64" would be twice as high as those in the corresponding structural elements of the ANT-42. The nation was at war, and it was unrealistic to expect to get new materials and intermediate products, modern equipment, and hardware in the short term. At best, our designers could only see in aeronautical magazines all those things that were available to American engineers when they were designing the B-29.

Notwithstanding all the difficulties, the ADP of the new bomber was ready in August 1944. According to this project, the

"aircraft 64" was intended for daytime missions in the enemy's rear, far behind the front lines. High cruising speeds and flight altitudes coupled with potent defensive armament enabled the aircraft to perform missions in areas where the enemy's anti-aircraft defense was at its strongest. The aircraft could carry a bomb load of up to 18,000kg (39,690lbs) while using bombs weighing up to 5,000kg (11,025lbs) apiece. The provision of pressurized cabins for the crew enabled the "64" to reach altitudes up to 10,000m (32,810ft).

Structurally, the aircraft was an all-metal mid-wing monoplane with a twin-fin tail unit. The circular section fuselage was of monocoque construction with a thick stressed skin and incorporated two capacious bomb bays fore and aft of the wing center section. The high aspect ratio wings with slight leading-edge sweep on the outer sections were of two-spar construction, featuring efficient high-lift devices to improve field performance. The aircraft had a tricycle undercarriage with single wheels on all units. As for the powerplant, five engine types came into consideration: the 2,000-hp Mikulin AM-42TK, the 2,300-hp AM-43TK-300B,

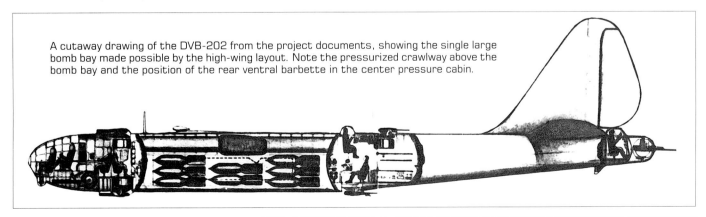

A cutaway drawing of the DVB-202 from the project documents, showing the single large bomb bay made possible by the high-wing layout. Note the pressurized crawlway above the bomb bay and the position of the rear ventral barbette in the center pressure cabin.

the 1,900-hp Charomskiy ACh-30BF diesel (all of them liquid-cooled Vee-12 engines driving three-blade propellers), the 1,900-hp Shvetsov ASh-83FN 14-cylinder two-row radial, or the monstrous 2,500-hp Dobrynin/Skoobachevskiy M-250 liquid-cooled 24-cylinder engine featuring power recovery turbines. Design studies and performance calculations were made for versions with the five engine options. The defensive armament, consisting of 20-mm or 23-mm (.78 or .90 calibre) cannons, was to be installed in four electrically actuated twin-cannon barbettes featuring remote control (two dorsal and two ventral, including one in a chin position) and a tail turret with one or two cannons. Navigation, radio communications, and radio engineering equipment were to comprise the most up-to-date items that could be produced by Soviet industry. All the main systems of the aircraft were electrically actuated, with the exception of units designed to sustain especially heavy loads, which were hydraulically actuated.

In the course of design work, several variants of the "64" were considered. They differed in overall dimensions, tail unit design (in particular, one version had conventional tail surfaces with a large vertical tail of similar shape to the B-29's), internal fuselage layout, equipment complement, weapons fit, and crew accommodation. The original extensively-glazed parabolic nose similar to that of the B-29 gave way to individual "bug-eye" teardrop canopies for the two pilots (similar to those of the Douglas XB-42 Mixmaster bomber) and a glazed workstation for the navigator/bomb aimer in the extreme nose.

In August 1944, proceeding from the preliminary work on the "64" aircraft, the Air Force issued a specification for a high-altitude long-range bomber powered by four AM-42TKs, which eventually could be replaced by Mikulin AM-46TK liquid-cooled Vee-12 engines. In the long-range bomber version, the aircraft was expected to operate at altitudes of 9,000-10,000m (29,530-32,810ft), have a combat radius not less than 2,000km (1,243 miles), and carry a bomb load of at least 18,000kg (39,690 lb). The potent defensive armament comprised twelve 20- or 23-mm cannons; the tail turret was to be fitted later with 45- to 57-mm (1.77 to 2.2-in) cannons.

By September 1944, the full-size mock-up of "aircraft 64" had been completed and presented to the Air Force's mock-up review commission (an obligatory stage to make sure any major deficiencies were identified and eliminated before the first metal was cut

Basic design specifications of "aircraft 64" with four AM-46TK-3PB engines	
Length	30.0m (98ft 5 in)
Wing span	42.0m (137ft 10½ in)
Height on ground	7.2m (23ft 7½ in)
Wing area	150.3m² (1,618 sq ft)
Normal take-off weight	38,000kg (83,790lbs)
Maximum speed at normal take-off weight	650km/h (404mph)
Service ceiling	11,000m (36,090ft)
Maximum range with a 4,000-kg (8,810-lb) bomb load	6,500km (4,040 miles)
Cannon armament	10 x 20-mm (.78 calibre)
	Berezin B-20
Crew	10

on the prototype). This inspection resulted in a large number of critical remarks; in particular, the military demanded that a bomb-aiming/navigation radar with a 360° field of view be installed, as was the case with the B-29. On 7 April 1945, the specification for the aircraft was endorsed. The Air Force wished the industry to produce a bomber possessing the following performance:

- maximum speed at nominal engine power rating, 610km/h (379 mph);
- at combat power rating, 630km/h (392 mph);
- service ceiling, 11,000m (36,090ft);
- range with a 5,000-kg (11,025-lb) bomb load, 5,000km (3,108 miles);
- tactical range with a 14,000-kg (30,870-lb) bomb load, 2,000km (1,243 miles);
- take-off run, 800m (2,625ft).

The aircraft's powerplant was expected to comprise supercharged Mikulin AM-43TK-300B or AM-46TK-300 engines.

The mock-up was approved on 27 April 1945, just a few days before the capitulation of Germany. After the end of the war with Germany, OKB-156 started issuing detail drawings of the machine; the OKB's experimental production facility began preparations for the construction of the first prototype. The airframe and the powerplant (with the exception of the turbosuperchargers) presented virtually no problems, but fitting the aircraft with up-to-date equipment meeting the Air Force's requirements within the specified time limits became a real stumbling block. Subcontractor enterprises were unable at that stage to provide "aircraft 64" with the necessary navigation and radio equipment and the automated remote control system for the cannon armament. The reason was plain: during the wartime years, all efforts had been directed at mass production (in keeping with Stalin's slogan "Everything for the front lines, everything for the Victory!") and there was no time or capacity to develop new equipment.

It became obvious that the work on the creation of the first Soviet aircraft intended for the strategic weapon carrier role was reaching a deadlock. Worried by the state of affairs in the "aircraft 64" program, the nation's leaders took the decision to reverse-engineer the Boeing B-29 and put it into production in the USSR; this was now possible, since four examples of the B-29 had ended up in the Soviet Far East (see Chapter 1). Andrey N. Tupolev's OKB was tasked with the job of copying the B-29, and all the resources of the design bureau were concentrated on this top-priority task.

Gradually, all the work on "aircraft 64" wound down, although it was still present in the OKB's plans in the course of the following two years. One more version of the "64" project was prepared; it was reworked completely to incorporate the knowledge gained when studying the B-29. The bomber was now a low-wing monoplane, which simplified its adaptation for the airliner role. It was to be powered by AM-46TK-3PB engines delivering a maximum of 2,300hp apiece, and the equipment complement was to include a Kobal't (Cobalt) bomb aiming radar, etc. As for the designation Tu-10, it was later reassigned to an experimental derivative of the production wartime Tu-2 tactical bomber (NATO reporting name *Bat*).

Chapter 1

"Presents" from America

As noted in the introductory section, the USA persistently turned down Soviet requests for the delivery of the latest B-29 bomber. Yet, despite this discrimination on the part of its ally, the Soviet Union had three flyable Superfortresses at its disposal by the beginning of 1945. Where did they come from? The origin of this windfall, very timely for the USSR, is as follows.

In the summer of 1944 USAAF units of the 20th Bomber Command equipped with B-29s began systematic raids against territories occupied by Japan, followed by bombardment of targets on the Japanese islands themselves. In the course of these raids four damaged machines made forced landings in the Soviet Far East. It should be noted that during the first raids the B-29 crews ran a fairly high risk of not returning to base. Quite apart from the danger posed by the Japanese anti-aircraft defences, the fate of the

crews depended on the aircraft's reliability – which left a lot to be desired in the B-29s of the early production blocks. The Wright R-3350 Duplex Cyclone engine was a particular source of problems; the early R-3350s powering B-29s *sans suffixe* and the first B-29As were extremely troublesome and engine life was initially a mere 15 hours. There were frequent failures of the defensive armament and so on. These were the usual teething troubles afflicting a new and highly complex machine that was hastily pressed into combat service with all its vicissitudes and thus the first crews could not use the aircraft's potential to the full. This was especially true for long-range flights; not until 1945 did the USAAF master the technique of cruising at the aircraft's service ceiling while heading for the target – a technique which substantially reduced fuel consumption, enabling the bomber to reach its design range.

Still in USAAF markings, Boeing B-29-5-BW 42-6256/'32' *Ramp Tramp* (crew No.K-59) sits at Tsentrahl'naya-Ooglovaya airbase in the Soviet Far East after force-landing there on 20th July 1944, with Soviet Naval Aviation/Pacific Fleet pilots in the foreground.

The second Superfortress to land at Tsentrahl'naya-Ooglovaya AB, B-29-15-BW 42-6358, prior to the ferry flight to Moscow in early 1945, with typical rugged local scenery in the background. Note that the Soviet serial '358 Black' is already applied to the fin.

The first B-29 to fall into Soviet hands made an emergency landing at Tsentral'naya-Ooglovaya, a Soviet Navy/Pacific Fleet Air Arm airbase located 30 km (18.6 miles) east of Vladivostok, on 20th July 1944 – less than two months after the Superfortress had received its baptism of fire. The machine, a B-29-5-BW built by Boeing at Wichita, Kansas (USAAF serial 42-6256, construction number 3390), belonged to the 462nd Bomb Group (Heavy)/770th Bomb Squadron and was christened *Ramp Tramp* because of its lengthy periods of maintenance downtime; it also wore the tail code '32' and the crew designator K-59 on the nose. It was part of a 100-ship formation that had taken off from Calcutta, India, to bomb the Showa Steel Works at Anshan in Japanese-occupied Manchuria. Crew K-59 comprised pilot Capt. Howard R. Jarrell, co-pilot 1st Lt. John R. Kirkland, navigator 1st Lt. Frank C. Ogden, bombardier 1st Lt. Edward J. Golden, flight engineer Sgt. Mike

Losik Jr., radar operator Cpl. Herbert A. Bost, radio operator Sgt. Jerome C. Zuercher, left gunner TSgt. Lewis A. Earley, right gunner Sgt. George Hummel and tail gunner SSgt. Roy Price. Some sources state that two of the four engines had been hit by Japanese AA fire while others say an engine failed and the propeller could not be feathered. Anyway, unable to make it back to base, the pilot did as he had been instructed, making an emergency landing in Soviet territory. Yakovlev Yak-9 fighters of the Pacific Fleet Air Arm intercepted the bomber and escorted it to Tsentral'naya-Ooglovaya AB, where the crew and the aircraft were interned.

An important note must be made at this point. Whatever Western authors may claim about it being 'contrary to agreements with the United States', the B-29, and various other American aircraft that force-landed in the USSR at that time, were interned on perfectly legal grounds. In April 1941, two months before the German

The same aircraft shortly after arrival at Izmaïlovo airfield, Moscow, in July 1945. The motorcycle is one of the numerous Harley Davidson WLA42s supplied to the USSR under the Lend-Lease Agreement.

Head-on view of B-29-15-BW '358 Black' at Izmaïlovo AB. This aircraft became the pattern aircraft for B-4 production.

This side view of '358 Black' shows the small Soviet red star on the fin, the open bomb bays and the open entry door for the center/rear pressure cabins. Fighters are parked on the flight line beyond.

Another view of the same B-29 at Izmaïlovo AB. Note the flight deck access ladder and the co-pilot's open direct vision window.

aggression, the Soviet Union and Japan had concluded a Pact of Neutrality with a five-year term of validity; in April 1945 the USSR denounced this pact, declaring war on Japan and joining the action in the Far Eastern theatre of operations. For four years the Soviet Union abided strictly by this pact. The reason was simple: virtually throughout the four-year Great Patriotic War the possibility of a Japanese aggression in the Soviet Far East was a threat to be reckoned with. Should Japan invade as Nazi Germany wanted it to, this, coupled with German onslaught, could place the USSR on the brink of total disaster – the nation would be unable to wage war on two fronts, at least in 1942-43. Therefore the Soviets diligently complied with all the clauses of the Neutrality Pact, including those

Close-up of the tail unit of B-29 '358 Black'. One can only guess why the red star was applied to the fin but not to the rear fuselage.

providing for the internment of combat materiel and personnel of the Allies in the Pacific theatre of operations – which included any and all American aircraft and their crews that landed in the Soviet Far East. The first such internee aircraft was a B-25B from the famous Group led by Lt.-Col. James H. Doolittle which attacked the Japanese islands on 18th April 1942 after launching from the aircraft carrier USS *Hornet* with effectively a 'one-way ticket'. Later, a considerable number of USAAF B-24s and B-25s, US Navy Lockheed PV-1 Ventura and Lockheed PV-2 Harpoon bombers made emergency landings in Soviet territory. Some of them were in airworthy condition; they were eventually taken on strength by the VVS and even took part in combat against Japan in 1945. The slightly damaged B-29s that had landed in the USSR in 1944 were interned on the same legal basis.

A month later, on 20th August 1944, another crippled 'Superfort' – a B-29A-1-BN built by Boeing at Renton, Washington – crossed the Soviet border over the Amur River. This was a 40th BG(H)/395th BS aircraft serialled 42-93829 (c/n 7236) and christened *Cait Paomat II* ('St. Catherine II' in Gaelic). The aircraft, which was part of a 75-ship formation, had been damaged during the first daylight raid against the Imperial Steel & Iron Works in Yawata (Fukuoka Prefecture, Kyushu, Japan), flying from Chengdu, China. With the radar inoperative, the B-29 lost its

way in the clouds where it had taken cover from Japanese fighters, straying into Soviet airspace as a result. The crew of 11 – pilot Maj. Richard M. McGlinn, co-pilot 1st Lt. Ernest E. Caudle, navigator 2nd Lt. Lyle C. Turner, bombardier 1st Lt. Eugene C. Murphy, flight engineer 1st Lt. Almon W. Conrath, radar operator Sgt. Otis L. Childs Jr., radio operator SSgt. Melvin O. Webb, senior gunner SSgt. William T. Stocks Sr., left gunner Sgt. Louis M. Mannatt, right gunner Sgt. John G. Beckley and tail gunner SSgt. Charles H. Robson – parachuted to safety and the uncontrolled B-29 crashed into the foothills of the Sikhoté-Alin' mountain range east of the city of Khabarovsk. Although the bomber was a total loss, the wreckage was salvaged and delivered to Moscow for close examination; part of it found practical use (see Chapter 9).

(Here, mention must be made of a controversy. According to the American aviation researcher Joseph F. Baugher, not just one but *two* B-29s diverted to Soviet territory on 20th August 1944. One was 42-93829, and Baugher maintains that the bomber crash-landed but was restored to flyable condition. The other aircraft, whose unit is not stated, was allegedly 42-93839 (c/n 7246), another B-29A-1-BN, and it was this machine which was purportedly named *Cait Paomat II* and had been damaged during the Yawata raid, crashing near Khabarovsk and being destroyed. However, other western sources – and Russian sources, for that matter – do not corroborate this.)

On 11th November 1944 B-29-15-BW 42-6365 (c/n 3499) belonging to the 468th BG (Very Heavy)/794th BS and piloted by Capt. Weston H. Price was flying as the pathfinder aircraft for a 27-ship package out of Pengshan, China, due to make a night raid on an aircraft factory in Omura on Kyushu Island, Japan. The 11-man crew included co-pilot 1st Lt. John E. Flanagan, navigator 1st Lt. Melvin E. Scherer, bombardier 2nd Lt. Edwin Morrison, flight engineer 2nd Lt. Eugene P. Rutherford, radar operator SSgt. Henry J. Stavinski, radio operator TSgt. David Pletter, senior gunner SSgt. Frank A. Weed, electronics specialist/gunner MSgt. Donald J. Larkin, powerplant specialist/gunner SSgt. John Bardunias and mechanic/gunner SSgt. Millard S. Cook. This B-29 was christened *General H. H. Arnold Special* to commemorate an inspection of it by Maj.-Gen. Henry H. 'Hap' Arnold, the Commander of the USAAF during the Second World War, during his visit to the Wichita plant on 11th January 1944; a commemorative plaque affixed in the aircraft's cockpit bore witness to this event. It was no coinci-

B-29-5-BW 42-6256 at an early stage of its Soviet test career with the USAAF serial removed but the Soviet serial '256 Black' not yet applied. This was the only example to carry the red star across the fin/rudder joint line.

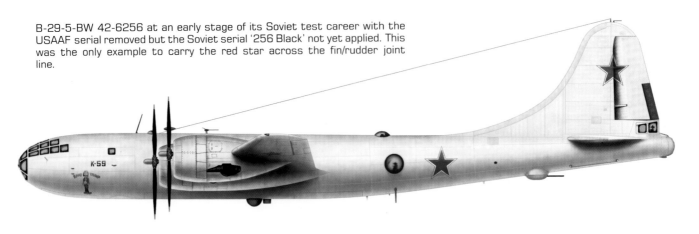

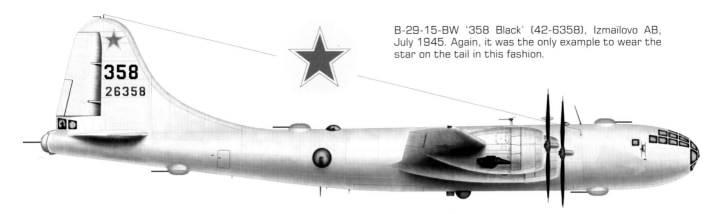

B-29-15-BW '358 Black' (42-6358), Izmaïlovo AB, July 1945. Again, it was the only example to wear the star on the tail in this fashion.

dence that Maj.-Gen. Arnold, who had backed the B-29 program since its inception, chose this particular aircraft, the 175th off the Wichita assembly line; it was the final machine he needed to get the newly-established 20th Bomber Command up to full strength and launch strategic bombardments of Japan.

En route to the target the aircraft went smack into the eye of a typhoon that had crossed the path of the formation. The storm was so violent that the huge bomber was tossed about like a dry leaf and was severely damaged, losing an engine, the radar and the central fire control system. When 42-6365 broke out of the storm, it had been blown far off course to the north and another engine on the same wing was running hot. The crew chose to head for the Soviet Far East. As soon as the B-29 crossed the border it was intercepted by Soviet fighters, which forced it to land at Tsentral'naya-Ooglovaya AB, where the bomber and its crew were interned. This was the aircraft's tenth combat sortie, and at the end of it 42-6365 had logged 563 hours 30 minutes total time since new.

The fourth Superfortress – another B-29-15-BW of the same 468th BG(VH)/794th BS – was interned ten days later. It was 42-6358 (c/n 3492), named *Ding Hao!* ('very good' in Chinese – a phrase that was common in the lingo of US servicemen deployed in China; the name is occasionally rendered as 'Ding How'). The crew comprised pilot 1st Lt. William J. Mickish, co-pilot 2nd Lt. John K. Schaefer, navigator 2nd Lt. Jack A. Diamond, bombardier 1st Lt. James R. Rutledge, flight engineer 1st Lt. James W. Ward, radar operator 2nd Lt. William R. Arentsen, radio operator TSgt. William P. Mann, senior gunner SSgt. Fred D. Brownwell, left

gunner SSgt. Edward J. Mertz, right gunner SSgt. Herman K. Sigrist and tail gunner SSgt. Therman Hassinger. On 21st November 1944 the aircraft was attacked by Japanese fighters during a further raid on the Omura aircraft factory. One of the engines was put out of action; on the power of three engines the aircraft reached the Soviet coastline where it was met by Soviet fighters; they escorted it to an airfield where the bomber made a safe landing.

Considerably later, on 29th August 1945, a fifth and final B-29 briefly fell into Soviet hands. By then Japan had ceased resistance but the Japanese Instrument of Surrender was not yet signed (this took place on 2nd September), and the Pacific Fleet's fighter element was on combat alert, ready to engage the enemy at a moment's notice. The B-29 passed over the airfield at Kanko, Korea, where the Pacific Fleet Air Arm's 11th IAP (*istrebitel'nyy aviapolk* – Fighter Regiment, ≈ Fighter Wing) was stationed. Two pairs of Yak-9Ms scrambled to intercept. Without hesitation, hot-tempered naval pilots made a firing pass, damaging the bomber's No.1 engine, whereupon the B-29 force-landed at Kanko. NKAP filed a report about the incident, suggesting that the bomber be flown to the USSR. However, in April 1945 the USSR had denounced the Pact of Neutrality and declared war on Japan, joining the action in the Far Eastern TO. Hence the B-29 at Kanko was repaired and promptly returned to the Americans together with its crew. (Another account gives a different story: it was Lt. Filimonov in a *14th* IAP Yak-9M who shot up the bomber, setting the *No.4* engine alight, after the B-29's pilots ignored orders to land at Kanko. Most of the crew bailed out and the captain landed the bomber single-

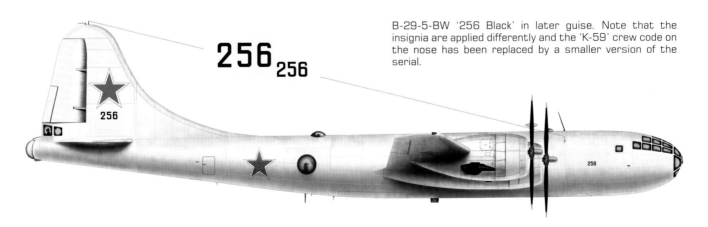

B-29-5-BW '256 Black' in later guise. Note that the insignia are applied differently and the 'K-59' crew code on the nose has been replaced by a smaller version of the serial.

handedly. After arriving at Kanko and inspecting the bomber, USAAF representatives concluded that repairing it on site was impossible and shipping it to the States was uneconomical, so the B-29 was abandoned.)

All the interned crews of Allied aircraft were sent to a special camp of the NKVD (*Narodnyy komissariaht vnootrennikh del – People's Commissariat of the Interior*, that is, the Soviet police-cum-secret service) near Tashkent in Uzbekistan. Before this, the airmen were allowed a meeting with the US Consul who promised them to send a message to the USA informing their families that they were alive. When they asked when they would be set free, the consul could only answer that they were interned in accordance with the Geneva Convention and that they would have to wait. The camp was monitored by Japanese officials, and the Soviet administration had to be extra careful not to give cause for accusations that they were breaking the rules.

It has been written that the conditions in the internment camp were comfortable by Soviet standards of the day. They could be compared neither with the Soviet camps for prisoners of war from the Axis powers, nor with the detention camps for former Soviet POWs who had been liberated from Nazi concentration camps, nor

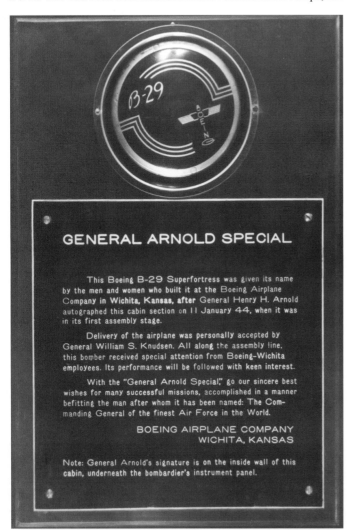

The control wheel 'hub cap' and the commemorative plaque from B-29-15-BW '365 Black' (42-6365).

even with the conditions for soldiers of the regular Red Army units in reserve regiments (there new units were formed before being sent to the frontlines). The rooms accommodated 15 to 20 persons; the Americans got decent food, they could practice sports, play games etc. Still, 'a rose is a rose is a rose', and a prison is a prison is a prison; the American airmen recalled that the conditions in the camp (and en route to it) were horrendous. Therefore there was no love lost between the Americans and their Soviet keepers. One of the B-29 crew members who had to spend some time in this camp recalled how a Soviet general came to the camp on Christmas day in 1944 and began to deliver a speech; at that moment one of the internees flung a pickled cucumber at him, hitting him squarely in the eye. Naturally, every one of the internees believed they all would be promptly shot for this, but the matter was sorted out without any disastrous consequences.

The camp was populated by some 200 internees at any one time; people came and went. The Allies were perfectly well informed about all these incidents involving their personnel and materiel but did not interfere, being well aware that they ought to avoid provoking Japan and that the appearance of one more front in the Far East could not only lead to the Soviet Union's debacle but also compromise the success of their own struggle against Japan. From time to time Soviet authorities allowed the internees to 'vanish'. They were transported in tightly shut Studebaker US6 trucks to Iran, part of which was occupied by the US Army; there they took a bath, changed into fresh clean uniforms and were shipped off to the United States. Before boarding the ships the airmen were warned that they had better keep their mouths shut, or else Stalin would not let the remaining Americans go. The interned B-29 crews travelled along that route, too, going to Iran in January 1945.

The return scheme was operated on condition that the American side would observe the strictest secrecy. The USAAF Command made the returning airmen sign a document stipulating that they would not breathe a word about their 'visit' to the Soviet Union for several years to come; no entries were made in their personal files about their sojourn in the USSR. Upon returning to the US the former internees were no longer posted abroad. The personnel officers of the units where they continued their active duty were advised that these men had served in some top secret 'unit D' and that it was categorically forbidden to ask them any questions about their previous service. Nevertheless, a leak did occur, and some information found its way into the press – apparently someone from these crews had blabbed. News of this was brought to Stalin's notice. 'Uncle Joe' was furious and immediately put a stop to the process of 'vanishing'. Rumours were rife among Americans that Stalin had ordered all the internees executed, and they were readily given credence by those who knew the true nature of the Stalin regime. Only a personal intervention on the part of the US Ambassador in the USSR helped hush up the affair.

As for the interned aircraft, they should have been returned to the USA once the Soviet-Japanese Neutrality Pact was denounced, but the Americans did not bother much about this problem. Their reasoning was that three flyable B-29s would not be a decisive factor; they would not dramatically boost the power of the Soviet Air Force. As for the Russian aircraft industry copying the latest US technology and even putting the B-29 into production at Soviet aircraft factories, the Americans simply dismissed this possibility. Even when the first three production examples of the Tu-4 – a copy of the B-29 – flew over Moscow's Tushino airfield during the 1947 Aviation Day flypast, for a long time people in the USA remained under the impression that these were the very B-29s interned by the Russians in 1944.

'To Moscow, to Moscow!'

Thus, by the beginning of 1945 the USSR had at its disposal three intact B-29s that could be restored to airworthy condition. The machines were formally taken on charge by the Pacific Fleet Air Arm but were not flown, sitting in storage at Tsentral'naya-Ooglovaya AB. Their original USAAF 'stars and bars' insignia were replaced by Soviet Air Force red star insignia and Soviet serials based on the original ones were applied; thus, B-29-5-BW 42-6256 became '256 Black', while B-29-15-BWs 42-6358 and 42-6365 similarly became '358 Black' and '365 Black' respectively.

People's Commissar of the Navy Admiral Nikolay G. Kuznetsov issued an order calling for a careful examination of the American bombers. Accordingly, Air Marshal Semyon F. Zhavoronkov, Commander-in-Chief of the Soviet Naval Aviation, summoned the Naval Aviation's Vice-Chief of Flight Inspection Lt.-Col. Semyon B. Reidel (some sources state his first name as Solomon) and aircraft operations engineer Nikolay A. Kravtsov, sending them on a mission to the Far East with the purpose of ferrying the B-29 to Moscow as soon as possible. The men had only one day to look through the available information on the B-29 at the Central Aero- and Hydrodynamics Institute named after Nikolay Ye. Zhukovskiy (TsAGI – *Tsentrahl'nyy aero-ghidrodinamicheskiy institoot*), receive intelligence data on the aircraft and pack their bags for the trip. A North American B-25J was picked for the two-day flight to Vladivostok – simply because it would give the crew some practice of flying and servicing American bombers.

Reidel had prior experience of testing and developing Naval Aviation materiel, having served in the Independent Naval Detachment (OMO – *Otdel'nyy morskoy otryad*) of the Red Banner State Research Institute of the Air Force (GK NII VVS – *Gosudarstvennyy Krasnoznamyonnyy naoochno-issledovatel'skiy institoot Voyenno-vozdooshnykh sil*; the 'Red Banner' bit means that the institute was awarded the Order of the Red Banner of Combat) and then in the Naval Aviation Research Institute in Sevastopol'. He spoke English and, by virtue of his professional skills, was the right person for the task of restoring the Superfortresses to active status.

Additionally, Maj. Vyacheslav P. Maroonov and one more pilot were seconded from the Black Sea Fleet to assist him, as both of them had flown Douglas A-20G Havoc bombers supplied under the Lend-Lease Agreement (these aircraft were used by the Soviet Navy as torpedo-bombers after appropriate modifications). The reason for this secondment was that the Pacific Fleet Air Arm had no American aircraft and hence no airmen qualified to fly them. However, the Pacific Fleet Air Arm did provide two engineers – A. F. Chernov and M. M. Krooglov. (Jumping ahead of our story, we may say that both Maroonov and Chernov were subsequently employed by the Tupolev OKB, flight-testing a number of aircraft, including the Tu-4. Both of them were involved in the crash of the first prototype of the Tu-95 four-turboprop strategic bomber (the

The manufacturer plate on a US Navy Bureau of Ordnance Mk 3 Mod. 9 bomb shackle removed from '365 Black'.

'95/1') on 11th May 1953; co-pilot Maroonov parachuted to safety while flight engineer Chernov was killed in the crash.)

From January 1945 onwards the B-29s became the object of systematic study. As of 1st January, two B-29s were on charge with the Pacific Fleet Air Arm's HQ and the third was operated by the 35th Independent Long-Range Bomber Squadron formed specially for testing the B-29; eventually this squadron came to include two B-29s and one B-25 used as a 'hack'.

Now, here comes a major discrepancy in the various accounts of how the B-29s were studied. One story has it that Reidel mastered the B-29 on his own, making use of his knowledge of English and the flight and maintenance manuals discovered in one of the aircraft. (Later these manuals were transferred to the Tupolev OKB and used by its personnel when working on the Tu-4; several books from this set of documents are now preserved in the company museum of the present-day Tupolev Joint-Stock Co.) Other accounts say that on arrival it turned out that the American crew had thrown a spanner in the works: before landing they had disposed of all technical manuals, removed and dumped the classified Norden bomb sight and identification friend-or-foe (IFF) transponder and further sabotaged the aircraft by cutting various wires. The Soviet specialists had to work almost by feel, using the labels in the cockpit and on various equipment, the red lines on the gauges – and, of course, their intuition. (It is possible that the sabotage bit applied to the *second* flyable B-29, 42-6365, whose crew may have been aware that the previous two B-29 crews diverting to the USSR had not returned; but then, why should the Soviet engineers have to 'work by feel' if there were manuals in the *first* B-29?)

Next, the Soviet specialists set to work checking the operation of the aircraft's engines, systems and equipment in preparation for a first low-altitude check flight. Predictably, the unfamiliar aircraft gave a few problems. Only one of the four engines of 42-6365 ran as it should; of the other three, one had insufficient manifold pressure, another would not go to maximum rpm and the third was suffering ignition problems. The manifold pressure problem on the No.1 engine was eventually traced to a defective vacuum tube in an electronic module controlling the electrically operated shutter in the manifold. However, it took a while to find a suitable equivalent among Soviet vacuum tubes.

More complications arose when it came to landing gear checks: there were no suitable jacks and no ground power unit (GPUs were non-existent in the Soviet Union at the time). Again, the engineers had to improvise, make do and mend. Stock screwjacks used for trestling IL-4 bombers were brought in and modified by fitting a geared drive and a suitable electric motor; the B-29's own generators provided the electric power for them via jumper cables. Next, as the jacks were too short and had insufficient lifting capacity, makeshift podiums were constructed from railway sleepers and welded steel connecting beams were fabricated, allowing the jacks to work in pairs. This allowed the bomber to be lifted and the gear to be cycled, which was a pretty hair-raising experience. The screwjacks were extended as far as they could go, and the aircraft was in a rather unstable position (with the engines running, mind you); should it topple the jacks and fall, suffering damage, the mission would go down the drain. The men succeeded in measuring the gear retraction/extension time and the strength of the current in the electric gear actuation system, but it was definitely not a moment they cared to recall.

There were lots of other difficulties associated with the ground checks. The hydraulics had to be tested without the benefit of a hydraulic power cart; refuelling and defuelling to determine the aircraft's fuel capacity had to be done using small fuel bowsers (no larger ones were available because there had been no need for them, Tsentrahl'naya-Ooglovaya being a fighter base) and hence was a time-consuming procedure, etc.

An 11-man crew was assigned to the B-29 for the first check flight, but its composition was somewhat different from the standard one. There was no radar operator and no bombardier; on the other hand, there were three pilots instead of the usual two (Reidel, Morzhakov and Chernyagov), one of them assisting the navigator (Filatov), and there were two flight engineers instead of one (Kravtsov and Morozov). Finally, there were the radio operator (Korotayev), three midship gunners (Ivanov, Yerioma and Khasanov), one of which acted as a back-up for the radio operator, and the tail gunner (Frolov).

Once the aircraft's systems had been ground-tested and all malfunctions eliminated insofar as possible, B-29 '365 Black' made several taxi runs at Tsentral'naya-Ooglovaya AB. One of these nearly ended in an accident. Everything had been OK while Reidel was in command, but when Maj. M. F. Morzhakov was in the captain's seat he started applying the brakes intermittently (apparently to avoid locking up the wheels and provoking a skid). This turned out to be ineffective; the aircraft accelerated and could only be brought to a halt a few feet from an earthen revetment at the end of the runway.

The high-speed runs were followed by brief hops and, finally, by the first real flight. This took the bomber to Nikolayevka AB (likewise located in the Primor'ye Region of the Soviet Far East) where it was safer from prying eyes.

Before the flight the aircraft was fully fuelled and the crew decided to have a good night's rest. However, their plans were foiled by the unexpected arrival of Lt.-Gen. Pyotr N. Lemeshko, Commander of the Pacific Fleet Air Arm, who summoned Reidel and Kravtsov and read to them a telegram from Naval Aviation C-in-C Zhavoronkov. It transpired that the US Ambassador in the

Here, B-29 '256 Black' is shown on the snowbound airfield of the Flight Research Institute (LII) in Zhukovskiy together with four initial production B-4s (Tu-4s). This particular aircraft was used by LII for various test and development work.

Soviet Union, William Averell Harriman, had visited Vyacheslav M. Molotov (who was then People's Commissar of Foreign Affairs) in Moscow, warning him that the B-29 was an extremely complex aircraft and if Soviet airmen attempted to fly it without prior training it was bound to end in disaster. This was not the most heartening message, to say the least, and Zhavoronkov wanted to hear the pilots' opinion. Addressing Reidel, Kravtsov said: *'Know what, Senya* (the informal form of the name Semyon – *Auth.*), *let's sleep on it. We'll give the answer tomorrow... after the flight.'*

It should be noted that the B-29 did show its temper during that first flight. The bomber took off with the flaps at the appropriate setting (which the crew had read on a plate in the flight deck) but the pilots quickly discovered that the aircraft would not climb until it had gathered speed – and the B-29 would not accelerate with the flaps down! The pilots had no way of knowing this beforehand, not even having anyone's advice to fall back on. Taking off in an easterly direction, they passed the city of Vladivostok, then passed Russkiy Island ('Russian Island') in Peter the Great Bay (part of the Sea of Japan) half a mile from Vladivostok, leaving the rescue launches in the bay far behind, and the B-29 was still reluctant to climb, 'crawling on its belly', as the crew put it afterwards. Only when the flaps were retracted – slowly and cautiously, in several steps – did the aircraft begin to gain altitude. On the way out it was escorted by fighters – perhaps because the command was worried about the possibility of the B-29 going to Japan! Some 45 minutes later the bomber touched down at Nikolayevka AB and the elated crew sent a telegram to Zhavoronkov: *'Take-off from [Tsentral'naya-]Ooglovaya, landing at Nikolayevka, all OK. Reidel, Kravtsov.'*

From Nikolayevka AB the B-29 made a number of practice flights so that the crew could explore the bomber's handling and master the piloting techniques. There was also a closed-circuit flight from Nikolayevka AB in order to determine the fuel consumption and pick suitable alternate airfields. Next, the B-29 moved to Romanovka AB, which was the largest airfield in the area and best suited for heavy aircraft in the B-29's class, permitting maximum gross weight take-offs. Also, unlike other airfields airfield in the Soviet Far East, Romanovka was not surrounded by hills.

This stage also posed its fair share of problems and exploration. Suffice it to say that for the first time in the Soviet Union such a heavy aircraft had flown with pressurized cabins (previous Soviet experiments with indigenous pressurized cockpits were concerned with fighters). The operation of the pressurisation system (including depressurisation), the heating and ventilation system, the emergency descent procedure etc. had to be tested. Remember, the pilots had no manuals, only the cockpit labels to provide any clue.

Apart from his share of the test work, Reidel had to coach Vyacheslav P. Maroonov and A. F. Chernov, who were mastering one of the other B-29s. They had only two days to get acquainted with the machine. Their English being far from perfect, they spent these two days mainly exploring the nooks and crannies of the bomber, armed with a thick English-Russian dictionary. On the third day Reidel made them take an exam. On 9th January Maroonov already flew as a trainee captain, the co-pilot's seat being occupied by Reidel. Two days later Maroonov performed a solo flight and received his unofficial type rating.

The final assignment at Romanovka AB before the ferry flight to Moscow was to check the defensive armament in case Japanese fighters should attempt an attack – after all, the border to Japanese-occupied Manchuria was quite close, and provocations from the Japanese did occur. As the bomber formated with a Polikarpov Po-2 *Mule* utility biplane towing a sleeve-type target, Khasanov took aim, pressed the trigger and... Nothing happened – the Browning M2 machine-guns were silent, even though the ammunition belts had been properly connected. The gunner tried the pneumatic cocking system but to no avail – the machine-gun barrels jerked, the weapons fired a single round each and no more. This went on until the air supply in the machine-gun cocking bottle ran out. The Soviet engineers tried charging the bottles with ordinary compressed air, then with nitrogen, but it was no good – the problem persisted. By trial and error they found out that the system had to be charged with carbon dioxide to work normally, enabling automatic fire.

The testing of the B-29s in the Far East went on until 21st June 1945. In the course of these tests the bomber's basic performance was recorded, proving to be somewhat inferior to the official figures stated by the Boeing Company. This is understandable: the machines were not brand-new, showing a good deal of operational wear and tear. Several high-altitude flights were undertaken to check the maximum range in a circuit and to perform bombing. The naval pilots established the following basic performance of the aircraft:
- empty weight, 32,200 kg (71,000 lb);
- normal all-up weight, 47,600 kg (104,960 lb);
- maximum all-up weight, 54,500 kg (120,170 lb);
- normal payload, 15,400 kg (33,960 lb);
- maximum payload, 21,700 kg (47,850 lb);
- maximum range without bombs, 6,760 km (4,201 miles);
- range with a 4,000-kg (8,810-lb) bomb load, 5,310 km (3,300 miles);
- maximum speed at normal AUW, 580 km/h (360 mph);
- climb time to 5,000 m (16,400 ft), 16.5 minutes;
- take-off run, 960 m (3,150 ft).

Finally, everything was ready for the flight to Moscow, and '365 Black' was the first to go. The machine was captained by Semyon B. Reidel, with M. F. Morzhakov as co-pilot and M. M. Krooglov as flight engineer. With a full fuel load the B-29 was capable of flying from Romanovka to Moscow non-stop; yet, to be on the safe side, the airmen decided to stage through Chita (eastern Siberia), Krasnoyarsk (western Siberia) and Taïncha (Kazakh SSR). This turned out to be a wise decision because the bomber promptly started acting up again. En route to Chita the entire group of barometric flight instruments – the airspeed indicator, the altimeter, the vertical speed indicator – started giving funny readings. Actually, for the crew flying a still-unfamiliar aircraft, this was no fun at all – especially with no ASI readings. Checking the air ducting from the pitots inch by inch as the aircraft cruised on, the crew reached a large air filter; its inner cartridge made of corrugated paper was completely clogged with red dust that must have come all the way from Calcutta. Attempts to clean it by blowing gave no result and the crew simply replaced it with a handy piece of AST-100 grade Soviet aviation fabric, restoring the instruments' operation.

There was trouble at the other staging points, too. In Krasnoyarsk, one of the engines suffered a starter failure – to be precise, the electric motor spinning up the flywheel of the inertia starter failed. As a result, the inertia starter had to be cranked up manually, using a tall and rickety ladder, which was most inconvenient. Anyway, on 23rd June 1945 B-29 '365 Black' arrived at Izmaïlovo AB located immediately east of Moscow, which was home to the Naval Aviation's 65th Special Mission Air Regiment, a unit performing both transport duties and research work. The runway at Izmaïlovo being rather short, the bomber had to circle for a while, burning off fuel to reduce the landing weight. (Today, Izmaïlovo is a residential area within the city limits; the base is long since closed and the land redeveloped. The airfield used to be where Sirenevyy Blvd. ('Lilac Boulevard') now is, between the present-day tube stations Pervomaïskaya and Shcholkovskaya.)

The heads of several People's Commissariats, including Vyacheslav M. Molotov, came to Izmaïlovo to examine the aircraft, the ferry crew acting as their 'tour guides'. Of course, Andrey N. Tupolev and several other high-ranking designers of OKB-156 (including Leonid L. Kerber, a leading equipment specialist, and Aleksandr V. Nadashkevich, an armament specialist) came too. Tupolev personally inspected everything he could, but when he tried the crawlway above the forward bomb bay connecting the forward and center pressure cabins, he stuck, being rather too portly. *'Humph!* – he snorted as he extricated himself – *This is only good for the skinny Americans, they must be malnourished under capitalism...'*

Afterwards, Kerber recalled that Tupolev voiced the same opinion as himself: the airframe as such was nothing out of the ordinary and the OKB would have no great trouble building such an aircraft. *"But what do we do with the armament and the equipment?"* – he said next – *"Frankly, I've no idea. The gunner sits almost in the tail, but he works the forward machine-guns: imagine how reliable the remote controls need to be! What's more: all these radars, rangefinders, automatic co-ordinate measuring systems – will our industry reproduce them properly? OK, let's assume it will; but look, they are all connected by hundreds, thousands"* – the *Old Man* (Tupolev's nickname within the OKB which, of course, was never mentioned to his face – Auth.) *was getting truculent –* *"millions of wires. How will Kerber and Nadashkevich sort them out, how will they understand where this particular wire comes from and where it goes?"* He tugged at one of the wires. *"What if there's a damn short circuit – what do we do?"* He spat with disgust, losing his temper, and raised his voice: *"Well, Kerber L'vovich and Aleksandr Vasil'yevich, why are you keeping mum? Tell me how we are to sort out this puzzle!"* (L'vovich and Vasil'yevich were Leonid Kerber's and Aleksandr Nadashkevich's respective patronymics; the irate Tupolev used Kerber's last name instead of his given name – Auth.) *What could we say? He was right.* *"All right"* – Tupolev said – *"back to the OKB. Let's get to grips with this Tower of Babel."*

In June and July 1945 B-29s '358 Black' and '265 Black' were likewise ferried to Moscow. The first machine was piloted by Semyon B. Reidel, with M. F. Morzhakov as co-pilot and Nikolay

A. Kravtsov (some sources say M. M. Krooglov) as flight engineer. So far it has not been possible to establish the names of the airmen piloting the other machine. Some sources claim Vyacheslav P. Maroonov was the captain; however, by then he had been transferred to one of the bomber regiments equipped with the Tu-2, taking part in the combat against Japanese Army units in Manchuria shortly thereafter. (According to one account, Andrey N. Tupolev demanded that three B-29s be placed at his disposal – one as a pattern aircraft, one for disassembly and study, and one for static testing – because he opposed the idea of copying the B-29, so he placed a demand which he believed could not be met. The reaction was simple: you want 'em – you've got 'em.) Later

'265 Black' was flown to Chkalovskaya AB near the town of Shcholkovo (Moscow Region), the airfield of GK NII VVS.

The second and third bombers were flown along the same route as the first one. Now that the Soviet pilots had some experience with the type, the process was easier but by no means trouble-free. On the third aircraft a fire broke out in the No.4 engine on the Krasnoyarsk-Taïncha leg of the journey. Luckily this occurred immediately before landing and the crew managed to land the burning bomber in one piece, whereupon the fire was extinguished, but the aircraft was unflyable for the time being. The situation was resolved by removing an engine from B-29 '365 Black' and delivering it to Taïncha in a C-47; once the engine had been replaced and

Above: B-29-15-BW '358 Black' at Moscow's Central airfield (Khodynka), with the fuselage of the Tu-70 – an airliner derivative of the Tu-4 – in the foreground as the aircraft awaits reassembly after delivery by road from the Tupolev OKB's experimental plant (MMZ No.156) in Moscow.

Right: B-29-5-BW '256 Black (42-6256) at LII, sporting revised Soviet insignia (the star on the tail is larger and positioned entirely on the fin, no longer overlapping the rudder hinge line). Note the protective canvas cover on the observation blisters.

Left: Another view of 42-6256 at LII at an earlier date, with a Tu-2 bomber and an IL-10 attack aircraft in the background. The nose art has been retouched away on this photo used for a test report.

the fire damage to the surrounding structure had been repaired, the bomber continued its journey to Moscow. As for the damaged engine, it was delivered to Arkadiy D. Shvetsov's OKB-19 engine design bureau in the city of Perm' for detailed study.

For fulfilling the important government assignment of ferrying the B-29s to Moscow Semyon B. Reidel and Nikolay A. Kravtsov were awarded the Lenin Order – the highest Soviet decoration. The other participants of the operation were not left out either, receiving various government awards.

At the request of Air Marshal Aleksandr Ye. Golovanov, Commander of the 18th Air Army, on 1st July 1945 B-29 '256 Black' was transferred to the 890th BAP (*bombardirovochnyy aviapolk* – Bomber Regiment, ≈ Bomb Wing) based at Balbasovo AB near the town of Orsha, Belorussia, and commanded by Endel K. Puusepp. Apart from nine Pe-8s, the regiment operated twelve ex-USAAF B-17Fs and B-17Gs restored to operational condition plus 19 B-25Js; this mixed bag was a result of the scarcity of flyable bombers forcing the Soviet Air Force to repair and use every available aircraft. The Superfortress remained with the 890th BAP only briefly. As soon as the decision to copy the B-29 was taken at the highest level, the machine was flown back to Moscow – to be precise, the airfield of the Flight Research Institute (LII – *Lyotno-issledovatel'skiy institoot*) near the town of Ramenskoye, Moscow Region. The positioning flight was performed by 890th BAP pilot N. A. Ishchenko and LII test pilot Mark L. Gallai, the latter mastering the aircraft in the process – sort of on-the-job training. (Actually the town of Stakhanovo was even closer to the airfield. This town was renamed Zhukovskiy in early 1947 to celebrate the 100th birthday of Nikolay Ye. Zhukovskiy, the founder of Russian aviation science, and now the LII airfield is commonly known as the one in Zhukovskiy.)

Interestingly, '256 Black' retained its 'Ramp Tramp' nose art throughout its flying career. So did many other ex-USAAF bombers in the 890th BAP – diligent propaganda officers did not seem to mind, demanding the removal only of nude beauties.

Thus, as the year of 1945 set in, the most horrendous war in the history of humanity was drawing to an end. The victors and the vanquished alike had in store for them peacetime years during which the former allies would soon be divided by a new front – the front of the Cold War. In the East this was clearly understood, and preparations for this situation were made in advance. The setbacks suffered by Tupolev's "64" project made it imperative that alternative ways be sought. In a gesture of despair, someone in the Air Force command suggested reinstating production of the Pe-8 in 'as-was' condition, without even modernisation. The idea was rejected outright. Stalin was furious; as was his wont, he tried to seek out those who were to blame for the Soviet Union's lagging behind the West in the field of heavy bomber technology (which was actually due to perfectly objective circumstances). Thunderous accusations were hurled at those who were responsible for the current condition and technical level of Soviet aviation. While the war was in progress, he did not touch anyone, but a year later he vented his wrath on the then People's Commissar of Aircraft Industry Aleksey I. Shakhoorin, VVS C-in-C Air Marshal Aleksandr A. Novikov and other high-ranking aircraft industry officials and Air Force commanders, who were removed from office and faced trial.

All this 'emotional' situation characteristic of the final months of the war induced many top-ranking Air Force officers and aircraft industry officials to ponder over the solution of the 'long-range high-speed high-altitude bomber' problem. At present it is hard to say who was the first to suggest solving the problem by copying the B-29 – the aircraft industry or the Air Force. According to Gen. Vasiliy V. Reshetnikov, the former Commander-in-Chief of the Russian Air Force's Long-Range Aviation, it was Air Marshal Aleksandr Ye. Golovanov who dared to approach Stalin with this proposal. According to Golovanov, Stalin's reaction was more than positive; it was coined in his succinct remark: '*All our aircraft that have fought the war must be scrapped, and the B-29 must go into production!*'

According to another account of the events, it was Vladimir M. Myasishchev who first floated the idea to NKAP in a letter to soon-to-be-dismissed People's Commissar Aleksey I. Shakhoorin on 25th May 1945, suggesting that his OKB and the OKB of Iosif F. Nezval' should take on the job. The letter stated, among other things: '*...all the work on issuing production drawings can be effected by the OKB led by Comrade Nezval', by the production engineering section and by some designers of our OKB residing at Plant No.22 which has been relieved of the work on production [Petlyakov] Pe-2 [dive] bombers...*' Myasishchev proposed that the Soviet copy of the B-29 be powered by indigenous Shvetsov ASh-72 radials and indigenous 20-mm (.78 calibre) Berezin B-20 cannons be installed instead of the American machine-guns. According to Myasishchev, the work should take place at aircraft factory No.22 in Kazan', the capital of the Tatar Autonomous SSR (which had a history of building bombers, including heavy ones). It was Shakhoorin's newly appointed successor Mikhail V. Khroonichev who reported this proposal to Stalin.

Gradually this mosaic of opinions on the problem transformed itself, in the minds of the Soviet leaders, into a firm conviction that it was necessary to initiate work on creating a Soviet copy – or rather analogue – of the B-29. This was a case of 'quantity transitioning into quality', as a philosophic maxim has it. Unperturbed by this unanimous opinion, Andrey N. Tupolev continued his work on the "64", without giving any thought to the possibility of copying the American machine. But Stalin, being well aware and appreciative of Tupolev's energy and thrust, selected none other than him for the implementation of this work.

During the early summer of 1945 Khroonichev summoned Tupolev and one of the latter's aides, Aleksandr A. Arkhangel'skiy, to a meeting in the Kremlin presided by Stalin. They surmised that the conversation would be concerned with the work on the "64" bomber, so they came armed with an album featuring a graphic presentation of this machine in colour (the artwork of Boris M. Kondorskiy, the Tupolev OKB's newly appointed chief of the projects team). However, virtually right from the start Stalin began discussing the B-29 and the conversation ended in explicit instructions to Tupolev requiring him to copy the B-29 within the shortest possible time frame. Shortly thereafter an appropriate directive of the State Defence Committee (the wartime Soviet legislative body) and an appropriate NKAP order were issued. The ball was set rolling.

Chapter 2

B-29 turns Tu-4: The reverse-engineering effort

As noted earlier, by the beginning of 1945 the leaders of the Soviet aircraft industry, as well as the nation's political leaders (including Stalin), fully realised that development of indigenous equipment for the "aircraft 64" by dozens of design bureau and research institutions subordinated to various People's Commissariats (reorganised into ministries by April 1946) would take a lot of time. This meant that in the short term the USSR would not field an aircraft capable of carrying a nuclear bomb which was so necessary for ensuring the nation's security, whereas the USA already

had such a delivery vehicle – the B-29. Stalin took the decision that was the only right one under the circumstances: to organise series production of a copy of the B-29 bomber – lock, stock and barrel – within the shortest time possible, using the four aircraft which had fallen into Soviet hands as recounted in Chapter 1.

It was OKB-156 that was tasked with copying the B-29 and putting it into series production. Initially the aircraft received the Soviet designation B-4 (for *bombardirovshchik* – bomber) – coined by Andrey N. Tupolev – and the product code *izdeliye* R. (*Izdeliye*

The team which was responsible for copying the B-29. Left to right: B-4 chief project engineer Dmitriy S. Markov, OKB-156 chief structural strength engineer Aleksey M. Cheryomukhin, Chief Designer Andrey N. Tupolev, his deputy Aleksandr A. Arkhangel'skiy and general arrangement group chief Sergey M. Yeger.

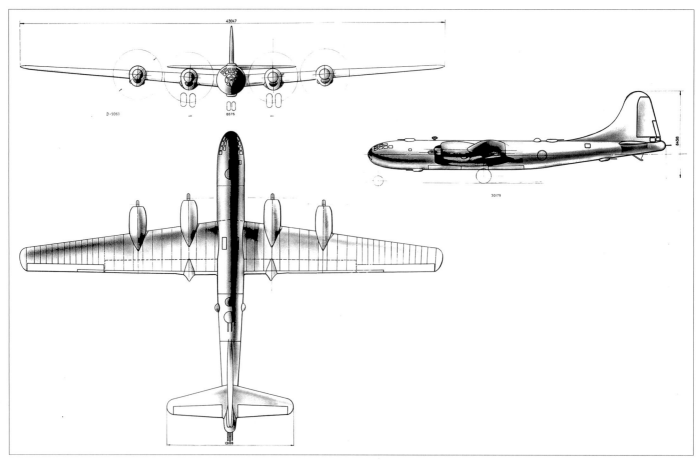

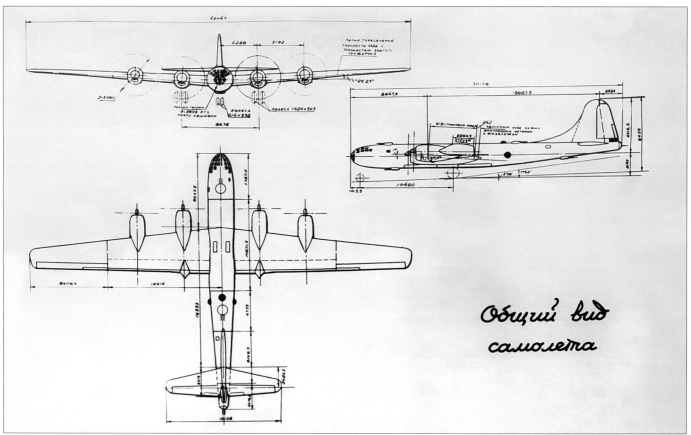

Общий вид
самолета

Opposite page:

A three-view drawing of the B-4 from the ADP documents.

Another, more detailed three-view of the B-4 from the ADP documents featuring more detailed dimensions. The legend reads 'Overall view of the aircraft'.

This page:

Above: A wooden scale model of the B-4 manufactured by the Tupolev OKB at the reverse-engineering stage.

Right: A side view of the same model.

Below: A cutaway drawing of the B-4 from the project documents. Note the different calibres of bombs in the two bomb bays.

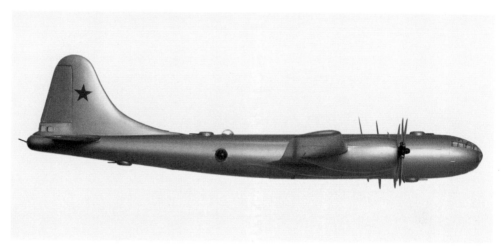

(product) such-and-such is a code for Soviet/Russian military hardware items commonly used in paperwork to confuse outsiders.) In fact, the designation B-4 denoted simply 'four-engined bomber' because Tupolev (who was not enthusiastic about copying someone else's design) did not want to assign a designation in the OKB's normal series to the B-29 copy. On 6th June 1945 the State Defence Committee issued directive No.8934; it was followed by the appropriate NKAP order No.263 issued on 22nd June 1945. In accordance with this order A. I. Kuznetsov (chief of NKAP's 10th Main Directorate and Vice People's Commissar of Aircraft Industry responsible for series production of heavy aircraft), Vasiliy A. Okulov (Director of aircraft factory No.22) and Chief Designer

Andrey N. Tupolev were required to organise production of the B-4 bomber. The schedule was extremely tough; the Tupolev OKB was to prepare a full set of manufacturing drawings within a year and transfer them to plant No.22, while the latter was to manufacture a trials batch of 20 aircraft within another year. Interestingly, no one worried much about copyright issues and such; military parity with the West was more important than licences (which would have been unobtainable anyway).

The NKAP order stated: *'Chief Designer Comrade Tupolev shall start immediately the work on preparing technical drawings, templates and technical documentation for the B-4 aircraft; this work shall be considered a priority task for the design staff and*

Left: B-29-15-BW '365 Black' in the post-mortem room... sorry, the TsAGI hangar at Moscow-Khodynka where it was taken apart completely for the purpose of studying the design and taking the dimensions of all parts and components. Here the forward and rear fuselage have been detached and the center fuselage is being worked on, with the upper skin already removed; note the detachable portion amidships for inserting the wing carry-through box. the port wing has been stripped down to the front spar, while the leading-edge fairings on the starboard wing are still in place. Workers are swarming all over the machine like scavenging ants.

Left: The wing center section of '365 Black' is readied for removal by means of a gantry crane.

Below left: One of the bomber's ailerons with the skin and the trim tab removed. Workers are measuring the ribs and spar of the aileron.

Opposite page:
Top and top right: Two starboard side views of the center pressure cabin of '365 Black' propped up on trestles, showing the pressure doors in both pressure domes; the forward door is closed (note the dished shape), with the opening for the crawlway above it. The skin on the port and upper sides is already gone.

Center and center right: Parts of the wings' internal structure are laid bare. Again, workers are meticulously taking measurements and entering the results in special tables.

Right: A more general view of the disassembly area. Note the outer wing leading edge fairing on the left, placed in felt-lined supports to avoid damage.

Tupolev OKB workers measuring the components of the wings of '365 Black' (note the star on the wing skin panel).

production personnel of Plant No.156 [...] For the purpose of speeding up the work on the B-4 aircraft as much as possible, an experimental design bureau (OKB) for the B-4 bomber shall be set up to aid the main OKB led by Comrade Tupolev; this new OKB is to take over the whole personnel of the OKB led by Comrade Nezval', Comrade Myasishchev's OKB at plant No.22 and the experimental workshop of plant No.22.' (Vladimir M. Myasishchev had

been transferred to Kazan' in order to supervise the production and upgrading of the Pe-2 bomber after Chief Designer Vladimir M. Polikarpov's death in an aircraft crash in 1942.)

The text of the order spelled out, item by item, the whole range of tasks assigned to different organisations and enterprises of the aircraft industry. The Tupolev OKB was entrusted with breaking down the B-29 into separate units and assemblies, reverse-engineering the theoretical outlines of all parts, removing the equipment items and transferring them to appropriate enterprises for reproduction. The All-Union Institute of Aviation Materials (VIAM – *Vsesoyooznyy insti**toot** aviatsionnykh materi**ahl**ov*) was to analyse all the structural materials used in the B-29's design; it was also to issue orders to factories for the production of materials which had not been produced by the Soviet industry prior to that. The Central Aero Engine Institute (TsIAM – *Tsen**trahl'**nyy insti**toot** aviatsionnovo motorostroyeniya*) was to study the power-plant and conduct the necessary test and development work that would make it possible to equip the new aircraft with indigenous Shvetsov ASh-73 engines (a derivative of the ASh-72) and special turbo-superchargers. TsAGI was entrusted with studying the aircraft's aerodynamics and strength. The Special Equipment Research Institute (NISO – *Na**ooch**nyy insti**toot** spetsi**ahl'**novo obo**roo**dovaniya*) was to study the B-29's equipment and prepare specifications that would be issued to Soviet enterprises for the purpose of putting the various equipment items into production.

These huge tables in the workshop at Moscow-Khodynka were used for making templates of the B-29's components.

One of the B-29's main gear doors (dished to accommodate the wheel) is weighed at Khodynka.

All parts of the B-29, down to the last nuts and bolts, were weighed as shown here.

Meanwhile, as already mentioned, preparations for ferrying the three flyable B-29s to Moscow were getting under way in the Far East; in June and July 1945 all three bombers arrived at Izmaïlovo AB. As a first step in the work on copying the B-29, the fate of the individual machines was decided. As already mentioned, after a brief stint in the 890th BAP B-29-5-BW '256 Black' (42-6256) was transferred to LII for personnel training and for preparing flight and maintenance manuals. B-29-15-BW '365 Black' (42-6365) was to be disassembled for the purpose of studying the airframe and preparing working drawings. Finally, B-29-15-BW '358 Black' (42-6358) was to be left intact as a reference specimen, remaining at Izmaïlovo AB. It made only one more flight, but it was periodically examined by various specialists, including representatives of the Air Force Engineering Academy who made

Copying the B-29 involved a lot of 'crawlwork' on the tables while the templates were being made.

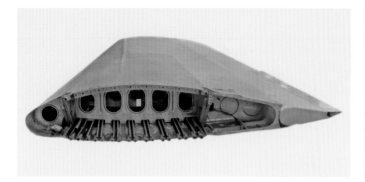

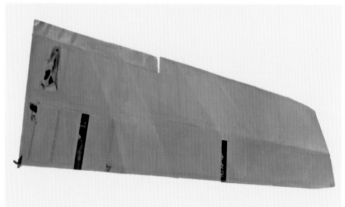

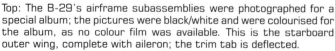

Top: The B-29's airframe subassemblies were photographed for a special album; the pictures were black/white and were colourised for the album, as no colour film was available. This is the starboard outer wing, complete with aileron; the trim tab is deflected.

Above: The fin and the fin fillet of B-29 '365 Black', showing the lightening holes in the fin's rear spar. The USAAF serial 42-6365 has turned grey due to weathering.

Top: The port aileron seen from below, with the inspection hatch covers removed.

Center: The rudder, with some skin damage evident on the starboard side.

Above: The starboard stabiliser (minus elevator).

detailed drawings of the powerplant installation. A few years later this aircraft was scrapped.

On 5th July 1945 NKAP issued order No.278 placing a hangar at the Central airfield named after Mikhail V. Frunze (better known as Moscow-Khodynka) at the Tupolev OKB's disposal; it was this hangar that '365 Black' was placed in after making the short hop from Izmaïlovo on the night of 10th-11th June. This was the only hangar in Moscow that was big enough for the B-29. It had been built for TsAGI before the war and was thus originally at the Tupolev OKB's disposal because the OKB was part of TsAGI until

1936; however, when Andrey N. Tupolev (like many of his associates) was arrested on false charges in 1937 and forced to work in the TsKB-29 'central design bureau' under the NKVD umbrella, the hangar had been transferred to the OKB-51 missile design bureau of Vladimir N. Chelomey. Thus, Tupolev (being, by some accounts, a 'petty owner' sort of person) reclaimed what was 'rightfully his'.

Tupolev assessed the scope of the forthcoming work as equivalent to three years. To substantiate this term, he pointed out that American technology and production methods differed from those of the Soviet industry not only in aircraft construction but in other branches as well. In response to Tupolev's arguments Stalin granted him carte blanche and support from Lavrentiy P. Beria, the influential and feared People's Commissar of the Interior, into the bargain, but the time frame for launching series production of the B-4 was limited to two years. The first Soviet-built bombers were to take part in the traditional air display at Tushino in the summer of 1947. Actually, the original plans required the first B-4 to be submitted for testing in a year – in June 1946, in the hope that the most complex units and subassemblies for the first machines – the Western Electric AN/APQ-13 'Mickey' X-band (centimetre-waveband) bomb-aiming/navigation radar, the BC-733 instrument landing system (ILS), the Hamilton Standard 6526A-6 variable-pitch propellers, the engine starters, the landing gear wheels etc – would be purchased in the USA. However, it soon became apparent that in the atmosphere of the incipient Cold War the Americans would not sell military hardware to the Soviets, and time limits had to be extended.

The work on the new aircraft encompassed some 900 enterprises and organisations subordinated to several People's Commissariats; in fact, some enterprises were created specially for the B-4 program. In particular, several new design bureau were set up within the NKAP system; they were tasked with copying various items of equipment, including the electronics, electrical equipment and instruments, and mastering production of same.

Tupolev started his preparations for the work by establishing a 'think tank' comprising the leaders of the OKB teams, each of which was responsible for this or that part of the work. Dmitriy S. Markov was appointed chief project engineer. This choice was not a matter of chance; before his arrest and incarceration in TsKB-29 in 1938 Markov had some experience with American aircraft, brilliantly managing the licence production of the Vultee V-11GB light bomber at plant No.1 in Moscow as the BSh-1 attack aircraft and PS-43 mailplane.

In the autumn of 1945 Tupolev and his associates undertook another inspection of the B-29. After the inspection Tupolev reiterated what he had said after examining the machine for the first time: *'That's a normal aircraft, I see nothing unusual in it. But, frankly, I cannot imagine how you will sort out all this tangle of wiring covering the entire machine, how you will ensure the linkage between the numerous gunsights and the weapons' remote control system, how you will tackle the flight control and navigation system – it beats me. A number of other questions are unclear, too. The defensive weapons and bombs will be indigenous, of course. The instrument dials will be converted to the metric system, but what about the IFF system, crew equipment, parachutes and so on? All these issues must be thoroughly worked out and the results incorporated into one more directive – lest we should debate these questions endlessly.'*

A while earlier, in the summer of 1945 (immediately after the arrival of the B-29 at Moscow-Khodynka), a special task force was set up at Tupolev's order for preparing outline drawings of the aircraft's main assemblies. An album with these outline drawings was ready at the beginning of August; it became a graphic visual basis for discussing the problems associated with copying the B-29. Preparation of these drawings revealed that it would be impossible to reproduce this aircraft in the USSR without radical technological changes in aviation metallurgy. The vast majority of technical features and materials used by the B-29 were new for the Soviet aircraft industry. Many intermediate products necessary for the B-4 (sheet metal and metal plates of the required size, stamped and extruded metal profiles, special nuts, bolts and fasteners), as well as up to 90% of the materials, had never been produced by the Soviet aircraft and metal industry until then. Equally new were the production techniques involved in the manufacture and assembly of the bomber's airframe.

The B-4 was included into NKAP's prototype aircraft construction plan for 1946, but it was not until 26th February 1946 that a Council of People's Commissars directive stipulating the future bomber's basic performance appeared. This document specified the normal take-off weight as 54,500 kg (120,150 lb), while the maximum TOW in overload configuration was not to exceed 61,250 kg (135,030 lb). Top speed was stipulated as at least 470 km/h (292 mph) at sea level and 560 km/h (348 mph) at 10,500 m (344,450 ft). With a normal TOW and a 1,500-kg (3,310-lb) bomb load, range was to be not less than 5,000 km (3,105 miles), increasing to 6,000 km (3,728 miles) with a 51,250-kg (112,990-lb) TOW and a 5,000-kg (11,020-lb) bomb load. In June 1946 the specifications were revised, the bomb load and range being actually downgraded somewhat.

Meanwhile, on 15th March 1946 NKAP was transformed into the Ministry of Aircraft Industry (MAP – *Ministerstvo aviatsionnoy promyshlennosti*). On 20th March 1946 Minister of Aircraft Industry Mikhail V. Khroonichev wrote to Soviet Air Force Commander-in-Chief Col.-Gen. Konstantin A. Vershinin in letter N-27/1176: *'Referring to the Air Force's letter No.864038s dated 1st March 1946, I hereby inform you that in accordance with the government's decision the B-4 aircraft to be put into production shall conform to the American B-29 original completely as regards not only the airframe but also the entire equipment suite. In accordance with the abovesaid, and the need to accelerate the launch of production, I deem it necessary to check [the conformity of] all equipment items at MAP [research] institutes by comparing the items produced by our design bureau with the ones installed on the B-29. The B-4 will undergo state acceptance trials at GK NII VVS together with its entire equipment suite. To this end, one of the three B-29s is retained [in original condition] for comparison purposes.'*

Three days later MAP issued order No.151s, which included the following item:

'In order to ensure that the government assignment concerning the B-4 four-engined bomber is fulfilled on schedule and with appropriate quality,

1. NISO Director [Maj.-Gen. Nikolay I.] Petrov shall organise and conduct tests of all the B-4's equipment to ensure conformity with the [American] samples and the specifications; [...]

5. LII Director [Prof. Aleksandr V.] Chesalov shall organise and conduct tests of all the B-4's equipment items on the B-29 at LII in accordance with a special list compiled by NISO and endorsed by Vice-Minister of Aviation Kuznetsov.'

In accordance with the MAP order dated 23rd March 1946, all items of equipment manufactured for the B-4 were checked against

Left: A huge 'exhibition' was staged on two floors of the TsAGI building, showing the components of the B-29, with specific R&D institutions and plants being assigned responsibility for each of them. The equipment items have already been allocated Soviet designations. Here, right to left, are the stands of OKB-something-or-other; MAP's experimental plant No. 110 responsible for the AP-5 autopilot; NII-something-or-other; and NII-17 which, among other things, was responsible for the Bariy IFF transponder (*not* a copied item, by the way).

Below: Another section of the display on the ground floor.

Opposite page:

Top left: A mock-up of the fuselage cross-section, showing three bomb cassettes; the center cassette is attached to U-shaped cross-pieces curving around the crawlway which passes over the bomb bay.

Top right: One of the remote-controlled dorsal barbettes with the fairings removed, showing the ammunition boxes for the 12.7-mm machine-guns.

Center: This stand showed how a cylindrical piece of metal was stamped into shape, step by step, to become one of the B-4's mechanical components.

Bottom left: An improvised discussion room with a large-scale cutaway drawing of the B-4 conveniently pinned on the wall for reference purposes.

Bottom right: An overall view of the display on the ground floor. The stands on the right are for the OKB's laboratories, including Section 963. A large wind tunnel model of the B-4 is suspended from the ceiling, with a full-size EMD mock-up of the bomber at the far end; a guard appears to be posted at the entrance to the mock-up.

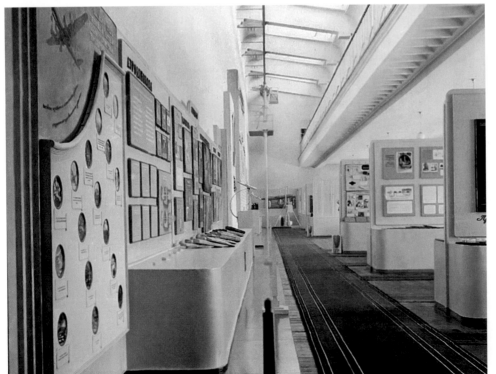

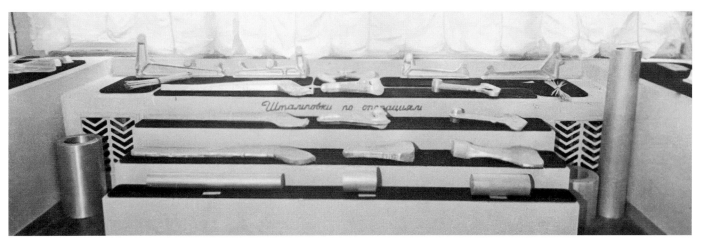

The main gear and nose gear struts are the centerpiece in this part of the display, which also includes tools and new manufacturing technologies mastered by plant No. 156.

the reference aircraft, B-29-15-BW '358 Black' (or 'No.358'), as proposed in Khroonichev's letter. This saved a lot of time and spared the design personnel for a lot of stress, ruling out unpleasant surprises in the process of fitting the equipment to the actual aircraft.

On 27th March MAP let loose with one more order, No.159ss. Among other things, it said: *'In its directive No.472-191ss dated*

A closer look at the wind tunnel model, with armament components arranged under it.

26th February 1946 the Council of People's Commissars has formulated the principal assignments to the Chief Designers [of the various aviation OKBs] and tasked NKAP with concentrating on the most important assignments whose aim is to ensure a major improvement in the performance of prototype aircraft and the creation of new aircraft types.

Pursuant to the CoPC directive, Chief Designer and Director of plant No.156 A. N. Tupolev shall:

1. Complete and deliver to plant No.22 a full set of technical documents for the B-4 four-engined bomber powered by ASh-73 supercharged engines and equipped with pressurized cabins, ensuring the following performance:

- *maximum speed at sea level at combat power, 470 km/h [292 mph];*
- *maximum speed at 10,500 m [34,450 ft], 560 km/h [348 mph];*
- *normal all-up weight, 54,500 kg [120,150 lb];*
- *maximum all-up weight, 61,250 kg [135,030 lb];*
- *range:*
- *at normal AUW with 1,500 kg [3,310 lb] of bombs, 5,000 km [3,105 miles];*
- *at maximum AUW with 5,000 kg [11,020 lb] of bombs, 6,000 km [3,728 miles];*
- *at maximum AUW with 8,000 kg [17,640 lb] of bombs, 4,900 km [3,044 miles];*
- *armament – remote-controlled barbettes:*
- *upper hemisphere, 4 x 20 mm in two barbettes;*
- *lower hemisphere, 4 x 20 mm in two barbettes;*
- *rear, 3 x 20 mm;*
- *internal bomb load: normal 1,500 kg, maximum 8,000 kg.*

The delivery deadline of the technical documents is 15th March 1946.'

It deserves mention that item 2 of the same MAP order tasked the Tupolev OKB with developing an indigenous bomber powered by four Mikulin AM-46TK engines that was to be superior in performance to the B-4. The performance included a maximum speed of 497 km/h (308 mph) at sea level and 605 km/h (376 mph) at 9,600 m (31,500 ft), a normal AUW of 50,000 kg (110,230 lb), a maximum-AUW range of 5,470 km (3,399 miles) with 5,000 kg of bombs and 1,000 km (621 miles) with a maximum bomb load of 18,000 kg (39,680 lb). The defensive armament was to comprise ten 23-mm cannons in two dorsal barbettes, two ventral barbettes and a tail turret. However, this project never materialised due to the B-4 enjoying higher priority.

In accordance with Stalin's instructions no deviations from the American patterns were tolerated in any of the parts reproduced. For this reason the OKB's structural strength specialists also had to tackle a 'reverse engineering' task: they were expected to issue specifications to the metal industry regarding the new alloys to be created, proceeding from the actual dimensions of the parts and strength characteristics of their materials, not vice versa. Of course, the dimensions of all components of the B-29 were expressed in feet and inches; when converted to the metric system, they were approximated to the standards adopted within this system. Later, detractors cracked a joke that Andrey N. Tupolev had been awarded the Hero of Socialist Labour title (the civilian equivalent of the Hero of the Soviet Union title awarded to military personnel) and

Right: This part of the display featured a chunk of the wing with an engine nacelle; note the propeller spinner – a feature that was never fitted to the actual B-29 (or the B-4, for that matter) – and the radial struts in the engine inlet/oil cooler air intake. One of the ventral machine-gun barbettes is displayed in the foreground, with the winged logo of OKB-43 which was responsible for reproducing it.

Below: These four stands are those of the Plastics Research Institute named after Mikhail V. Frunze, its experimental plant and the Vladimir Chemical Plant, which were responsible for replicating and producing the various plastics used in the B-29's systems and equipment.

Below right: Another view of the wind tunnel model, showing the high gloss finish.

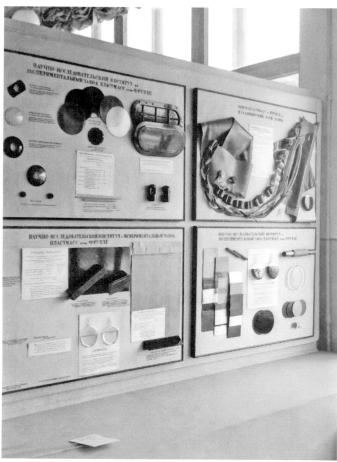

the Stalin Prize for converting the screw threads from inches to the metric system. In reality, however, Tupolev and his team had to shoulder the extremely heavy burden of co-ordinating the activity of different industry branches involved in the program and bringing them up to the state-of-the-art level of technology and production processes. After analysing the B-29's airframe Aleksey M. Cheryomukhin, the OKB's chief specialist in structural strength issues, came to the conclusion that the aircraft had been designed with substantially lower ultimate loads in mind than those prescribed by the Soviet structural strength standards (SSS) then in force. As a result, the SSS were reworked and approximated to those adopted in the USA; subsequently this facilitated the design work on Soviet heavy aircraft.

One phrase in the American manuals baffled the Soviet engineers: *'Start the putt-putt'*. None of the engineers could recall coming across this term before; the team sifted through loads of aviation literature but to no avail. The mystery was solved by pure chance – much later, when the first Soviet-built B-4 was being prepared for flight. Someone cranked up the M-10 emergency generator/auxiliary power unit in the rear fuselage, which was driven by a two-stroke engine, and the latter started emitting an unmis-

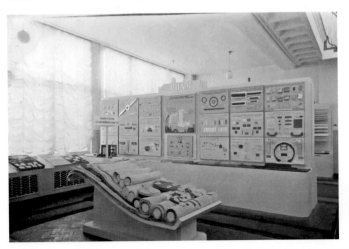

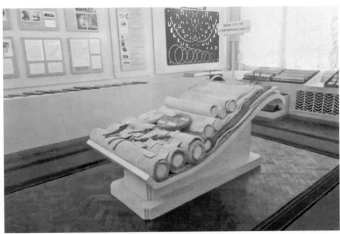

Opposite page, anti-clockwise from top:

More views from the B-29 reverse-engineering display at TsAGI. This is the stand of the Ministry of Rubber Industry, showing various rubber seals; a complete mainwheel tyre and half-finished versions of the main-wheel and nosewheel tyres are in the foreground.

A stand showing the B-29's radio navigation equipment. The faired loop aerial of the Bendix SCR-269G automatic direction finder can be seen, among other things.

Another stand of NII-17, showing the Kobal't radar copied from the Western Electric AN/APQ-13.

The stand of the All-Union Institute of Aviation Materials (VIAM) responsible for various structural materials.

The stand on the other side of the same bay was that of the Ministry of Light Industry (that is textiles, shoes and the like), which was to reproduce the special fabrics used in the B-29.

This page:
Top right: Pictures from the the abovementioned album showing some of the B-29's equipment items copied and produced by plant No.327: the BC-357 marker beacon receiver (this is the original American designation, BC standing for 'basic component') and the BC-733 blind landing system receiver. As on the other sheets, product specifications are given in a box. The BC-357 had a fixed 75-MHz operating frequency and a horizontal dipole aerial; the BC-733 had six wavebands ranging from 108.3 to 110.3 MHz and a strake aerial.

Above right: More copied B-29 components from the same album The SCR-269G ADF was copied by plant No.528 as the *Aist* (Stork).

Right: Still more copies (this time manufactured by plant No.197), namely the Collins AN/ART-13 communications radio (called *Berkut*, Golden eagle, in the USSR). Left to right: the extension coil, the trailing wire aerial control panel, the transmitter and another control panel. The 90-kW transmitter operated in 200-1,500 kHz and 2.0-18.1 MHz wavebands, in telegraph and voice modes, and had 11 preset channels.

ПРИБОРЫ СЛЕПОЙ ПОСАДКИ
ЗАВОД 327

МАРКЕРНЫЙ ПРИЕМНИК
ТИПА ВС-357

ПРИЕМНИК СЛЕПОЙ ПОСАДКИ
ТИПА ВС-733

Сигнализирует прохождение самолетом маркерного передатчика.
Рабочая частота - 75 мгц.
В полете н е управляется.
Антенна-горизонтальный диполь.

Диапазон - 108,3 ÷ 110,3 мгц.
Число фиксированных волн - 6.
Управление для смены волн в полете - дистанционное.
Антенна - шлейф-диполь, установлена на фюзеляже.

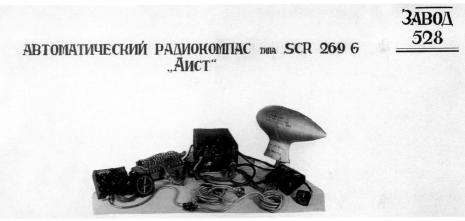

АВТОМАТИЧЕСКИЙ РАДИОКОМПАС ТИПА SCR 269 6
„Аист"
ЗАВОД 528

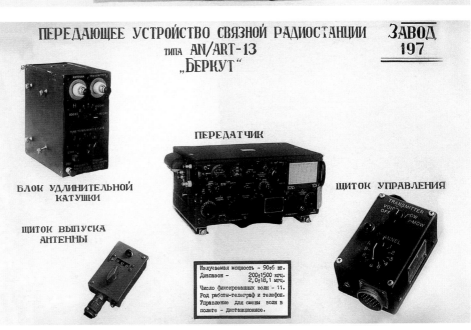

ПЕРЕДАЮЩЕЕ УСТРОЙСТВО СВЯЗНОЙ РАДИОСТАНЦИИ
ТИПА AN/ART-13
„БЕРКУТ"
ЗАВОД 197

ПЕРЕДАТЧИК

БЛОК УДЛИНИТЕЛЬНОЙ КАТУШКИ

ЩИТОК УПРАВЛЕНИЯ

ЩИТОК ВЫПУСКА АНТЕННЫ

Излучаемая мощность - 90±6 вт.
Диапазон - 200±1500 кгц.
2,0±18,1 мгц.
Число фиксированных волн - 11.
Род работы-телеграф и телефон.
Управление для смены волн в полете - дистанционное.

Above: The aforementioned scale model of the B-4 in TsAGI's T-101 wind tunnel. The electrically-driven propellers are not fitted here.

Below: The Shvetsov ASh-73TK 18-cylinder radial engine.

Below: The TK-19 exhaust-driven supercharger copied from the General Electric B-11.

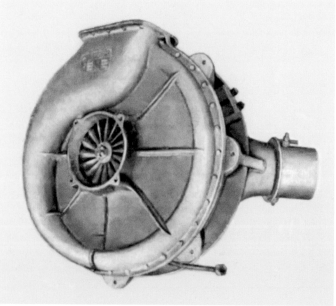

takable *putt-putt-putt-putt-putt...* The Soviet engineers may have been well versed in English aviation terminology, but nobody had counted on aviation *slang*!

The preparation of manufacturing drawings for the Kazan' aircraft factory was preceded by a systematic disassembly of B-29 '365 Black'; this work proceeded under the guidance of the Tupolev OKB's chief technologist Semyon A. Vigdorchik. The airframe was dismantled into major subassemblies, to each of which a special team of designers and production engineers was assigned. Each subassembly was placed in a purpose-built jig lined with felt to prevent damage, weighed, measured, photographed and given a detailed written description. Then all detachable equipment was removed, revealing the basic framework, but not before all piping and wiring runs had been duly documented; each piece of equipment was likewise weighed, measured, photographed etc. Finally, the skin panels were removed, leaving only the internal structural members inside the jig. All parts were subjected to spectral analysis in order to determine the material. 40,000 A4-size sheets of technical drawings were issued on the basis of that research. Preparation of the drawings was completed in March 1946.

Almost all items of equipment removed from the airframe, totalling 350 units, were sent for study and copying to specialised design bureau. The plaque in the flight deck commemorating Gen. Henry H. Arnold's visit to this aircraft was removed and kept as a souvenir by Leonid L. Kerber; later on he gave it to

A ground test rig for the ASh-73TK with a complete engine nacelle and a section of the wing. The V3-A3 propeller is feathered.

Maximilian B. Saukke, another Tupolev OKB designer who also took part in the work on the Tu-4.

Here it is worth mentioning that a major problem arose at this stage: which enterprise should this or that item be given to for reproduction? It was not the OKB's concern to pick them –

Two views from a LII test report showing B-29-5-BW 42-6256 (still with 'Ramp Tramp' nose art and the crew number K-59) during tests of the ASh-73TK, which was installed in the starboard inboard nacelle. Most of the armament has been removed.

otherwise the OKB would be swamped in bureaucracy. However, Andrey N. Tupolev was not only a talented aircraft designer but also a brilliant organiser. He went to Vice-Chairman of the Council of Ministers Anastas I. Mikoyan (the elder brother of aircraft designer Artyom I. Mikoyan), suggesting that someone with personal responsibility for the B-4 program be appointed in each of the ministries involved. Mikoyan agreed. Thus was born – probably for the first time in Soviet practice – an informal inter-department body whose job was to sort out the problems arising between the ministries and cut through the red tape. The 'curators' from the various ministries started visiting the Tupolev OKB, accompanied by representatives of specific plants; the latter were given specimens of equipment, together with their detailed specifications, and a delivery schedule for the copied items. Furthermore, Tupolev established a special bureau whose job was to keep track of the items subcontracted out; as a result, not a single piece of

equipment was lost. This sort of co-operation continued successfully after the B-4 program.

The B-4's powerplant issue was resolved fairly easily. There was no need to copy the B-29's Wright R-3350-23A Duplex Cyclone engines outright because Arkadiy D. Shvetsov's OKB-19 had a suitable engine available. In the early 1930s the Shvetsov OKB had organised licence production of the Wright R-1820-F3 Cyclone nine-cylinder radial as the 635/700-hp M-25, and from then on further development of the 'Soviet Cy-Clone' proceeded independently. Back in 1937 OKB-19 had brought out the 1,400/1,500-hp M-25D18 – an 18-cylinder two-row radial based on M-25 components, hence the D18 (for *dvookhryadnyy* – two-row). Soon redesignated M-70, this engine remained in prototype form but was progressively developed further into the 1,700/2,000-hp M-71 built in small numbers (plus uprated versions delivering up to 2,200 hp), the supercharged M-72 and, finally, the ASh-73 developed in

Two views of the first B-4 (c/n 220001) at Kazan'-Borisoglebskoye during initial flight tests.

1944-45. In its original naturally aspirated form the ASh-73 delivered 2,400 hp for take-off and 2,000 hp at maximum continuous rating. This engine, which entered production at plant No.19 in Perm', was thus a 'cousin' of the Wright R-3350, having been evolved independently from the original Cyclone in the same fashion. The two engines had an identical bore of 155.6 mm (6⅛ in), however, the ASh-73 had a longer stroke – 169.9 mm (6¹¹⁄₁₆ in) versus 160.2 mm (6⁵⁄₁₆ in) – and an accordingly bigger displacement of 58.122 litres (3,546.8 cu in) versus 54.86 litres (3,347 cu in).

However, when the R-3350 became available, the initial ASh-73 *sans suffixe* was developed further into the supercharged ASh-73TK, which differed in being fitted with twin TK-19 exhaust-driven two-speed superchargers (**toor**bo**kompres**sor – turbosupercharger) and a single-speed geared centrifugal supercharger to improve high-altitude performance. The TK-19 with its RTK-46 control system (*regoolyator **toor**bokom**pressora*), was copied from the General Electric B-11 supercharger fitted to the American engine; the magnetos and the heat-resistant main bearings were copied as well. This Soviet-American symbiosis resulted in the rapid creation of a powerplant which fully met the requirements of the Tupolev OKB. The ASh-73TK delivered 2,400 hp for take-off, with a combat rating of 2,200 hp up to the altitude of 8,000-8,700m (26,250-28,545 ft) and a nominal rating of 2,000 hp at 2,400 rpm up to 8,600-9,300 m (28,215-30,515 ft). (It may be mentioned at this point that the engine was put through its paces on B-29 '256 Black'; more will be said about it in Chapter 4.)

In contrast, copying the equipment proved to be a far greater challenge. By the day's standards the B-29 was chock full of sophisticated equipment, much of which had no Soviet analogues. This included the B-11 superchargers with all-mode regulators, the use of electronics (amplifiers) in the engine control and cockpit pressurisation systems, electrically actuated Fowler flaps and

Two more views of the first Kazan'-built B-4. Unlike subsequent examples, this aircraft did not carry the construction number visibly.

landing gear, automated fuel system controls, the complex defensive machine-gun installations with a central fire control system. The avionics included an electric autopilot, the AN/APQ-13 bomb-aiming/navigation radar, the rear warning radar, the ILS, high-range and low-range radio altimeters etc. Also, the capacious pressurized cockpits, two of which were connected by a crawlway, were quite unlike anything seen on Soviet bombers until then.

The leaders of some design bureau responsible for specific units and systems of the B-4 persistently shied away from copying the American prototypes, claiming that the equipment developed in their own OKBs was in no way inferior and was already series-produced into the bargain. However, it was perfectly clear to Tupolev that, should one relinquish the principle of strict reproduction, this would entail a host of co-ordination problems, jeopardising the completion of the work on schedule. Both Stalin and the government shared Tupolev's opinion. Tupolev resorted to an unusual step: at his initiative an exhibition was arranged in the mock-up hall of the TsAGI building in Moscow. Virtually all the units and equipment items of B-29 '365 Black' were on display, accompanied by notice boards specifying the ministries, enterprises, stipulated delivery dates and names of the officials directly responsible for reproducing this or that item. Members of the Politbureau (Political Bureau – the top policy-making organ of the Communist Party's Central Committee – *Auth*.), ministers, chief designers and factory directors were invited to visit the exhibition. It became a peculiar kind of tool for putting pressure on the enterprises responsible for the equipment, making it possible to take swift punitive measures against those chief designers or factory directors who displayed negligence or reluctance – failure to meet the component delivery deadline would be reported to Stalin. Thus, the exhibition resulted in a decision to remove from his post chief designer M. Oikher who had consistently criticised the American prototype of a certain instrument and refused to copy it, proposing that an instrument developed by himself be used instead.

Subsequent events proved this policy to be the right one, as it boosted the aviation industry and the subcontractor industries involved in the program to a qualitatively new level. The pressure brought to bear on the subcontractors yielded the desired result; deliveries of equipment for production B-4 bombers were made on schedule and, most importantly, the Soviet-produced copies of subassemblies and equipment items were not overweight compared to their American prototypes. When the first B-4 was completed, its empty weight was 35,270 kg (77,770 lb) versus 34,930 kg (77,020 lb) for the B-29; the weight penalty did not exceed 1%, which was a remarkable result.

Major difficulties stemmed from the need to switch from the Imperial units of measurement used in the USA to the metric system used in the Soviet Union. Changing nuts and bolts and other threaded connectors was easy, but skin panel thicknesses and wire cross-sections were a real problem – the Soviet metallurgical industry had no equivalents to the American items. For example, one of the duralumin sheet gauges used on the B-29 was 0¹⁄₁₆ in, which equals 1.5875 mm. The Soviet industry was not prepared to provide that kind of accuracy; the nearest local equivalent was 1.75 mm (0.06889 in). Using standard 1.75 mm gauge duralumin sheet meant incurring a weight penalty, with an attendant reduc-

tion in flight performance, whereas reducing the gauge to 1.5 mm (0.059 in) was justifiably vetoed by the OKB's structural strength department, as it would compromise structural integrity. Now, why on earth is it possible to manufacture rolled duralumin sheet if the thickness is called 0¹⁄₁₆ in, but impossible to do it if the same thickness is called 1.5875 mm? This kind of nonsense could send you up the wall.

Even the crew parachutes proved to be an issue. The American ones were designed so that the packed parachute acted as a padded backrest; in contrast, the standard Soviet ones acted as a cushion fitting into a dished seat pan. Which model of parachute and seat should be used? Copying the American parachutes would be nonsense, as it meant creating an unnecessary logistical headache; as Leonid L. Kerber put it, *'everyone agreed that this was stupid, but demanded a signed document authorising them to be clever'*.

There were only two instances where Stalin's order to copy everything strictly was not observed. Firstly, this was the case with the IFF equipment – which was quite natural, since it was the American air defence system that the B-4 was to penetrate. Actually, this was a case of even greater nonsense; everyone was so afraid to challenge Stalin's explicit instructions that the B-29 be copied in full that at first the electronics plant responsible unthinkingly copied... the American SCR-695 IFF transponder and supplied it to the OKB. Imagine a Soviet bomber (with American IFF!) being interrogated by a Soviet fighter (with Soviet IFF) which, not getting a 'friendly' response, would regard the bomber as hostile and attack it! Eventually, of course, common sense won.

Fortunately, even then there were people in the industry who were cannier than the 'machines for fulfilling instructions' around them and realised that even Stalin's assignments, however important they be, should not be fulfilled mindlessly. It turned out that the B-25J bombers delivered under the Lend-Lease Agreement featured an SCR-522 VHF command radio with automatic tuning which was more up-to-date than the ubiquitous Aircraft Radio Corporation SCR-274-N HF command radio fitted to the B-29. Logic suggested that it was the newer model that should be copied and fitted to the B-4 – but, again, logic was up against *ipse dixit*. Kerber had a talk with Vice-Minister of Radio Industry I. G. Zoobovich, suggesting that the SCR-522 be copied because a switch to VHF communication was inevitable anyway. After pondering for a few seconds, Zoobovich said: *'As far as my close associates are concerned, I know them, and I have no fear of them. But there's bound to be some son of a bitch from outside who will report on us, writing to the Big House* (the NKVD headquarters in Moscow, which was accommodated in a huge building – *Auth*.) *that we are disobeying the Leader's orders, and we'll all be jailed – that's for sure. No, that's not the way to do it; what we need to do is enlist the support of a high-ranking and upstanding man in the military who will explain if summoned to the higher command – even to Stalin if necessary – that the changes were made for the success of the matter, not out of selfish opposition.'* They succeeded in finding such a supporter – Sergey A. Danilin, the chief of a military research institute, who had flown a long-distance record-breaking mission in a Tupolev ANT-25 before the war. This was no vain precaution because, ever so often, even in the most close-knit collective there is a rat – especially under a totalitarian regime, and if someone

These three views show B-4 c/n 220001 on the hardstand at the LII airfield in Zhukovskiy. B-29 '256 Black' is visible beyond.

The pilots who test-flew the first Tu-4s. Top row, left to right: Nikolay S. Rybko (captain of Tu-4 c/n 220001), Aleksandr G. Vasil'chenko (captain of Tu-4 c/n 220101) and Mark L. Gallai (captain of Tu-4s c/ns 220002 and 220405). Bottom row, left to right: Kazan' aircraft factory test pilot Nikolay N. Arzhanov (captain of Tu-4 c/n 220402), Vyacheslav P. Maroonov (captain of Tu-4 c/n 220102) and Aleksey P. Yakimov (captain of Tu-4 c/n 220103).

reported this act of disobedience the consequences could be fatal. It was the same story in the abovementioned case of the parachutes, when a general in the Air Force's parachute service agreed without hesitation to sign a paper authorising the use of Soviet parachutes, and in numerous other cases.

Secondly, the B-4's armament differed from that of the B-29. The Soviet military and the defence industry were unanimous that the entire armament should be indigenous – for procurement reasons if nothing else. The 12.7-mm (.50 calibre) Browning M2 machine-guns were initially substituted with indigenous Berezin UB-12.7 machine-guns, with the intention of replacing them with 23-mm (.90 calibre) Nudelman/Rikhter NR-23 cannons; on the other hand, the central fire control system (CFCS) of the B-29 was

retained. This decision was justified: a more potent cannon armament considerably enhanced the B-4's defensive capabilities compared to those of the B-29. Yet, again, one thing leads to another: the ballistic parameters of Soviet and American bombs and projectiles were different; the American bomb shackles were incompatible with Soviet bombs and had to be replaced; the same applied to the ammunition boxes and sleeves etc. Nobody had thought about it because the directives concerning the copying of the B-29 had been prepared in great haste. It took a lot of arguing and wrangling before the required papers were signed – and a lot of signatures there were!

(As an aside, subsequently the Korean War where Soviet-produced MiG-15 fighters clashed with B-29s gave indirect proof that the decision to switch to an all-cannon defensive armament for the

B-4 was correct; the MiGs could often make a firing pass at a B-29 and inflict great damage while staying outside the effective range of its machine-guns. Had the B-29s been armed with cannons, the fighter pilots' mission would have been far more difficult.)

Speaking of avionics, the old SCR-274-N radio was copied after all – at any rate, early-production Tu-4s had it; even the original designation was retained, albeit in a Russian transcription. The AN/APQ-13 radar was copied under the designation *Kobal't* (Cobalt); the work was undertaken by NII-17, an avionics house headed by Chief Designer Viktor V. Tikhomirov (A. I. Korchmar was the project chief), and the design office of the Leningrad Electromechanical Plant No.283. The Norden telescopic bomb sight linked to a vertical gyro was copied as the OPB-4S (*opticheskiy pritsel bombardirovochnyy, sinkhronnyy* – optical bomb sight, synchronised).

It proved to be an arduous task for the Soviet industry to master production of American materials, especially the non-metallic ones: plastics, high-quality rubber and synthetic fabrics. This was due to the Soviet chemical industry generally lagging behind the world standards of the day. As a rule, factories and research institutes persistently proposed that the American materials be replaced by traditional ones well-established in Soviet production, such as bakelite, compressed impregnated cellulose fibre, Plexiglas and so on. Yet Tupolev and his deputies were adamant; they understood full well that failure to use new materials would result in a substandard aircraft, and insisted that the subcontractors strictly abide by the specification.

By mastering the whole range of the B-29's design features, materials and technologies the Soviet aircraft industry and associated industry branches created a potent material basis which enabled the Soviet aviation to attain world standards by the early 1950s, and then to surpass them on some counts. The work effected then was so wide-ranging and anticipatory in its scope that to this day instruments or equipment items that were put into production in 1946-47 and were either modelled directly on those of the B-29

or represented their further modernisation can be found in Soviet/Russian-built aircraft.

B-29-5-BW '256 Black' (referred to in paperwork as 'No.256'), which was in airworthy condition, was test-flown to check out the aircraft's controllability, powerplant operation and obtain more precise performance data. In the autumn of 1945 it made a visit to Kazan'-Borisoglebskoye, the factory airfield of plant No.22, giving the employees an opportunity to examine the aircraft firsthand. At the same time the factory flight crews that were expected to test the first B-4s started their conversion training; this work was supervised by test pilot Maj. Vyacheslav P. Maroonov, who by then had been transferred from the Naval Aviation to the Tupolev OKB as a test pilot.

OKB-156 was expanded to suit the needs of the B-4 program. Not only did it obtain authority over the hangar at Khodynka where the dismantling and examination of B-29-15-BW '365 Black' took place; it came to include the personnel of OKB-22 – the Kazan' aircraft factory's design office led by Iosif F. Nezval', which became a branch office of the Tupolev OKB. The latter also came into possession of a supplementary production facility in central Moscow near Paveletsky Railway Station.

The Kazan' branch of the Tupolev OKB undertook a considerable portion of the work. Making use of the components of '365 Black' delivered from Moscow, the personnel prepared manufacturing drawings of the landing gear doors, bomb bays, crew seats, outer wing panels, the trailing-edge portion of the wing center section, the stabilisers, elevators, fin, rudder and the gunners' pressurized cabin. The drawings were completed by 1st January 1946; in all, the Kazan' branch issued 1,383 drawings and 657 specifications. It also prepared theoretical outline mold lofts for the B-4.

In December 1945 the Kazan' branch began reassembling B-29 '365 Black'. Yet, this work was not completed; the aircraft never became its former self again because a decision was taken to use its wings, engine nacelles, tail surfaces and main gear units

Clad in his Lieutenant-General's uniform and wearing the Gold Star medal that went with his Hero of Socialist Labour title, Chief Designer Andrey N. Tupolev has a discussion concerning the B-4 with Long-Range Aviation Commander Air Marshal Aleksandr Ye. Golovanov.

for building the prototype of an airliner derivative of the B-4 – the Tu-12 ('aircraft 70'), which is described in Chapter 9.

The actual bombers available to the Soviet designers were not the only source of information used for cloning the Superfortress. Good old intelligence was as important as ever, and priceless technical information on the B-29 was furnished by Soviet undercover agents abroad – and the local talent they recruited. These included first and foremost N. M. Gorshkov, a Soviet diplomat posted in Rome, who was chief of the Soviet intelligence network in Italy. In the mid-1990s the KGB of the Republic of Tatarstan (yes – the security/counter-intelligence service of this constituent republic of the Russian Federation is still called KGB) and the Russian Federal Security Service (FSB) declassified certain documents shedding light on how the technical documents on the B-29 were obtained and featuring Gorshkov's name.

Once he had taken the decision that the B-29 should be copied, Stalin summoned Lavrentiy P. Beria (who, apart from being People's Commissar of the Interior, co-ordinated several key military programs) to the Kremlin. He ordered Beria to task Soviet intelligence agents abroad with obtaining whatever information on the B-29 it was possible to get. This is how N. M. Gorshkov recalled this assignment:

'I had to think about this one real hard. After all, I could not go stealing blueprints personally, could I? And without them it was impossible to understand the design, even if you had the actual aircraft. To top it all, Boeing had no aircraft production facilities in Italy. So, I had meetings with all my Italian informers – and then I hit the jackpot! One of the Italians had a brother, an aircraft engineer, who had emigrated to the US and got a job with Boeing. So I asked my informer, "Can you go to America and visit him?" "Gladly – he told me – but this is gonna cost a lot of money." So it boiled down to money. Now, bureaucracy is the same all over, and the intelligence community is no exception. I had the required sum, but in order to pay out such a large amount legitimately I had to file a report, stating the purpose, and obtain permission from the bosses in Moscow. That would take up precious time, which I could not afford to. So I just took the money from my funds and paid the man, and filed the report afterwards, putting a different date on it. A while later the man came back from America with two suitcases chock full of blueprints and manuals. My friends in the trade, who knew of this operation, joked afterwards: "See, Andrey Nikolayevich [Tupolev] has three Hero of Socialist Labour stars? One of them should rightfully be yours!" (The star in question is the Gold Star Medal that goes with the Hero of Socialist Labour title – *Auth.*)

Construction of a full-size engineering & manufacturing development (EMD) mock-up of the B-4 began in mid-1946. All design faults and discrepancies that came to light were rectified and appropriate corrections were immediately made to the working drawings in the presence of a supervisor; the corrections were checked out on the mock-up and corrected drawings were sent to the production plant No.22 in Kazan'. In this way it proved possible to reveal most of the mistakes which were caused by the slam-bang tempo of the work and by fatigue suffered by the personnel who had only one day off per month. When the mock-up was completed, all of its hatches were provided with locks and no one could access the inside of the mock-up without permission from Andrey N. Tupolev or Dmitriy S. Markov, the chief project engineer. This was done to preclude any attempts to undertake modifications without putting them on record in the production drawings.

At the same time a large-scale wind tunnel model of the B-4 was produced for TsAGI; it was used for studying the peculiarities of the aircraft's aerodynamics in the T-101 wind tunnel. The model had a wingspan of 18 m (59 ft $0^{21}/_{32}$ in) – only a fraction less than a full-size Tu-2 bomber (18.86 m/61 ft $10^{33}/_{64}$ in) – and was equipped with four electric motors driving the propellers.

On 29th June 1946 MAP issued yet another order, No.413ss. This basically confirmed the speed and range targets stipulated in the earlier orders but stated that range was to be 3,000 km (1,863 miles) with a maximum bomb load of 7,120 kg (15,700 lb). More could not be carried because then-current M-44 series (1944 standard) bombs were too bulky. The order also mentioned that production B-4s were to be capable of carrying new, shorter M-46 series bombs of the same calibres, which would allow the maximum bomb load to be increased to 9,000 kg (19,840 lb) – that is, three FAB-3000 bombs (*foogahsnaya aviabomba* – high-explosive bomb).

Shortly thereafter, on 9th July, the Council of Ministers issued directive No.1534-682ss, setting the B-4's production schedule. The Kazan' aircraft factory was to manufacture the first three bombers on 30th September, 30th October and 30th November 1946 respectively. Apparently this means the completion/rollout dates.

Meanwhile, the Kazan' aircraft factory was gearing up to build the B-4. By 5th March 1946 (ten days ahead of the deadline) the Tupolev OKB had completed the manufacturing drawings and documents, and by 1st April most of them had been issued to the plant. The B-4's airframe manufacturing technology was based on the use of mold lofts and was completely new for plant No.22; quite different production methods had been used earlier at Kazan' for the Pe-2 (NATO reporting name *Buck*) and Pe-8 bombers. Moreover, a number of workshops had to be rebuilt and re-equipped, as much of the equipment and machine tools was obsolete and worn out, and numerous new production technologies had to be mastered.

This immense, thorough and carefully planned work culminated in the completion of the first prototype B-4 (c/n 220001 – that is, Plant No.22, Batch 00, 01st aircraft in the batch) in February 1947 – ten months after the factory had received the manufacturing drawings. Since the type had the product code *izdeliye* R, this particular aircraft was known as R-01. Strictly speaking, despite this batch number, there was no prototype because the first Kazan'-built B-4 was completed to production standard and thus can be regarded as the first production machine. Bearing no serial, the aircraft was rolled out at Kazan'-Borisoglebskoye on 28th February. Adjustment and development work went on until mid-May, whereupon the aircraft was ready for the maiden flight. A mixed crew was assigned to the bomber on 19th May; R-01 was captained by LII test pilot Nikolay S. Rybko, with Kazan' factory test pilot Aleksandr G. Vasil'chenko as co-pilot and V. N. Satinov (LII) and Ya. S. Osokin (plant No.22) as engineers in charge of the tests. The crew included flight engineers Shestakov (LII) and A. G. Andreyev

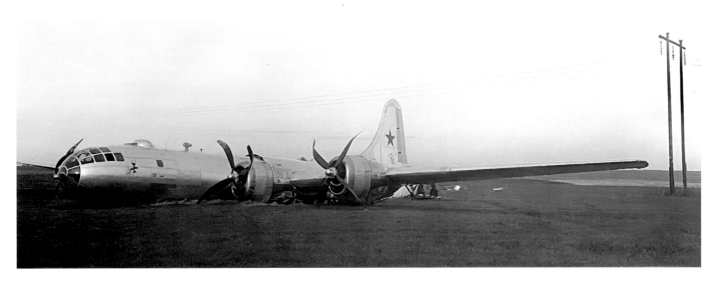

(plant No.22), mechanics Antonov (LII) and M. I. Nigmetzyanov (plant No.22) and radio operator A. V. Krasnov (plant No.22).

No matter how the local counter-intelligence officers tried to conceal the date of the first flight, spreading deliberate misinformation, the correct date was circulated on the grapevine all around the factory. On 21st May 1947, when the first B-4 became airborne at 1506 hrs Moscow time, there was not a soul in the workshops

Top and center: Tu-4 c/n 220101 after crash-landing in a field near Kolomna on 18th September 1947 due to a fire in the No.3 engine. The aircraft broke its back on impact and was a write-off. Note how the bomber narrowly missed a high-voltage power line.

Above left: This photo shows how the fuselage broke in two aft of the wings.

Above right: The twisted remains of the No.3 engine nacelle; the engine is gone, having broken away in flight.

– the entire workforce was out to watch the event. The rooftops of the factory buildings, the lawns of the factory airfield, even the roads leading to the airfield were packed with crowds of cheering people who were proud – and rightly so – to see the results of their hard work.

The maiden flight lasted 34 minutes. The R-01 made a few more development flights at Borisoglebskoye. Two weeks later the aircraft was ferried to the LII airfield at Ramenskoye for further testing.

At the end of June the second production B-4 serialled '202 Black' (aircraft R-02, c/n 220002) was rolled out; during its first flight it was captained by LII test pilot Mark L. Gallai, with factory test pilot Nikolay N. Arzhanov as co-pilot/engineer in charge. The first flights of the first two B-4s took place in the presence of senior OKB-156 staff, including Andrey N. Tupolev. Later, Gallai flew the R-02 (with Tupolev and Kerber aboard,

among the 'passengers') to Moscow. This flight featured a funny episode. En route the heating system created such a tropical climate inside the aircraft that the occupants had to strip to their underwear (it proved impossible to reduce the temperature), and the ceremonial meeting at Ramenskoye was spoiled – the men needed time to dress before they could deplane!

'When the Kazan' factory turned out the first batch of Tu-4s – Gallai wrote later in his memoirs – *the first machine was accepted by Rybko and Shoomeyko (sic – Auth.). I was assigned to the second aircraft, and before departing to Kazan' I made one more flight in the very same "Ramp Tramp" to refresh my skills. Even though Tu-4 No.2 was a copy of the B-29, there was a difference – and, unfortunately, not in the Tu-4's favour. The view from the cockpit, which was hampered by the closely spaced glazing framework as it was, was further spoiled – and quite a lot – by the glazing which distorted everything you looked at through it. Dur-*

Two views of '202 Black' (c/n 220002), the second Kazan'-built B-4.

ing the approach, you would look through one "window" and it seemed you were too high [above the glideslope]; looking through another, you got the impression you were too low. Jumping ahead of the story, I may say that the last aircraft of the development batch – the "twenty" ('20 Black', c/n 220205 – Auth.), which my crew likewise tested – suffered from the same deficiency; our industry had not yet mastered the knack of manufacturing curved transparencies that did not distort the view. The Tu-4's other deficiency as compared to the B-29 was the heavy controls, particularly in the roll channel. It's either that the tension of the cable runs was not optimum or that the seals where the cables exited the pressure cabin differed in some way [from those of the B-29] – I don't know. Later, I had the opportunity to fly the B-17, and I was surprised to find that the B-29 was a step backward, not forward, as far as stability and handling were concerned. The B-17 had a better view from the cockpit and was easier to fly. I guess the

transition to pressurized cockpits [on the B-29/Tu-4] had something to do with it.'

In August 1947 the unserialled third production B-4 (aircraft R-03, c/n 220101) was handed over to a crew captained by Aleksandr G. Vasil'chenko; that same month a crew captained by Vyacheslav P. Maroonov took charge of the (likewise unserialled) fourth production aircraft (c/n 220102). In the course of the summer and autumn of 1947 a further ten or so machines were manufactured, test flown and ferried to Ramenskoye.

The fourth production machine (aircraft R-04), again wearing no serial, introduced an important change. The first three aircraft had been armed with eleven UB-12.7 machine-guns (two in each of the dorsal and ventral barbettes and three in the tail turret). The fourth B-4, however, featured a PV-20 defensive weapons system (**push**echnoye vo'oruzheniye – cannon armament) developed under the guidance of the Tupolev OKB's chief armament special-

Two more aspects of the same aircraft.

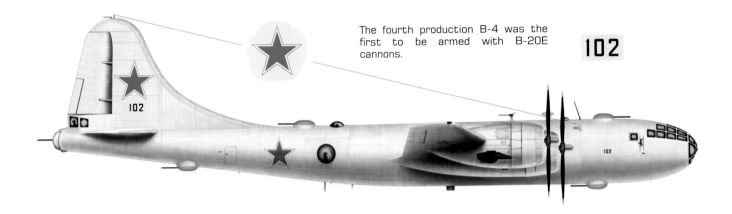

The fourth production B-4 was the first to be armed with B-20E cannons.

102

ist Aleksandr V. Nadashkevich. The original intention was to use the brand-new NR-23 (*izdeliye* 150P) cannons developed by Aleksandr E. Nudelman and Aron A. Rikhter at OKB-16 of the Ministry of Armament based in Tula. The NR-23 was a faster-firing derivative of the proven Nudelman/Suranov NS-23 designed in 1944. Andrey N. Tupolev had suggested using these cannons back when the directive concerning the manufacture of the trials batch of B-4s was being drafted, but the NR-23 was not yet cleared for installation when B-4 c/n 220102 was completed. As the next-best thing, the aircraft was fitted with 20-mm (.78 calibre) Berezin B-20E cannons, which were similar to the UB machine-guns in size and weight.

As mentioned earlier, the government directive required the B-4 to be immediately submitted for state acceptance (≈ certification) trials involving a total of 20 aircraft. Holding the trials on this

Left and below: Tu-4 c/n 220102 – the first example with the PV-20 defensive weapons system – on the flight line of the LII airfield in the winter of 1947-48. The crew entry door is open a crack.

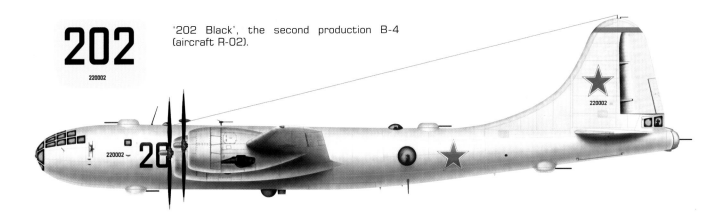

'202 Black', the second production B-4 (aircraft R-02).

scale was expected to speed up obtaining the necessary data on B-4 operation, yielding more comprehensive information which would then be summed up; thus, the test results would be obtained within the shortest possible time, including information on the operational reliability of the aircraft and its entire equipment complement. However, certain difficulties arose in ensuring such a large-scale trials program, unprecedented in the USSR. First of all, it was necessary to select the required number of pilots and engineers and form flight and ground crews. A stock of spare parts had to be provided for the aircraft; repair teams had to be formed; it was necessary to ensure efficient co-ordination with production plants delivering equipment and engines.

A special headquarters, headed by Air Chief Marshal Aleksandr Ye. Golovanov representing the Air Force and Deputy Minister Pyotr V. Dement'yev representing the Ministry of Aircraft Industry, was formed to supervise the trials. Permanently present in the headquarters were representatives of the organisations and plants responsible for the equipment and of the ministries involved in the B-4 program; Dmitriy S. Markov was the permanent representative of the Tupolev OKB. The trials were expected to be completed in 1948, whereupon the aircraft would enter service with the Soviet Air Force.

The best test pilots of MAP and the VVS were entrusted with the testing of the B-4. In addition to the crews captained by Rybko, Gallai, Vasil'chenko and Maroonov, the trials involved GK NII VVS and DA pilots B. G. Govorov, Aleksey P. Yakimov, Stepan F. Mashkovskiy, Fyodor F. Opadchiy, Aleksandr I. Kabanov, Pyotr M. Stefanovskiy, Valerian I. Zhdanov, Vasiliy I. Alekseyenko, Mikhail A. Nyukhtikov, Ivan P. Piskunov, Andrey G. Kochetkov, Afanasiy G. Proshakov, Aleksey G. Koobyshkin, Vladimir Ye. Golofastov, V. V. Ponomarenko, Aleksey D. Perelyot, Konstantin K. Rykov, I. Sh. Vaganov, Mikhail V. Rodnykh and others. The trials took place at the LII airfield.

Before the work on determining the performance of the aircraft and its equipment could be started, much time had to be spent on adjustment and development work. Gradually the aircraft forming the initial production lot of twenty machines were brought up to the full complement of equipment and started flights in accordance with the test program.

Stage A involved three aircraft – '202 Black' (c/n 220002), the unserialled B-4 c/n 220102 and '1 Black' (c/n 220201). The trials

showed that the B-4's normal all-up weight of 47,850 kg (105,490 lb) and maximum AUW of 54,430 kg (120,000 lb) did not meet the figures stipulated in the government directive. Flight performance at maximum continuous power fell short of the target; maximum speed was 420 km/h (261 mph) at sea level and 550 km/h (341 mph) at 9,500 m (31,170 ft), which was the service ceiling at 75% maximum continuous power. On the other hand, range was greater than expected; with a 1,500-kg (3,310-lb) bomb load the range was 5,200 km (3,231 miles). With a maximum bomb load of 7,120 kg (15,700 lb) and 12,300 litres (2,706 Imp gal) of fuel the range was 3,060 km (1,901 miles).

Static tests had shown that the bomber could be flown with a normal all-up weight of only 47,630 kg (105,010 lb); therefore the military demanded that the airframe be reinforced to permit a normal AUW of 54,500 kg (120,150 lb) and a maximum AUW of 61,250 kg (135,030 lb). The crew's armour protection was considered poor, being adequate only against 12.7-mm bullets; the military demanded that the armoured seat backs be made thicker – for instance, the pilot's seat backs should be 15 mm ($0^{19}/_{32}$ in) thick instead of 6.5 mm ($0^1/_4$ in). Considering the number of crew members, this would incur a sizeable weight penalty.

At the initial stage the testing of the first B-4s looked more like development work on certain systems, individual units and equipment items – such as the powerplant (primarily the ASh-73TK engine itself but also the propeller governors), landing gear and flap actuators (faulty functioning of screwjacks was common in the early days and was cured by selecting the correct type of lubricant), electric motors, generators etc. There were many complaints concerning the radar equipment and the synchros in the CFCS.

Despite these faults, the crews which test-flew the B-4 had a fairly high opinion of the aircraft – they recognised its good stability and benign handling, advanced equipment, comfortable and well-designed crew workstations, the provision of service devices which facilitated piloting and eased the crew workload. The B-4 presented no problems for the averagely-skilled bomber pilot versed in night and adverse-weather flying. In the event of a single or dual engine failure, prolonged level flight was possible with an AUW of up to 47,850 kg (105,490 lb); in this case the bomber showed no tendency to roll sharply.

The military also noted such an indisputable advantage as the Kobal't radar – the first of its kind on a Soviet bomber. The radar could detect large cities at a range of at least 90 km (56 miles), smaller towns at 60 km (37 miles) range, railway junctions and bridges at 30-45 km (18.6-28 miles) range, and large lakes and rivers at 45 km range.

However, the pilots also voiced a number of complaints: they were not quite happy with the curved flight deck glazing which distorted the view, the rather heavy controls of the first machines and the lack of a de-icing system on these aircraft. The main drawbacks were eliminated during the flight testing of the initial lot of B-4s.

The first three B-4s took part in the traditional Aviation Day flypast at Moscow-Tushino airfield on 3rd August 1947. Aircraft R-01, R-02 and R-03 were captained by Rybko, Gallai and Vasil'chenko respectively; the co-pilot of the lead bomber captained by Vasil'chenko was none other than Air Chief Marshal Aleksandr Ye. Golovanov. The three B-4s made a dramatic low-level pass over Tushino, producing the desired impression on the government officials and the attending foreign military attachés. Curiously, for a long time the West was under the impression that it was the refurbished B-29s that had been shown at Tushino, being certain that the USSR was unable to copy and put into production such a complex aircraft. Yet, the fact was there: at the price of unbelievable exertion the Soviet aircraft industry had succeeded in mastering technologies of utmost complexity within 18 months, furnishing a first-rate bomber for its Air Force. Later, when the

West learned the truth, the NATO's Air Standards Co-ordinating Committee (ASCC) allocated the reporting name *Bull* to the Tu-4.

After the Aviation Day performance the testing resumed but was soon interrupted by an accident. At 1512 hrs Moscow time on 18th September 1947 B-4 c/n 220101 (aircraft R-03) captained by Aleksandr G. Vasil'chenko departed the LII airfield on a routine test flight; the mission was to check the powerplant temperature at 5,000 m (16,400 ft) and 3,000 m (9,840 ft). Suddenly the No.3 engine caught fire; flight engineer N. I. Filizon activated the fire suppression system and attempted to feather the propeller, but to no avail. At the captain's orders ten of the 12 crewmen bailed out, including co-pilot I. P. Piskunov, navigator S. S. Kirichenko and engineer in charge of the tests D. F. Gordeyev; only Vasil'chenko and Filizon remained on board. Shortly afterwards the No.3 engine fell off as the fire burned through the engine bearer and spread to the leading edge of the wing center section. In this critical situation, realising the wing might fail catastrophically, Vasil'chenko managed a belly landing in a farm field 7 km (4.3 miles) south-west of the town of Kolomna in the south-east of the Moscow Region – barely missing a high-voltage power line in so doing. Scrambling to safety, Vasil'chenko and Filizon promptly attacked the blaze with hand-held fire extinguishers taken from the flight deck; yet it took the assistance of local peasants and a fire engine that appeared on the scene to put out the fire completely. The entire

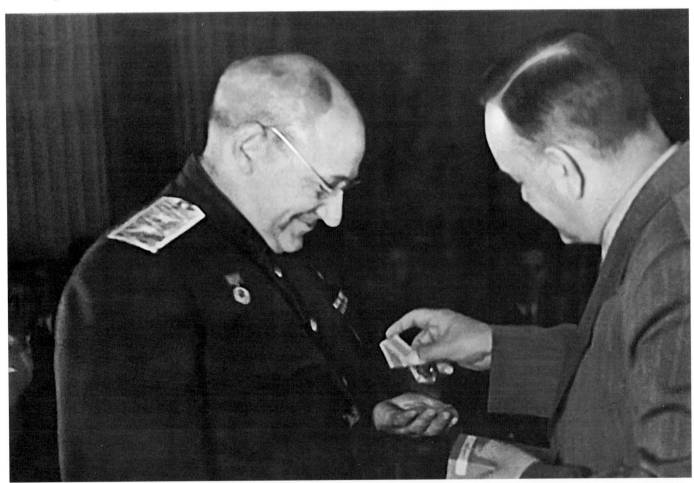

Nikolay M. Shvernik, Chairman of the Presidium of the USSR Supreme Soviet, hands the Order of Lenin to Andrey N. Tupolev. The latter received this award for copying the B-29.

crew was unhurt, but the aircraft was a write-off, having broken its back ahead of the wings; it was eventually stripped for spares.

The second, and far more serious, accident occurred on 5th November 1947 when the 13th production B-4 (c/n 220303) was being ferried from Kazan' to LII's airfield in Zhukovskiy. It was to be part of a large bomber formation participating in a flypast over Moscow's Red Square on 7th November during the annual military parade marking the anniversary of the October Revolution; this was one of the most important public holidays in the Soviet Union, especially because in 1947 it was the 30th anniversary. (Here it should be explained that on 26th January 1918, three months after the revolution, Soviet Russia switched from the outdated Julian calendar to the Gregorian calendar; hence the apparent discrepancy – 25th October, the 'old style' date of the revolution, equals 7th November according to the current calendar.) The crew was captained by Lt.-Col. Mikhail V. Rodnykh (HSU) and included co-pilot Lt. (SG) G. I. Kondrat'yev, navigator Capt. V. Ya. Malykin, engineer in charge of the tests Lt.-Col. B. N. Fedoseyev, propulsion group technician Lt. P. V. Mezentsev, flight engineer Capt. N. I. Bannikov, radio operator MSgt. V. P. Boolatov and six others. (It deserves mention that Rodnykh had been shot down and captured while flying a bombing sortie over enemy territory in 1943, ending up in a German POW camp from which he was liberated in April 1945. Former POWs were generally treated with distrust in the Soviet Union, so it was a small miracle that Rodnykh had been trusted with testing the top secret B-4 bomber.)

B-4 c/n 220303 took off from Kazan'-Borisoglebskoye at 1350 hrs Moscow time. For the next 15-20 minutes it circled over the airfield while the crew checked that all systems were go before the long flight to Moscow; then, satisfied that everything was OK, Rodnykh reported this to the control tower and was cleared to proceed along the planned route. However, 23 minutes after take-off the captain sensed vibration in one of the engines; immediately afterwards the navigator reported a fire in the No.2 engine. As had been the case with Vasil'chenko's aircraft, attempts to extinguish the fire and feather the propeller gave no results. Fedoseyev suggested making a forced landing rather than bailing out; Rodnykh concurred, and at 1415 hrs the crew radioed: *'We have an engine fire, we're landing!'*

Rodnykh chose to land in a field which lay ahead, to the left of the bomber's course. Flying at 400-450 m (1,310-1,480 ft), the aircraft began a landing pattern, extending the landing gear on the downwind leg. However, shortly after the aircraft turned onto the base leg the navigator reported that the No.1 engine was on fire as well. The engineer in charge told him to shut that engine down, not informing the captain of this action. When the aircraft was on short finals the pilots suddenly saw a gully and a farm building straight ahead and applied full power to clear the obstacles. However, because (unbeknownst to the captain) both engines on the port side were dead, the bomber yawed and rolled sharply to the left when the Nos. 3 and 4 engines were revved up. Touching down on the port main gear unit, it swung to the left and crashed into trees at the side of the field, the port outer wing and No.1 engine breaking away. Still under momentum, the aircraft continued to roll forward and struck another tree, collapsing the nose gear unit and shedding the No.2 engine; the forward fuselage failed exactly where the

navigator and the radio operator sat, killing both of them, and broke away from the rest of the airframe. As the aircraft came to rest after swinging 140-150° to the left, its starboard wing smashed by the impact, the fuel from the inboard wing tanks poured onto the hot engines and a huge fire ensued which destroyed the aircraft almost completely, burning for eight hours. Only the tail unit remained intact. This was the only fatal crash of a B-4 (Tu-4) during the tests. Should one doubt the magic power of the number 13 after that?

It should be noted that powerplant problems figured most prominently at the initial stage of the Tu-4's flight test program. For a long time the first B-4s were plagued by engine failures, malfunctioning of engine accessories and failures of the propeller feathering system. In one of the flights a fire erupted in the port inboard engine of a B-4 flown by a GK NII VVS crew captained by Semyon M. Antonov. Cutting off the fuel supply, the crew activated the fire suppression system, then tried to put out the flame by making a sharp side-slip, but to no avail. As the burning bomber limped towards the airfield, the starboard inboard engine started running roughly, while fuel and oil poured from its engine nacelle. Luckily Antonov managed a safe landing; it turned out that the No.2 engine had suffered a failure of the master con-rod, while one of the No.3 engine's TK-19 superchargers had disintegrated, the fragments rupturing the fuel line and the oil tank.

Much trouble was caused by two cases of propeller overspeeding which occurred during descent from maximum altitude. On aircraft R-02 captained by Mark L. Gallai the crew eventually managed to feather the runaway No.4 propeller and the consequences were limited to a wrecked engine. About two weeks later, Vyacheslav P. Maroonov had a far more serious case on aircraft R-04: the reduction gearbox shaft failed and the runaway propeller broke off at great speed, striking the fuselage and the neighbouring engine as it departed – fortunately without causing catastrophic damage. Later, when the bomber was already in wide-scale operational service, there were several cases of the engine mounts failing due to metal fatigue. Step by step these faults were eliminated by the aircraft and engine designers; the engine mounts were reinforced and alterations were introduced into the engines' shock absorption system.

In the autumn of 1947 the testing of the first B-4s for maximum range commenced, two machines (c/ns 220002 and 220102) being used. The two bombers took off simultaneously from the LII airfield and followed each its own route (Zhukovskiy – Sevastopol' – Zhukovskiy for Gallai's aircraft and Zhukovskiy – Sverdlovsk – Chelyabinsk – Zhukovskiy for Maroonov's machine). The first aircraft, flying under adverse conditions (heavy icing and extra drag created by bomb bay doors opened inadvertently due to a crew member's negligence), covered a distance of 2,560 km (1,591 miles). The all-up weight was 49,100 kg (108,265 lb), including a bomb load of 1,500 kg (3,307 lb). The second machine covered a distance of 3,123 km (1,941 miles), staying airborne for 7 hours 39 minutes; the average speed was 400 km/h (249 mph). The flight was performed at an AUW of 50,000 kg (110,250 lb), including a 1,500-kg bomb load.

Another long-distance flight was performed on 25th January 1948 by a pair of B-4s, c/ns 220102 and 220201 (some sources say 220102 and 220103) which were captained by Maroonov and

The first B-4s' flight performance as per LII test report No.47-178	
Maximum speed at a 47,700-kg (105,180-lb) all-up weight:	
at sea level	427 km/h (265 mph)
at 1,000 m (3,280 ft)	442 km/h (275 mph)
at 3,000 m (9,840 ft)	471 km/h (293 mph)
at 5,000 m (16,400 ft)	500 km/h (311 mph)
at 8,000 m (26,250 ft)	543 km/h (337 mph)
at 9,600 m (26,980 ft)	557 km/h (346 mph)
at 11,000 m (36,090 ft)	543 km/h (337 mph)
Rate of climb at a 47,700-kg all-up weight:	
at sea level	4.6 m/sec (905 ft/min)
at 1,000 m	4.6 m/sec (905 ft/min)
at 3,000 m	4.5 m/sec (886 ft/min)
at 5,000 m	4.2 m/sec (827 ft/min)
at 8,000 m	3.3 m/sec (650 ft/min)
at 9,000 m	2.9 m/sec (571 ft/min)
at 11,000 m	1.5 m/sec (295 ft/min)
Time to altitude at a 47,700-kg all-up weight:	
at sea level	0 min
at 1,000 m	3.7 min
at 3,000 m	11.0 min
at 5,000 m	18.7 min
at 8,000 m	31.8 min
at 9,000 m	37.1 min
at 11,000 m	51.9 min
Service ceiling	11,500-11,600m (37,730-38,060 ft)
Range:	
with a 54,800-kg (120,834-lb) AUW and 1,500 kg (3,310 lb) of bombs	5,150 km (3,418 miles)
with a 58,000-kg (127,890-lb) AUW and 2,000 kg (4,410 lb) of bombs	6,000 km (3,729 miles)
with a 57,500-kg (120,830-lb) AUW and 9,000 kg (19,840 lb) of bombs	3,320 km (3,418 miles)
Take-off run/take-off field length at an AUW of 47,850 kg (105,510 lb)	980/2,200 m (3,215/7,220 ft)
Landing run/landing field length at a landing weight of 44,000-kg (97,020 lb)	870/1,970 m (2,850/6,460 ft)
Landing speed at a landing weight of 44,000 kg	170 km/h (105.7 mph)

Weight characteristics obtained during tests of the first production B-4s	
Empty weight	34,520 kg (76,100 lb)
Crew weight	990 kg (2,180 lb)
Cannon ammunition weight	1,058 kg (2,332.5 lb)
Normal all-up weight	47,600 kg (104,950 lb)
fuel load	4,080 kg (8,995 lb)
bomb load	6,000 kg (13,230 lb)
High gross weight version 1:	
All-up weight	54,500 kg (120,150 lb)
Fuel load/bomb load:	
option 1	15,300 kg/1,500 kg (33,730 lb/3,310 lb)
option 2	14,300 kg/2,500 kg (31,530 lb/5,510 lb)
option 3	13300 kg/3500 kg (29,320 lb/7,720 lb)
option 4	12,300 kg/4,500 kg (27,120 lb/9,920 lb)
option 5	11,300 kg/5,500 kg (24,910 lb/12,125 lb)
option 6	10,300 kg/6,500 kg (22,710 lb/14,330 lb)
High gross weight version 2:	
all-up weight	60,000 kg (132,280 lb)
fuel load	14,300 kg (31,530 lb)
bomb load	7,120 kg (15,700 lb)
High gross weight version 3:	
all-up weight	62,500 kg (137,790 lb)
fuel load	19,600 kg (43,210 lb)
bomb load	3,560 kg (7,850 lb)
High gross weight version 4:	
all-up weight	64,000 kg (141,095 lb)
fuel load	14,300 kg (31,530 lb)
bomb load	12,000 kg (26,455 lb)
High gross weight version 5:	
all-up weight	65,000 kg (143,300 lb)
fuel load	19,600 kg (43,210 lb)
bomb load	6,000 kg (13,230 lb)

V. V. Ponomarenko respectively. Their route took them from Zhukovskiy to Turkestan (Yany-Kurgan railway station) and back again. The first bomber covered a distance of 5,380 km (3,344 miles) at an average speed of 384 km/h (239 mph); the all-up weight was 58,200 kg (128,330 lb), including a bomb load of 2,000 kg (4,410 lb). The remaining fuel upon landing was 2,294 kg (5,058 lb). The second machine covered 5,090 km (3,163 miles) at an average speed of 390 km/h (242 mph), the all-up weight being 57,900 kg (127,670 lb), again with a 2,000-kg bomb load. The remaining fuel upon landing was 2,428 kg (5,354 lb). Design range, with the full use of fuel, was calculated to be 5,800-6,100 km (3,605-3,791 miles). The B-4s dropped 2,000 kg (4,410 lb) of bombs at a target range in Turkestan during the mission.

The maximum range testing of the first production B-4s marked the completion of Stage A of the state acceptance trials. MAP test pilots turned two machines over to GK NII VVS for further work within the trials program, followed by another three. Gallai's original aircraft (c/n 220002) was the first to be transferred; instead, Gallai received the 20th machine off the line (c/n 220405), the last in the initial tranche, for development and testing.

At the end of 1947, summing up the test results of the first B-4s, LII issued report No.47-178 which cited the performance figures given in the tables on this page. This performance basically met the requirements set forth in directive No.1282-524 issued by the Council of Ministers of the USSR on 29th June 1946, according to which the final performance characteristics of the Tu-4 (speed, range and payload) were to be endorsed after the completion of state acceptance trials of the first aircraft. At the same time it was stipulated that maximum speed at sea level should not be lower than 470 km/h (248.6 mph), maximum speed at altitude should be at least 560 km/h (348 mph), maximum range with a 1,500-kg (3,307-lb) bomb load should be at least 5,000 km (3,107 miles), maximum range with a 7,128-kg (15,717 lb) bomb load should be at least 3,000 km (1,864 miles).

One of the first production B-4s is refuelled from a BZ-35 fuel bowser on a ZiS-6 6x4 chassis at the LII airfield during trials.

Below: A very early B-4 on the hardstand at Zhukovskiy during trials, with a Lisunov Li-2 transport parked on the grass beyond.

Stage A of the state acceptance trials had turned up 82 defects and shortcomings, which had to be eliminated pronto. Rectifying 65 of them, the OKB and the Kazan' factory resubmitted the B-4 for Stage B of the trials; this involved ten aircraft, including c/ns 220202, 220204, 220205, 220301, 220401, 220405 and 220501. These aircraft could be operated with a normal AUW of 55,000 kg (121,250 lb); the maximum landing weight was 48,000 kg (105,820 lb). With the 65,000-kg (143,300-lb) maximum AUW the G load was reduced from 4.05 to 3.56, but even that was adequate for turbulent conditions. With ASh-73TK Srs 3 engines the bomber attained a maximum speed of 435 km/h (270 mph) at sea level at combat power and 558 km/h (346 mph) at 10,250 m (33,630 ft) at

maximum continuous power. The service ceiling of 11,250 m (36,910 ft) was attained in 58 minutes.

With a 54,430-kg (120,000-lb) AUW, including 1,500 kg (3,310 lb) of bombs and 13,500 litres (2,970 Imp gal) of fuel, the B-4 had a range of 5,100 km (3,170 miles); with a 61,500-kg (135,580-lb) AUW, including 3,000 kg (6,610 lb) of bombs and 26,700 litres (5,874 lb) of fuel, it rose to 6,580 km (4,088 miles). However, the B-4 could not carry FAB-3000M-46 bombs because they were too large to fit inside the bomb bays.

The state acceptance trials and special tests at GK NII VVS involving such a large number of aircraft (the initial 20 Tu-4s) required substantial time, continuing throughout 1948 and well

into 1949. This is comparable to the B-29's test cycle (the first prototype XB-29 flew on 21st September 1942 and the type was inducted by the USAAF on 8th May 1944). The program included performance testing, tests of the aircraft's offensive and defensive armament and verification of the Kobal't bomb-aiming radar. Various bomb load options were carefully tested because the Tu-4 was to be used as a strategic bomber carrying nuclear weapons; the weight of the first Soviet atomic bomb was expected to be within 6,000 kg (13,230 lb).

In the bomber's normal and overload configurations the 6,000-kg (13,230-lb) bomb was accommodated in the rear bomb bay. If M-43 and M-44 series (1943 and 1944 standard) HE bombs were used, the maximum bomb load was 7,120 kg (14,600 lb), increasing to 12,000 kg (26,460 lb) in the case of M-46 series bombs. In overload configuration, when the all-up weight exceeded 60,000 kg (132,300 lb), an additional 5,300 kg (11,690 lb) of fuel was carried in auxiliary tanks suspended in the forward bomb bay; in this case the aircraft's empty weight rose by more than 1,000 kg (2,205 lb) due to the additional weight of the fuel system.

At length the voluminous trials report was presented to the State commission, but the designers weren't out of the woods yet. As the report was discussed at progressively higher levels, the debates grew stronger and stronger. One of the professors on the State commission unexpectedly queried if the B-4 conformed to the Soviet SSS mentioned earlier. This immediately caused some other members of the commission to waver. Andrey N. Tupolev reasoned that for two years the B-29 had seen action in South-East Asia, where gale-force winds were common (unlike Europe), and the airframe structure had not succumbed to them. He argued that even if there was a discrepancy between Soviet and US structural strength standard, it had been taken into account when the aircraft was copied and its dimensions were converted from Imperial to metric units – but to no avail. At length, having reached the limit of his eloquence, Tupolev picked up the phone – a direct line to the Kremlin – and dialled a number. Everyone fell silent, wondering what this maverick was up to. *'Comrade Stalin, this is Tupolev reporting* – he said in his high-pitched voice – *Here there are some who believe the structural strength of the B-4 to be inadequate...'* He listened to Stalin's reply, signed off, hung up and said, addressing everyone: *'Comrade Stalin does not share your point of view and recommends that you'd better not delay the signing of the report'.*

In 1948 the official designation was changed from B-4 to Tu-4. The Tu-4 became the only aircraft in Soviet aviation history to have the final report on its state acceptance trials personally endorsed and its designation personally given by Stalin – or by any other Soviet head of state, for that matter. The way it happened is related by Pyotr V. Dement'yev, who at the time was first deputy to Mikhail V. Khroonichev, the Minister of Aircraft Industry:

'One evening Khroonichev received a telephone call from [Maj.-Gen. Aleksandr N.] Poskryobyshev, Stalin's secretary, who gave the message: "The Master (that is, Stalin – Auth.) asks for the final report on the trials [of the B-4] to be brought to him at his nearest country house". (That was the term used for Stalin's residence in a wooded area in Kuntsevo, now a residential area well within the boundaries of Moscow; the house is still there – *Auth.*)

The Minister invited me to accompany him, since I had been in charge of the testing of the aircraft at LII and knew all the details. On arrival we were met by [Lt.-]Gen. [Nikolay S.] Vlasik, the chief of Stalin's personal bodyguards, who led us to the terrace, telling us to wait there and dismiss our chauffeured limousine. About half an hour later the door opened without a sound and Stalin entered. Giving us a nod, he sat down at the table and, not saying a word, started working through the thick report. Then, puffing at his pipe, he uttered: "Precisely one year behind schedule" and, taking the report with him, left the terrace.

We sat and waited there for a long time – and, as you might imagine, we were ill at ease. At length Vlasik came out again and told us that Stalin had endorsed the report and we could go back to Moscow. Khroonichev made a move towards the phone, intending to summon the car, but Vlasik stopped him, saying: "The car is waiting for you".

Sure enough, a ZiS-110 (a limousine of the kind used by the Kremlin garage, a derivative of the 1942-model Packard 180 – Auth.) was parked near the entrance, but it was not our regular car. An officer carrying a pistol was sitting beside the chauffeur, holding a big Manila envelope sealed with sealing wax.

We climbed in and started off. The car passed Arbat Street, turned [left] towards Okhotnyy Ryad ('Game Row', places in the center of Moscow – Auth.) and was now going up the hill towards Lubyanka Square (notorious for the NKVD (later successively MGB/KGB/FSB) headquarters being located there – Auth.). Khroonichev and I kept silent. Stalin had greeted us so drily that we could only guess if the officer sitting in front was just another aide or our warden. The car skirted the square and went up Bol'shaya Lubyanka Street – but then, there were quite a few NKVD buildings in that street as well! Only when the car had driven up Sretenka Street and turned [right] into Dayev Lane leading up to the Ministry [of Aircraft Industry] did we breathe a sigh of relief.

At the main entrance the officer silently handed the envelope to Mikhail Vasil'yevich [Khroonichev]. It was 3 AM. We rose to the "ministerial" floor (the third floor where Khroonichev's study was – Auth.), entered the study and shut the door. Wearied by the harrowing appointment, by Stalin's hazing and by the worry caused by guessing at what he had written [in the report], we slumped into the armchairs. After a brief pause we broke the seal on the envelope and read: "Approved. Chairman of the Council of Ministers and Minister of Defence. (signed) I. Stalin" Stalin had signed the report in blue pencil and, in his own hand, amended the designation "B-4" to "Tu-4".

Khroonichev rose, wiped his brow, took a bottle of vodka from his safe [where he kept documents] and poured each of us a shot. "No, Pyotr Vasil'yevich, this way we won't last long", he said. We drained our glasses in a gulp without even eating something afterwards (as required by the Russian vodka drinking custom – Auth.) because there was nothing to eat in the study, but we were so nervous that the vodka did not have any effect on us...'

In April 1949 twenty Tu-4s underwent modifications in accordance with the State commission's findings, and in May four of them (c/ns 220702, 220703, 220802 and 220804) were designated as pattern aircraft for check-up trials.

Chapter 3

The Tu-4 in production

As already mentioned, aircraft factory No.22 in Kazan' was selected as the first plant to build the B-4 (Tu-4) in quantity. The reason was that the plant had a history of building bombers, including heavy ones – which, by coincidence, happened to be Tupolev types (the TB-3 and the TB-7; the latter, though designed by Vladimir M. Petlyakov, started life in the Tupolev OKB as the ANT-42). On 11th July 1945 the plant received an order from People's Commissar of Aircraft Industry Aleksey I. Shakhoorin to terminate production of the Pe-2U bomber trainer and start prepa-

rations for producing the B-4. This document marked the beginning of the Kazan' aircraft factory's seven-year involvement with the *Bull*.

Putting the B-4 into production was a tremendous task involving a complete change of airframe manufacturing technology. A mold loft department was set up at the factory; at the initial stage it received a lot of support from the Tupolev OKB's head office in Moscow and the Kazan' branch. A number of workshops had to be rebuilt or erected anew, more than 2,500 items of manufacturing

Early-production Tu-4s in the final assembly shop of the Kazan' aircraft factory No.22. Note the cloth covers preventing damage to the skin from the boots of the assembly workers.

The Tu-4's fin takes shape in the assembly jig.

equipment and machine tools had to be moved; also, the plant had to build additional housing for its workforce. Tooling up for B-4 production involved developing no fewer than 30,000 new manufacturing techniques, producing 2,700 pieces of jigs and tooling, 389,900 pieces of standard tools and 40,700 templates. The new equipment obtained by dismantling some factories in the vanquished Germany started coming in at the end of 1945; some of the new machine-tools came from the final deliveries under the Lend-Lease Agreement.

The new technologies introduced in the course of the preparations to build the B-4 included manufacturing and welding stainless steel structures, welding duralumin parts, electrochemical coating of duralumin parts, making pressurized rivet joints and

joints using large-diameter rivets etc. Many problems were associated with the manufacturing of flap and landing gear actuation screwjacks, the observation/gun-aiming blisters and other aircraft parts that were new for the plant. The B-4 introduced new complex navigation, electronic support measures and radio communications equipment systems, as well as the remote control system for the defensive armament; this necessitated the setting up of a whole network of laboratories for studying these systems at the factory. In these laboratories various items of equipment were subjected to bench testing and adjusted with assistance from specialists from subcontractor enterprises, prior to being installed in the aircraft. Personnel from the OKB and its Kazan' branch also took part in the process of final assembly of the airframe and installing the

The forward fuselage build-up area; Section F-2 of a Tu-4's fuselage displays the internal structure before the skin panels are attached. Note the section of the pressurized crawlway at the top of the pressure dome.

The shop where the Tu-4's fuel tanks were manufactured. As the photo shows, much of the Kazan' aircraft factory's workforce was female – in similar fashion to Boeing's 'Rosie the Riveter' workers, and for exactly the same reason.

Below: The newly fitted starboard main gear unit of a Tu-4; the engine nacelle is not fitted yet, exposing the gear fulcrum and actuation mechanism. Note also the wing ribs cut away at the rear to accommodate the trailing-edge flap.

Below right: An assembly worker busies himself with the nose gear unit of a Tu-4. Note the different nose-wheel tread pattern (squares instead of diamonds on the mainwheels). Note also the temporary structure fitted instead of the forward glazing frame (Section F-1).

Left: Two Tu-4s in the final assembly shop of plant No.22. The one on the right is the basically assembled fuselage (sections F-2 through F-6) mated with the wings (with the engine nacelles and inner wing leading-edge fairings yet to be fitted); the tail surfaces have been bolted together into a single subassembly and are sitting alongside. The other aircraft is at a much more advanced stage of construction.

Right: Here we have the rear fuselage (section F-5) and tail gunner's station (section F-6) with the stabilisers attached – but nothing else.

Left: Here, section F-2 is joined to the rest of the fuselage (already mated with the wings, with trestles under the outer wing panels).

Right: It seems the Kazan' plant was not particularly consistent about the order in which the major airframe subassemblies were joined together. Here, section F-6 is apparently fitted after everything else, including the tail unit (complete with control surfaces), has been installed.

Left: A substantially complete Tu-4 sits on its gear in the assembly shop. The apertures of the lateral sighting blisters are temporarily blanked off. Note the open bomb bay doors.

Right: Probably the same aircraft from a different angle, showing that the nacelles and engines are already installed.

systems and equipment, which helped avoid errors in the installation and adjustment of the latter.

As if this wasn't enough, the plant was under constant pressure to adhere to the schedule. Reports on progress in the preparations were filed to Stalin daily; the plant was expected to complete the first three bombers in the third quarter of 1946! This was at a time when much of the country was lying in ruins, the national economy was in shambles because of the war effort and the people were starving. As one western author put it, the hectic atmosphere at Kazan' in 1946 was much the same as at Wichita in 1942, when the first B-29s were being built.

To tackle this mountain of work, the factory sent a 106-man team of production engineers to the OKB's head office in Moscow where they worked with the drawings. By 1st April 1946 plant No.22 had received most of the manufacturing documents, and the top-priority jigs and tooling had been designed by 1st May; these were manufactured both by the preparatory workshops (whose job it was in the first place) and by the main assembly shops. In so doing the factory killed two birds with one stone, maintaining the production schedule and boosting the skills of the engineering staff and workforce.

Despite these strenuous efforts, the Kazan' plant failed to keep the schedule; by the end of 1946 only one airframe had been assembled and the major airframe components for a further two were prepared for mating. The delay was mainly due to late delivery of items supplied by subcontractor plants – though, in fairness, they can hardly be blamed either. Given Stalin's order that the B-29 should be copied exactly, the number of enterprises involved and the scope of new technology to be mastered all along the line meant that delays were inevitable. For this reason the Kazan' aircraft factory was unable to build up a stock of components for next year's B-4 production.

Knowing that the Soviet leader would not take kindly to his orders not being fulfilled and reprisal would be swift, the Tupolev OKB and the Kazan' factory did their utmost to accomplish the mission; the management and workforce alike literally took up residence in the workshops, toiling around the clock. This yielded results; as mentioned in Chapter 2, the first production B-4 (c/n 220001) was completed in February 1947. A festive meeting was held in Shop 8 (the final assembly shop) to celebrate the occasion, the plant's director Vasiliy A. Okulov cutting the red ribbon.

The first flight of this aircraft took place on 21st May. Andrey N. Tupolev was there for the occasion; thus was born a tradition – in subsequent years, health permitting, he would always come to Kazan' to witness the first flights of new Tupolev aircraft built there.

The second B-4 followed at the end of June and the third in August. About ten more aircraft had been completed by the year's end. As the state acceptance trials progressed, the Kazan' aircraft factory was launching full-scale production of the Tu-4.

With the exception of the first two production batches (Nos. 00 and 01), which consisted of two and three aircraft respectively, Tu-4s built by plant No.22 typically had five aircraft per batch, although not later than Batch 45 the number of aircraft per batch was increased to ten. At least one Kazan'-built Tu-4 had an alphanumeric c/n (226609B-07), the alphanumeric suffix probably

denoting the seventh example of some special version. Late batches had the batch number presented in a three-digit format (for example, 2207301 or 2207401), apparently in anticipation of 100 or more batches being built, although this was not consistent – Batch 75 featured both 227506 and 2207510. The highest known batch is No.84 (c/n 2208407), but as Kazan' production of the Tu-4 totalled 655 there should be another three and a half batches – or perhaps Batch 45 was not the first ten-machine batch.

Unlike the B-29, which did not display its c/n visibly, the Tu-4 had the c/n prominently stencilled on both sides of the fin (in the same size of digits as the USAAF serial on the B-29) and on both sides of the nose. In the Kazan' factory's internal paperwork an abbreviated version of the c/n (the batch number and the number of the aircraft in the batch) was usually used; thus, B-4 c/n 220101 was referred to as 'No.101'. With a few exceptions, early-production Kazan'-built Tu-4s had sequential serials reflecting their place in the production sequence; thus, c/n 220404 was '19 Black', c/n 220405 was '20 Black', c/n 220501 was '21 Black' and so on. However, a few aircraft had out-of-sequence serials – for example, c/n 220201 was '1 Black', although logically it should have been '6 Black', and c/n 220205 was '1000 Red' (!) rather than '10 Black'; some Tu-4s had no serials at all.

As an aside, the famous order that the B-29 should be copied exactly had one more consequence. During the preparations for B-4 production the Kazan' aircraft factory's Bureau of Rationalisation and Invention was closed down – lest some inventive engineer or technologist should make any changes to the design! The Russian acronym for such bureau, BRIZ (*Byuro ratsionalizahtsii i izobreteniy*), is pronounced like the word 'breeze'; well, the management sure was not going to let any fresh breezes in! Thinking outside the box was not encouraged; 'toe the line' was the catch phrase of the day.

Okay, production had started, but the factory's troubles were by no means over – in fact, the worst was yet to come. The Air Force's representatives (quality control inspectors) at plant No.22 flatly refused to accept the bombers because the latter had numerous defects – which was not unexpected during the learning curve. The chief source of complaints was the powerplant; initial-production ASh-73TK engines were notoriously unreliable and had a time between overhauls (TBO) of only 25 hours. There were cases when a brand-new Tu-4 did not manage to make a single pre-delivery test flight before the engines ran out of engine life. Of course this was no good for the military. More snags were caused by the late deliveries of items supplied by subcontractor enterprises, such as the V3-A3 propellers, electric starters, generators and so on.

Now, since the Air Force was not accepting the completed bombers, the Ministry of Defence was not paying for them. This put the Kazan' factory in a dire financial situation in 1947.

As if that weren't enough, there were personal relations and antipathies to contend with. Col.-Gen. Pavel F. Zhigarev, who had just been appointed Commander of the Long-Range Aviation (and Deputy C-in-C of the Soviet Air Force) in 1948, was one of the Tu-4's main opponents. He would not budge, no matter how Andrey N. Tupolev tried to reason with him. News of this standoff quickly reached Stalin, who summoned both men to the Kremlin. When they entered Stalin's study, he was alone in the room.

Right: The crash scene of B-4 c/n 220303 near Kazan' on 5th November 1947

Below: The forward fuselage lying on its side; note the collapsed nose gear.

Below right: The rear fuselage (Sections F-5/F-6) and tail unit were among the few parts remaining intact after the post-crash fire.

Center: The detached No.2 (foreground) and No.1 engines.

Center right: Lower rear view of the detached forward fuselage, with 'F-1' chalked on it.

Bottom right: The starboard outer wing that broke away on impact with the trees.

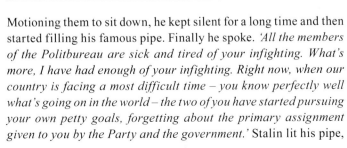

Motioning them to sit down, he kept silent for a long time and then started filling his famous pipe. Finally he spoke. *'All the members of the Politbureau are sick and tired of your infighting. What's more, I have had enough of your infighting. Right now, when our country is facing a most difficult time – you know perfectly well what's going on in the world – the two of you have started pursuing your own petty goals, forgetting about the primary assignment given to you by the Party and the government.'* Stalin lit his pipe,

A brand-new Tu-4 (c/n 220605) awaiting delivery at Kazan'-Borisoglebskoye.

rose from the table and started pacing behind the men's backs. *'If you don't want to work, then say so. We will convene the Polit-bureau and discuss your conduct. What do you think you are, hag-gling at a bazaar?'* Puffing at his pipe, Stalin slowly made a round of the table and stood facing Zhigarev and Tupolev. He fell silent for a long time, apparently making up his mind. Finally, after a nerve-racking pause, he took the pipe from his mouth and, point-ing the stem of the pipe at Tupolev, said: *'Comrade Tupolev, you have one month to rectify the faults of the Tu-4'.* Next, he nodded towards Zhigarev: *'You, Comrade Zhigarev, have one month to accept the machines. That's it. You may go.'*

Yet, this meeting in the Kremlin did nothing to resolve the situation; the Kazan' aircraft factory kept rolling out completed Tu-4 bombers and the military steadfastly refused to accept them.

Mid-production Kazan'-built Tu-4 c/n 225801 underwent check-up trials.

The allotted month was drawing to an end when Tupolev and Zhigarev met again at the OKB's flight test facility at LII's airfield in Zhukovskiy. *'Will you sign the Act of acceptance, Pavel Fyodorovich?'* – Tupolev asked; Zhigarev shook his head in rejection. Then Tupolev picked up a phone that was a hotline to the Kremlin, dialled Stalin's number and said: *'Comrade Stalin, this is Tupolev speaking. I have fulfilled the assignment'* – and handed the receiver to Zhigarev. Taken aback, the latter had no choice but to utter: *'Comrade Stalin, this is Zhigarev speaking. I accept the bombers'*.

Knowing all too well that this trick might cost him and Zhigarev dearly, Tupolev arranged a big conference attended by representatives of all the subcontractor industries involved in the Tu-4 program. Describing the disastrous situation, he asked them to take urgent measures aimed at eliminating the manufacturing defects. On 31st March 1948 the VVS formally accepted the first Kazan'-built Tu-4. That year the plant rolled out 59 Tu-4s but delivered only 17 instead of the planned 57, and the factory apron was getting crammed with incomplete and hence undeliverable aircraft. At one point, 14 bombers sat in storage at Borisoglebskoye minus engines; six others lacked landing gear actuation electric motors, ten Tu-4s lacked US-9 radios and a further ten lacked the M-10 APU (the famous 'putt-putt').

Thus, the Kazan' aircraft factory had failed to fulfil the production plan for three consecutive years (1946-48). The Ministry of Aircraft Industry drew its conclusions from this. On 11th January 1949 Vasiliy A. Okulov was fired from his post as Director of plant No.22 and replaced by L. P. Sokolov.

In 1949 the quality of the Tu-4 finally began to improve. An important contributing factor was that the worst of the ASh-73TK's teething troubles had been overcome. Plant No.19 in Perm' had increased the engine's guaranteed TBO from the initial 25 hours to 200 hours; some engines clocked as many as 800 hours. As a result, in 1949 the Kazan' plant finally met the production target, delivering 120 Tu-4s; these were followed by 177 bombers in 1950 (that is, seven more than stipulated by the plan), 191 *Bulls* in 1951 (again, with one extra aircraft) and 150 in 1952.

All the same, the troubles with subcontractors failing to observe the delivery schedule were far from over. In its year-end report for 1951 the Kazan' factory stated that plant No.283 in Leningrad was not supplying enough Kobal't radars; 44 Tu-4s had to be delivered minus radars and retrofitted with them in service. Due to the lack of parking space at the factory airfield the Air Force was obliged to accept the bombers in as-was condition, placing them in storage at the bases and formally taking them on charge only after the missing items had been installed. Moreover, the Kobal't radar was a source of trouble in its own right – it was rather unreliable and failed to provide the stated detection range.

The modernisation of the factory itself continued in the course of production. In late 1949 a new runway (11/29) measuring 2,400 x 100 m (7,870 x 330 ft) was commissioned at Kazan'-Borisoglebskoye, as were an additional paved hardstand and a compass base. The new runway was not only longer and wider than the original one (no longer in existence) but also featured a stronger surface. The old runway paved with hexagonal concrete slabs 100 mm ($3^{15}/_{16}$ in) thick had been adequate for the Pe-2 bombers but proved too weak for the Tu-4 in maximum gross weight configuration, forcing maximum-range test sorties to be flown from other airfields (in Kuibyshev or Zhukovskiy). Such sorties were mostly flown in a southerly direction (to Tashkent and back) or an easterly direction (to Novosibirsk and back).

At peak output the Kazan' factory would crank out 15 Tu-4s per month. This kept the factory's own test personnel and the military test pilots performing the pre-delivery tests plenty busy. To ease the workload, the factory management asked several experienced bomber crews from a DA unit based in Kazan' to lend a hand. For the crews this was an excellent deal, as they built up experience with the Tu-4 and were generously paid by the plant. One such test flight, however, ended in a tragic episode. On 17th December 1949 a Tu-4 flown by a DA crew was flying at 6,000 m (19,685 ft) near the town of Galich (the one in the Kostroma Region of Russia) when a weight at the end of the Berkut radio's trailing wire aerial started spinning in circles, eventually striking and shattering the starboard gunner's observation blister. Due to the resulting explosive decompression one of the crewmen, Capt. Moroz, who was not strapped into his seat, was blown out of the aircraft and fell to his death. In spite of an extensive search, his body was never found.

Since a single bull farm... sorry, aircraft factory obviously could not cope with the task of equipping the Air Force with a sufficient number of Tu-4 bombers (especially considering the initial problems at Kazan'), a decision was taken to widen the scope of production. As early as 1947 the Council of Ministers issued a directive requiring aircraft factory No.18 in Kuibyshev to commence production of the Tu-4. (The city has now reverted to its historic name of Samara, and the plant is now called Aviacor Joint-Stock Co.) The Kazan' factory, which retained its status as the chief manufacturer of the type, was obliged to assist the Kuibyshev factory in launching production. In 1948 it supplied three wing center sections, three forward fuselage sections, three center fuselage sections, three shipsets of Fowler flaps and some other components, as well as mock-up versions of the flight deck glazing and other transparencies, to Kuibyshev as 'starter sets'. Measurement tools and other necessary items were also supplied. A year later, in February 1949, the first production machine manufactured by plant No.18 took to the air from Kuibyshev-Bezymyanka, the factory airfield shared by plants No.1 and No.18.

Kuibyshev-built Tu-4s had two distinct construction number systems. Originally they, too, had six-digit c/ns but these were structured differently; for example, the first three machines assembled from components supplied by the Kazan' plant received the c/ns 184101, 184201 and 184301 – that is, plant 18, Tu-4, respectively 1st, 2nd and 3rd aircraft in Batch 01. This format was retained until Batch 34, with five aircraft per batch (except Batch 01). From Batch 35 onwards the number of aircraft per batch was increased to ten and the c/n was altered to a seven-digit format to accommodate this; for example, c/n 1840848 was the 08th aircraft in Batch 48. In 1952 the Kuibyshev factory introduced a completely new system in Batch 49 or 50 – for example, c/n 2805110, that is, year of manufacture 1952, plant 18 (the first digit was omitted to confuse hypothetical spies), Batch 051, 10th and final aircraft in the batch.

In 1948 the government took the decision to task one more aircraft factory – namely plant No.23 in Fili, then a western suburb of

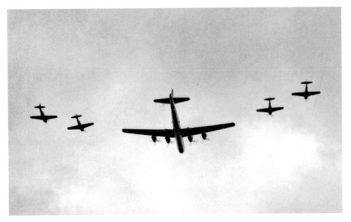

Above: Parades were part of the job. Here, a Tu-4 escorted by four Lavochkin La-11 fighters performs at a military parade in Moscow's Red Square.

Above right: This Tu-4 seen during a later parade is escorted by four Mikoyan/Gurevich MiG-15 fighters in very close formation.

Below: A Tu-4 with La-11 escort passes over Moscow's History Museum during a parade in Red Square. One more (unescorted) is coming up.

Moscow (now a residential area well within the city) – with manufacturing the Tu-4. Again, this factory had a history of building bombers, having manufactured the IL-4 and Tu-2 during the war. (As a point of interest, it used to be the premises of plant No.22 until the latter was evacuated to Kazan' in October-November 1941 and pooled with the existing plant No.124.) Series production of the Tu-4 at plant No.23 commenced by the beginning of 1950 but proceeded on a much smaller scale than in Kazan' or Kuibyshev.

Plant No.23 began preparations for Tu-4 production on 19th May 1949. It was much the same story as in Kazan' three years earlier – the assembly shops and the tooling shops had to be reconstructed at a slam-bang pace, new workshops, laboratories and storage facilities had to be built, new structural materials had to be mastered. Just one minor example: the pressurized cabins and the crawlway between the forward and center cabins were coated with heat insulation blankets known in the USSR as ATIM (*aviatsionnyy teplo'izolyatsionnyy materiahl* – aviation-specific heat insulation material). This material was a layer of wool 10-12 mm ($0^{25}/_{64}$ in to $0^{15}/_{32}$ in) thick, quilted with two layers of gauze and coated with waterproof fabric on the outer side and with fireproof fabric on the inner side. Until then the Soviet industry had not produced anything of the sort.

Again, Moscow/Fili-built Tu-4s had two construction number systems. The original one was similar to the early Kuibyshev system – for example, c/n 230217 (that is, plant 23, 02nd aircraft in Batch 17) – and was in use up to and including Batch 23. In Batch 24 or 25 this gave place to an 'inverted' seven-digit format – for example, c/n 2303201 (that is, plant 23, Batch 032, 01st aircraft in the batch). There were five aircraft per batch under both systems.

With three plants being engaged in Tu-4 production, this inevitably created a need for all units of the aircraft built in Kazan', Kuibyshev and Moscow to be interchangeable, which meant that the jigs and tooling at the three plants had to be standardised. To check component interchangeability, the Customer (as the Air Force was euphemistically referred to) arranged a trial joining of airframe subassemblies manufactured by different plants. This procedure, which took place at one of the plants, went smoothly, no adjustments being needed to any of the parts. The units thus checked comprised both the basic airframe subassemblies and the engine/propeller packages together with engine cowlings and engine mounts, as well as engine exhaust manifolds. The results corroborated the high technical level attained by production factories building the Tu-4 and eliminated a great number of questions concerning the repair of operational aircraft.

Almost immediately after the Tu-4 entered production, the Tupolev OKB set about updating the design. As a rule, the Kazan' aircraft factory, being the primary manufacturer of the type, was the first to introduce the design changes. Some of these were only temporary. Thus, an additional RSU-5 command radio – a Soviet copy of the American SCR-522 – was installed in order to facilitate air-to-air communication in a large group of aircraft near the base where communication channels were extremely busy. This radio was fitted to Kazan'-built aircraft in batches 19-61 (c/ns 221901 through 226110), Kuibyshev-built aircraft in batches 04-39 (c/ns 184304 through 1841039) and Moscow-built aircraft in batches 01-12 (c/ns 230101 through 230512). From aircraft c/ns 222101 and 184505 the SCR-274-N radio was replaced by the indigenous RSB-5 set with an RSI-6M-1 receiver; Moscow-built aircraft had this radio from the outset, since plant No.23 was the last to launch production.

From c/ns 222903 and 184107 onwards the old vacuum-type artificial horizon gave way to the AGK-47B electric artificial horizon; again, Moscow-built Tu-4s had this new item all along. From c/ns 223002, 184209 and 230102 onwards the pilots' and navigator's workstations were provided with tinted visors and anti-flash shutters, and the radio operator, radar operator and flight engineer had more comfortable seats. Also from c/ns 223002 and 184209 onwards, and on all Moscow-built examples, the NK-46 navigation computer (*navigatsionnyy ko'ordinahtor*) was installed – for the first time on a Soviet aircraft. This analogue processor based on vacuum tube technology made it possible to determine the distance covered by the aircraft and its co-ordinates

Above: A pair of early Kazan'-built Tu-4s in echelon starboard formation; the wingman is apparently serialled '93 Black'.

Below: A trio of Tu-4s seen on approach to Moscow-Tushino airfield during the Aviation Day flypast.

in real time. The 'navigation co-ordinator' cover name had to be invented because in those years cybernetics was branded as a 'false science' in the USSR, heresy and anathema to some high-ranking officials, notably the infamous Trofim D. Lysenko – a man whose actions were profoundly detrimental to Soviet science.

All Tu-4s had a camera port with a tilting camera mount and a KP (*komahndnyy pribor*) control panel for bomb damage assessment (BDA), but originally the bombers were delivered *sans* camera, the units installing whatever cameras were available. Not until c/ns 223002, 184209 and 230102 were built did the factories begin installing BDA cameras. One aircraft in each three was fitted with an AFA-33/100 (*aerofotoapparaht* – aerial camera) with a 100-cm focal length; the others came with AFA-33/75 cameras having a 75-cm focal length. Each aircraft came complete with an optional NAFA-3S/50 night camera (*nochnoy aerofotoapparaht*) with an Industar-A quad lens; its shutter speed was set automatically, triggered by the flash of FotAB flash bombs (*fotograficheskaya aviabomba* – 'photo bomb'). Also, each aircraft came complete with an AKS-1 cine camera for filming short routes.

Tu-4s c/ns 223002, 184209 and 230102 introduced yet another change – in anticipation of the new NR-23 cannons, which had a more powerful recoil than the B-20E, the fuselage structure was reinforced to prevent fatigue cracking. The NR-23 had a longer barrel, and when the forward ventral barbette was stowed the cannon barrels, which were pointing aft, made it impossible to open the forward bomb bay doors. Therefore a safety feature was added to disable door opening if the barbette was stowed; small spring-loaded flaps were built into the doors near the cannon barrels to permit opening, should this feature fail. To compensate for the apertures weakening the doors, additional fixtures securing the doors in the open position were introduced concurrently on Kazan'-built and Kuibyshev-built machines; for some reason plant No.23 did not add this feature until Batch 09 (c/n 230109). Actual installation of the new NR-23 cannons as standard commenced in late 1949 from Kazan'-built c/n 223201; the other two factories

were late in introducing it, starting with Kuibyshev-built c/n 1840136 and Moscow-built c/n 230104.

Several fatal accidents involving the Tu-4 were caused by fatigue failure of the engine bearers, leading the Tupolev OKB to take remedial action. From Kazan'-built c/n 223701 and Kuibyshev-built c/n 184115 onwards the original engine bearers with mounting lugs made of 30KhGSA grade Chromansil steel were replaced with new ones made of 40KhMA grade chromium/molybdenum steel; Moscow-built aircraft had the new design from the start.

From c/n 224506 onwards the *Magniy* (Magnesium) IFF transponder gave place to the Magniy-M version; this improved electromagnetic compatibility of the avionics, but IFF response range was reduced from 55 to 35 km (from 34 to 21.75 miles). Concurrently the *Bariy* (Barium) IFF interrogator was likewise replaced with the Bariy-M version. The number of IFF codes entered at a time was reduced from six to four, making the system easier to operate; a 'mayday' transmission mode was added, the duration of the response signal was reduced and other changes were made. The Kuibyshev factory started installing the new IFF system from c/n 184121 onwards; the Moscow factory fitted the Bariy-M interrogator from the very first aircraft, whereas the Magniy-M transponder had to wait until Batch 3 (c/n 230103).

Additional indicators for the SD-1 *Shipovnik* (Dog Rose) distance measuring equipment (*samolyotnyy [rahdio]dal'nomer* – aircraft-mounted DME) and the RV-2 radio altimeter (*rahdiovysotomer*) were fitted at the captain's workstation from c/ns 225303, 184133 and 230307 onwards. A more reliable and user-friendly SPU-14 intercom (*samolyotnoye peregovornoye oostroystvo*) was fitted, starting with c/ns 225501, 184430 and 230205. An improved Kobal't-M radar in a deeper semi-retractable radome was installed from c/ns 225502, 184133 and 230407 onwards. Actually the radome itself had been introduced a while earlier on c/n 225002, but the upgraded radar was still undergoing trials at the time, so the old version had to be installed for the time being.

Andrey N. Tupolev (in 'dress whites') and the famous Soviet aviatrix Valentina Grizodoobova watch the Aviation Day parade at Tushino.

Opposite page: Shots from Tushino as a massive formation of Tu-4s passes over the field in flights of three.

Tu-4 production (according to Russian State Archive of the Economy materials)							
Year	1948	1949	1950	1951	1952	1953	Total
Plant No.22	17	120	177	191	150	–	655
Plant No.18	–	41	108	150	166	16	481
Plant No.23	–	–	27	80	52	1	160
Grand total	17	161	312	421	368	17	1,296

As production proceeded, the Tu-4's constantly changing avionics and equipment suite required progressively more electric power. This need was addressed from c/ns 225401, 184132 and 230307 onwards by installing new GSR-9000 generators, which provided the same voltage as the old GS-9000M units at lower rpm.

The changes made in the course of production concerned the offensive armament as well. First, new KD3-547 bomb cassettes (*kassetnyy derzhahtel'* – cassette-type rack, Group 3, that is, for ordnance up to 500 kg/1,102 lb calibre, Model 547) were added in the autumn of 1949; Kazan'-built Tu-4s had them from Batch 50 onwards. Next, new bomb armament which was standardised with the Soviet Air Force's other new bomber types was introduced from c/ns 225701, 1840136 and 230109 onwards, allowing the Tu-4's payload capacity to be used more rationally. The number of maximum bomb load options was increased from two to four thanks to new bomb racks, including the BD5-50 (*bahlochnyy derzhahtel'* – beam-type rack, Group 5, that is, for ordnance up to 6,000 kg/13,230 lb calibre, Model 50). In order to facilitate the loading of large-calibre bombs into the rear bomb bay, the protruding blade aerial of the IFF interrogator was made detachable; this was replaced by a retractable version on Tu-4 c/n 225801 and on all aircraft from c/ns 226001, 1840140 and 230115 onwards.

From c/ns 226601 and 230219 onwards the designers incorporated the option of switching the RSB-5 command radio to the fixed wire aerial of the 1RSB-70 communications radio as a back-up. The reason was that the RSB-5's own trailing wire aerial was prone to breaking off in heavy icing conditions.

At the end of 1950 the original OSP-48 ILS (*oboroodovaniye slepoy posahdki* – blind landing equipment, 1948 model) gave way to the new SP-50 *Materik* (Continent) ILS. The RV-2 radio altimeter's accuracy left a lot to be desired, and various ways of tackling the problem were tried without success until the dipole aerials were placed farther apart on Kazan'-built Batch 71.

The last major revision to the airframe structure was made from c/ns 227209, 1840348 and 230123 when the horizontal tail spar, leading edge and tip fairings were beefed up. Some changes took a long time to implement. For example, it was not until c/ns 2207301, 2805001 and 2302501 that a seat was provided for the flight technician, even though the latter had been a member of the crew for years; previously the technician had been forced to sit on his parachute throughout the sortie. The final major change introduced on the production lines was the new V3B-A5 propellers which, unlike the earlier V3-A3, could be unfeathered after they had been feathered in flight. The new model was fitted as standard from c/ns 227506, 2805009 and 230101 onwards.

Available information concerning Tu-4 production figures is extremely contradictory. The late Dmitriy S. Markov (the Tu-4's project chief) quoted an overall production run of 'nearly 1,000 machines' in 1987. Referring to MAP sources, the Russian aviation historian Ivnamin S. Sultanov maintained in the early 1990s that only 847 Tu-4s had been built. The *Kazan'* magazine cited the production run at plant No.22 as 655 Tu-4s – a figure also found in other sources; however, in 2002 the Kazan' Aircraft Production Association named after Sergey P. Gorbunov (KAPO, as plant No.22 is now called) stated Tu-4 output as a single aircraft in 1947 (*sic*), 12 in 1948, 120 in 1949, 170 in 1950, 191 in 1951 and 150 in 1952, which amounts to 644 aircraft!

According to information received from Samara the Tu-4's production run at plant No.18 was 480 aircraft. An article by the Russian aviation writer Nikolay V. Yakubovich based on materials found in the Russian State Archive of the Economy cites the following Tu-4 production figures: 655 in Kazan' (1948-1952), 481 in Kuibyshev (1949-1953) and 160 in Moscow (1950-52), which totals 1,296! As for the odd fact that no Tu-4s are listed as built in 1947 (except in the abovesaid publication by KAPO), the reason must be that the years given are the years when the aircraft were accepted by the VVS, *not* when they were completed.

Whichever figure is correct, it is far smaller than the B-29's production run: four factories – the Boeing plants at Wichita and Renton, the Bell Aircraft Co. plant in Marietta (Georgia) and the Martin Aircraft Co. plant in Omaha (Nebraska) built a total of 3,943 Superfortresses.

From 1953 onwards all three plants switched to turbine-powered aircraft. Plant No.22 launched production of the Tu-16 *Badger* twin-turbojet medium bomber; plant No.18 started producing the Tu-95 *Bear* four-turboprop strategic bomber in 1955, while plant No.23 was transferred to Vladimir M. Myasishchev's new design bureau (OKB-23) and started turning out Myasishchev M-4 *Bison-A* four-turbojet strategic bombers in 1954. Successes scored in mastering quantity production of new jet aircraft in many respects had their foundation in the experience of modern aircraft technology that had been gained during the years of the Tu-4 production.

The Tu-4 was mostly produced in the standard bomber version suitable for conversion into a long-range reconnaissance aircraft. A few were modified into nuclear-capable bombers. Several dozen machines were converted into anti-shipping missile platforms. The Tu-4 served as a basis for various special mission aircraft: airborne command posts, transports, flying testbeds etc. (see Chapter 4); as often as not, the conversion job was performed by the original manufacturer.

In recognition of their efforts in creating the Tu-4 bomber and organising its mass production, the leaders and many employees of the Tupolev OKB and the production plants received Government awards and the title of a Stalin Prize laureate.

Chapter 4

Independent development: Tu-4 versions

In the Soviet Union the Superfortress spawned a number of versions, many of which had no US equivalent.

Tu-4 (B-4, *izdeliye* R) long-range bomber

The vast majority of Tu-4s produced by the three factories were built in the standard long-range bomber version. The aircraft was intended for delivering massive bomb strikes with conventional bombs against strategic targets located deep in the enemy's rear area, both singly and in groups, day and night, in any weather. As already mentioned, the Tu-4 (*izdeliye* R) was not a 100% carbon copy of the B-29, having a different powerplant (indigenous ASh-73TK engines) and heavier defensive armament.

In the course of production the bomber was progressively upgraded. As mentioned earlier, the first three Kazan'-built B-4s (Tu-4s) had a defensive armament of eleven 12.7-mm (.30 calibre) Berezin UBK machine-guns – two in each dorsal and ventral bar-bette and three in the tail barbette. From the fourth aircraft (c/n 220102) onwards the machine-guns were replaced with the PV-20 weapons system featuring eleven 20-mm (.78 calibre) Berezin B-20E cannons aimed by means of PS-48 sighting stations (*pritsel'naya stahntsiya*); these were a standard fit until April 1950. Developed in 1944, the B-20 was a 'king size' version of the UB machine-gun, differing from the latter mainly in having a new barrel of larger calibre; the B-20E version of 1946 had an electric charging mechanism, hence the suffix.

In keeping with Council of Ministers directives issued in July 1947 and June 1948, Tu-4s built from May 1950 onwards were armed with ten 23-mm (.90 calibre) NR-23 (*izdeliye* 150P) cannons. This weapon had been developed in 1947 by Aleksandr E. Nudelman and Aron A. Rikhter at OKB-16. The NR-23 was a derivative of the 1944-vintage Nudelman/Suranov NS-23 with a breechblock accelerator increasing the rate of fire from 550 to

Unserialled Tu-4 c/n 220202 sits on a snow-covered hardstand in the course of trials. This aircraft was later serialled '7 Black', being the seventh Tu-4 off the Kazan' production line.

850 rounds per minute for virtually no increase in weight. As compared to the B-20E, the NR-23 had more than twice the weight of fire – 2.66 kg/sec (5.86 lb/sec) versus 1.32 kg/sec (2.91 lb/sec) – and a higher rate of fire (800-950 rpm versus 600-800 rpm) but was slightly heavier, weighing 39 kg (86 lb) versus 25 kg (55 lb). Dmitriy F. Ustinov, who was then Minister of Armament (he is best known as the Soviet Minister of Defence in the 1970s), gave the NR-23 top priority status and assigned plant No.2 named after Vasiliy A. Degtyaryov in the town of Kovrov (Vladimir Region, central Russia) for series production of the cannon, which began in 1948. The delay in 'gunning up' the Tu-4 was caused not so much by the cannon's inevitable development problems but rather by the fact that the NR-23 was primarily a fighter weapon. Such aircraft as the Mikoyan/Gurevich MiG-15 *Fagot*, Lavochkin La-15 *Fantail* and Yakovlev Yak-23 *Flora* enjoyed priority, and it was not until plant No.525 in Kuibyshev joined in that enough NR-23s could be produced to cater for bombers as well.

Speaking of teething troubles, early NR-23s had an extremely strong recoil of some 5,200 kgf (11,460 lbf), which affected gunnery accuracy and could cause fatigue problems. OKB-16 cured the problem by introducing hydraulic recoil dampers which

Basic performance of a typical late-production Tu-4	
Empty weight	36,850 kg (81,240 lb)
All-up weight:	
normal	47,850 kg (105,490 lb)
maximum	55,600-63,600 kg (122,575-140,210 lb)
Top speed (with normal AUW):	
at sea level	435 km/h (270.0 mph)
at 10,250 m (33,630 ft)	558 km/h (346.5 mph)
Service ceiling	11,200 m (36,745 ft)
Climb time to 5,000 m (16,400 ft)	18.2 minutes
Range at 3,000 m (9,840 ft) with 10% fuel reserves:	
with a 55,600-kg (122,575-lb) AUW/1,500 kg (3,310 lb) of bombs	4,200 km (2,608 miles)
with a 63,600-kg (140,210-lb) AUW/3,000 kg (6,610 lb) of bombs	5,400 km (3,354 miles)
with a 63,600-kg AUW/9,000 kg (19,840 lb) of bombs	3,580 km (2,223 miles)
Range at 3,000 m (9,840 ft) until fuel exhaustion:	
with a 55,600-kg AUW/1,500 kg of bombs	4,850 km (3,012 miles)
with a 63,600-kg AUW/3,000 kg of bombs	6,200 km (3,850 miles)
with a 63,600-kg AUW/9,000 kg of bombs	4,100 km (2,546 miles)
Take-off run:	
with a normal AUW	960 m (3,150 ft)
with a 55,600-kg AUW	1,255 m (4,120 ft)
with a 63,600-kg AUW	2,000 m (6,560 ft)
Take-off field length:	
with a normal AUW	2,210 m (7,250 ft)
with a 55,600-kg AUW	2,585 m (8,480 ft)
with a 63,600-kg AUW	3,830 m (12,565 ft)
Landing run:	
with a 47,850-kg (105,490-lb) landing weight	1,070 m (3,510 ft)
with a 41,000-kg (90,390-lb) landing weight	920 m (3,020 m)
Landing speed:	
with a 47,850-kg landing weight	172 km/h (107 mph)
with a 41,000-kg landing weight	160 km/h (99 mph)

reduced the recoil force by 50% to 2,600 kgf (5,730 lbf). Moreover, early examples had a service life of only 700-1,000 shots; this was soon increased to 3,000 shots, but it was not until 1951 – a year after the NR-23 had been officially added to the inventory – that the service life reached the originally stipulated 6,000 shots.

The new defensive weapons system based on the NR-23 was designated PV-23 *Zvezda* (Star) and was developed under the overall guidance of Ivan I. Toropov, Chief Designer of the OKB-134 weaponry design bureau based at Tushino. The Tu-4's project chief Dmitriy S. Markov and the Tupolev OKB's chief armament specialist Aleksandr V. Nadashkevich were also actively involved. The PV-23 system included improved automatic charging mechanisms, more powerful electric drives, PVB-23 ballistic computers (*pritsel'no-vychislitel'nyy blok* – targeting computation module) and updated PS-48M sighting stations offering higher accuracy and greater flexibility. The PVB-23 had a habit of lagging behind the gunsight if the cannons were traversed at a high rate to track the target; still, no alternative design was available. Another disadvantage of the PV-23 system was that the Tu-4's overall ammunition capacity was reduced from 4,680 to 3,150 rounds because of the latter's larger calibre – the ammunition boxes were the same size, and you can't put a quart into a pint pot.

Several production Tu-4s, including c/n 220403, were modified for testing the PV-23 weapons system in 1949. The state acceptance trials took place in one of the Soviet Union's Central Asian republics, and the conditions were extremely tough; the airfield was unpaved, which meant loads of dust, and the temperatures ranged from extreme heat at ground level to –40°C (–40°F) at cruise altitude. The weapons system stood up to the challenge, functioning perfectly.

The tests of the PV-23 weapons system revealed a few bugs. For instance, when the NR-23 cannons were test-fired on Tu-4 c/n 223002 in January 1950, using fragmentation/incendiary/traced (FI-T) rounds, the ammunition belts broke in three of the five cannon installations (both dorsal barbettes and the tail turret). On inspection it turned out that the breeches of the NR-23 cannons did not permit unhindered passage of FI-T rounds and required modification.

It should be mentioned that other defensive weapons had been proposed for the Tu-4 as well. One was the TKB-481 ultra-fast-firing 12.7-mm machine-gun designed by M. N. Afanas'yev at the Tula-based KBP (*Konstrooktorskoye byuro priborostroyeniya* – Instrument Design Bureau) in 1949, with an impressive 1,400-rpm rate of fire. (However, this weapon had its share of problems as well, and the A-12,7 production version had a rate of fire reduced to 1,100 rpm.) The other was the Sh-23 cannon developed by the Ministry of Armament's OKB-15 under Boris G. Shpital'nyy. However, Andrey N. Tupolev opted for the NR-23 as soon as he first saw it. Well, if aircraft can have a fly-off, then the NR-23 (which was in production by then) and the Sh-23 had what you might call a shoot-off. Shpital'nyy claimed that the NR-23, which was recharged by recoil action and had a movable barrel, could jam when firing at right angles to the slipstream – unlike the Sh-23, which had a fixed barrel; however, tests on a ground rig showed otherwise. The final argument in favour of the NR-23 was that the Sh-23 failed the continuous fire test in which the cannon

The Tu-4K prototype (c/n 224203) at Bagerovo AB with two KS-1 missiles attached. Note the recess in the missile pylon for the cockpit canopy of the manned demonstrator version of the missile used for the initial trials of the guidance system.

was required to expend the entire ammunition complement in one long burst without jamming.

Tu-4A nuclear-capable bomber

A version of the Tu-4 designated Tu-4A (*ahtomnyy* – nuclear; in this context, nuclear-capable) was developed as a delivery vehicle for the first Soviet free-fall nuclear bombs. It differed from the basic Tu-4 primarily in having an electrically-heated, thermostabilised bomb bay in which the correct temperature was maintained right up to the aircraft's service ceiling. This was necessary because nuclear munitions are sensitive to sub-zero temperatures and the chain reaction might not go ahead after the nuke has had a cold soak at high altitude. The suspension system for the bulky nuclear bomb had to be designed anew, and a special system of cables removed the bomb's safety pins before a drop in detonation mode (the pins remained in place for an emergency drop in non-detonation mode).

The pressure cabins featured radiation shielding and special shutters over the glazing to protect the crew from the flash of the nuclear blast. The bomb-aimer's workstation was equipped with a modified optical bomb sight and a control panel for the electronic

system downloading the detonation data to the bomb via a special cable and connector plugged into the bomb's rear end. Interestingly, the Tu-4A's nuclear weapon control system was so top secret that even the Tupolev OKB's Deputy Chief Designer Leonid L. Kerber, who was chief of the design team, was denied access to it by the MGB – probably because he had been in detention in the 1940s; only a female employee of the OKB who had designed the circuitry knew all the details of the system.

Outwardly the Tu-4A was no different from the conventionally armed Tu-4 *sans suffixe*. It did not yet have the white anti-flash finish typical of nuclear bomb carriers; this feature was added later on its successor, the Tu-16A.

Tu-4K (Tu-4KS) missile strike aircraft

The issue of the bomber's survivability in the face of strong enemy air defences loomed large as the Tu-4 entered service; this was especially the case with maritime targets. The obvious solution was to use stand-off weapons against heavily protected ground and maritime targets. Hence the Soviet work on air-to-surface missiles, which had started in the late 1930s, was given new impetus. On 13th May 1946 the Council of Ministers issued the first

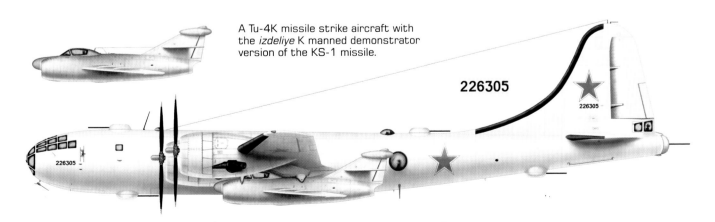

A Tu-4K missile strike aircraft with the *izdeliye* K manned demonstrator version of the KS-1 missile.

226305

Top left: An armourer hoists one of the four *izdeliye* K manned demonstrators (with the landing gear down) into position, using a BL-47 winch.

Top: The electrical connector and sway braces are checked.

Above left: The demonstrator on the Tu-4K's port pylon.

Above: A view of the demonstrator through the Tu-4K's starboard blister shortly after take-off from Bagerovo AB. The wingtip fairings housed the outrigger struts.

Left: The Tu-4K prototype in flight with two *izdeliye* K demonstrators.

directive to this effect ordering research and development work on anti-shipping cruise missiles (referred to in then-current Soviet terminology as *samolyot-snaryad* – lit. 'missile aircraft', or aircraft-type missile).

Development of the K-1 *Kometa* (Comet) weapons system began pursuant to a further Council of Ministers (No.3140-1028 issued on 8th September 1947) which outlined the basic opera-

tional requirements. The B-4 (Tu-4) was envisaged as the missile platform. Apart from the aircraft itself, the system included the Kometa-3 (K III) cruise missile, its Kometa-1 (K I) guidance system and the Kometa-2 (K II) targeting system installed on the missile platform. At an altitude of 1,500-4,000 m (4,920-13,120 ft) the Tu-4's radar was to ensure detection and tracking of a surface ship with a 10,000-ton displacement at a range of 100 km (62.1

miles); the missile would be launched at 60 km (37.2 miles) range and was to have a cruising speed of at least 950 km/h (590 mph). The advanced development projects of the missile proper, the guidance/targeting systems and the weapons system as a whole were to be completed in the second, third and fourth quarters of 1948 respectively.

Overall co-ordination of the Kometa project was performed by the Council of Ministers' 3rd Main Directorate, which was responsible for missile systems. The whole program was shrouded in utmost secrecy and only a few people had access to it on a 'need to know' basis.

The missile was originally developed by OKB-51, which was then headed by Vladimir N. Chelomey and specialised in missiles. In accordance with the abovesaid directive the design drew heavily on the 14Kh air-to-surface missile, itself a derivative of the 10Kh (a Soviet copy of the German V-1 'buzz bomb'), which had a range far in excess of 100 km. However, OKB-51 never had a chance to complete the project; with the envisaged 900-kgp (1,980-lbst) Chelomey D-7 pulse-jet engine the missile's speed would fall short of the stipulated figure. Therefore, on 2nd August 1948 the Council of Ministers issued a further directive (No.2922-1200) transferring responsibility for the Kometa-3 missile to OKB-155 under Artyom I. Mikoyan. Mikoyan's closest associate Mikhail I. Gurevich, who headed a section of OKB-155 tasked with designing missiles (this subsequently became a separate design bureau headed by Aleksandr Ya. Bereznyak), led the actual design effort.

The original project of 1948, designated simply KS (for *Kometa-snaryad* – 'Comet-missile'), was rather different from the missile that eventually emerged. The first version resembled a scaled-down MiG-9 *Fargo* straight-wing fighter with one 800-kgp (1,760-lbst) RD-20 axial-flow turbojet (a Soviet-built version of the captured BMW 003A) instead of two, minus cockpit canopy and with guidance antenna radomes on the nose and atop the fin. Version 2 was no longer based on an existing design; the wings swept back 35° at quarter-chord were positioned well forward on the fuselage, the swept tail unit (a T-tail or a cruciform tail) carried a cigar-shaped fairing with a dielectric rear end housing components of the mid-course guidance system. The engine was the same but the air intake was now located under the center fuselage, and the resulting long 'snout' housing the radar seeker gave the thing a cartoony appearance. The stipulated range at a launch altitude of 4,000 m (13,120 ft) was 195 km (120 miles). The ADP was submitted for review in December 1948; however, the project review panel discovered numerous faults in the project and axed it.

OKB-155 had to start from scratch; work on the second version of the project began on 25th March 1949. Unsurprisingly, the second version – the KS-1 – looked like a scaled-down version of the

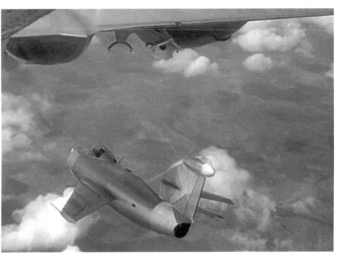

Top to bottom:
Another view of the Tu-4K with two 'manned missiles'.

The *izdeliye* K falls away from the port pylon at about 4,000 m.

One gone! The Tu-4K seen after the port demonstrator has been launched.

Here, the *izdeliye* K is launched from the starboard pylon.

Left: A KS-1 missile (most likely an inert test round) under the starboard wing of Tu-4K c/n 226305. The actual missile lacked the cockpit and the wingtip fairings.

Right: A full frontal of Tu-4K c/n 226305 with two KS-1 missiles.

Center right: Three-quarters front view of the same aircraft, showing the radome of the Kometa-M radar in fully extended position.

Below right: This side view shows the missiles' position relative to the propellers' rotation plane.

Bottom right: Three-quarters front view of the Tu-4K.

one-off MiG-15P*bis* (*izdeliye* SP-1) radar-equipped all-weather interceptor minus cockpit canopy. It had a cigar-shaped fuselage, mid-set swept wings, a swept cruciform tail unit and a nose air intake with a bullet-shaped guidance antenna radome on top of the vertical splitter. The wings had unusually strong sweepback (57°30' leading-edge sweep and 55° sweepback at quarter-chord versus 37° and 35° respectively for the MiG-15) and 5° anhedral. The fin was tipped by the same cigar-shaped fairing housing guidance system components. The KS-1 was powered by a 1,590-kgp (3,500-lbst) Klimov RD-500K centrifugal-flow turbojet – an expendable short-life version of the RD-500, which was a Soviet copy of the Rolls-Royce Derwent V. The missile was 8.29 m (27 ft 2³⁄₈ in) long, with a wing span of 4.722 m (15 ft 5²⁹⁄₃₂ in) and a fuselage diameter of 1.145 m (3 ft 9 in). It had a launch weight of 2,735 kg (6,030 lb) and carried an 800-kg (1,760-lb) high-explosive/fragmentation warhead. The ADP of this version was submitted for review on 3rd November 1949.

The Kometa-1 guidance system was a product of the SB-1 special bureau, which was created in 1949 (and renamed KB-1 in 1950), with overall responsibility for the Kometa weapons system. It was headed by Col. Pavel N. Kooksenko, with Engineer-Maj. Sergo L. Beria – the son of the infamous Lavrentiy P. Beria – as Chief Engineer. (After Stalin's death in March 1953 the new Soviet government exposed the crimes of the Stalin regime. Lavrentiy Beria was found guilty of high treason and executed, sharing the fate of many he had sent to death. Hence Beria Jr. was removed from office and arrested, being replaced by Konstantin Patrookhin)

The Kometa-2 targeting system started life in 1948 at the NII-17 avionics house headed by Chief Designer Viktor V. Tikhomirov (it is now known as the Vega-M Research & Production Association). However, pursuant to Council of Ministers directive No.1228-436 issued on 25th March 1949 SB-1 took over development of this system as well.

The task of the Tupolev OKB was to equip the Tu-4 bomber with the Kometa-2 targeting system and a pair of BD-KS pylons (**bahlochnyy derzhahtel'** – beam-type [weapons] rack) for carrying

the missiles. The aircraft was designated Tu-4K, the K referring to the Kometa system; Aleksandr V. Nadashkevich was the project chief. The pylons were mounted between the inner and outer engines, and the rear attachment point allowed the pylon's incidence to be varied from –1° to +7°. The targeting system included the Kometa-M 360° search/target illumination radar (based on the Kobal't radar but with a different directional pattern) housed in the usual semi-retractable radome; this radar was occasionally referred to as the Kobal't-N, the N standing for *nositel'* (carrier; in this context, missile platform). It was capable of detecting large surface ships at 250-300 km (155-186 miles) range. The KS-1's engine was to be started while the missile was still on the wing; in order to avoid depleting the missile's limited fuel supply it would be fed from a tank on the aircraft right up to the moment of separation. Since the RD-500K turbojet ran on kerosene, of course; this required special 1,158-litre (254.76 Imp gal) kerosene tanks to be installed near the missile pylons, reducing the Tu-4K's own fuel capacity by 2,316 litres (509.52 Imp gal). The ADP of the Tu-4K (sometimes called Tu-4KS) was completed in 1949.

The envisaged operational procedure was as follows. The Tu-4K flew towards the target at 3,000-4,000 m (9,840-13,120 ft); the bomber's radar acquired the target in search mode, whereupon the radar operator switched it to tracking mode, the radar generating a high-power directional beam which was scanned in a cone. The missile's engine was then started up five to ten minutes before launch, and the KS-1 was launched at 70-130 km (43.5-80.7 miles) range, depending on the launch altitude; as the missile was released its engine switched automatically to internal fuel and went to full military power. After launch the missile accelerated in level flight, controlled by the APK-5B autopilot, until it acquired the radar beam and headed towards the target at 1,060 km/h (658 mph). 40 seconds after launch the autopilot switched to remote control mode, using the triangulation method, and the KS-1 changed its flight path, following the radar beam's equisignal line to the target; the missile's flight level varied at this stage. When the radar echo from the target picked up by the Kometa-1 passive radar

seeker head reached a preset level, the guidance system switched to homing mode for terminal guidance; this was necessary because, as the distance between the Tu-4K and the missile increased, so did the width of the aircraft's radar beam cone and hence the error margin. When the KS-1 came within 15-20 km (9.3-12.4 miles) of the target, the missile's seeker head achieved a lock-on and guided the missile all the way to impact; when the range decreased to 500-700 m (1,640-2,300 ft) the autopilot put the missile into a terminal dive onto the target.

Because of the need to illuminate the target continuously the Tu-4K was forced to proceed towards the target until the missile hit; yet, in the early 1950s this course of action was deemed to be effective enough. After the launch the aircraft reduced speed from 360 to 320 km/h (from 223 to 198 mph) in order not to approach within 40 km (24.85 miles) of the target, staying out of range of the latter's air defences.

In 1950 the KS-1's guidance system underwent initial flight tests on two specially converted Li-2 transports, one of which emulated the missile and the other played the part of the missile platform equipped with the target illumination radar. By 1951 the Mikoyan OKB had verified the guidance system on the MiG-9L (*izdeliye* FK) and MiG-17SDK (*izdeliye* SDK-5, *samolyot-doo-blyor Komety* – stand-in aircraft for the Kometa [missile]) avionics testbeds – heavily modified versions of the production MiG-9 (*izdeliye* FS) and MiG-17 *Fresco-A* (*izdeliye* SI) fighters respectively.

As the next step in the program, the Mikoyan OKB brought out the *izdeliye* K demonstrator aircraft pursuant to a Council of Ministers directive dated 25th March 1949. This was a manned version of the actual KS-1 with an unpressurized cockpit (equipped with an ejection seat and a very basic instrument fit) instead of the explosive charge, a retractable bicycle landing gear and wing flaps to improve the landing performance. The *izdeliye* K was to be taken aloft and launched like the real thing; then, at the terminal guidance phase, the pilot would disengage the guidance system, take over the controls and bring the demonstrator back to base. Unlike the actual missile, the aircraft had a regular RD-500 with a normal throttle and fuel control unit instead of the single-mode expendable RD-500K. Four such aircraft (K-1 through K-4) were built.

Meanwhile, in early 1951 the Tupolev OKB completed the first prototype of the Tu-4K; the unserialled aircraft was converted from a Kazan'-built bomber (c/n 224203). The attachment lug for hooking the missile up to the BD-KS pylon was located amidships – immediately aft of the cockpit in the case of the *izdeliye* K; therefore, the pylons of the prototype had to feature recesses accommodating the cockpit canopy.

In 1951-52 the Tu-4K prototype passed manufacturer's flight tests at Chkalovskaya AB (the main facility of GK NII VVS at the time) and, starting in November 1951, at the 71st Test Range (Bagerovo AB) near Kerch on the Crimea Peninsula (see Chapter 7 for more details on the latter facility). Vyacheslav P. Maroonov (Tupolev OKB) and Capt. Nikol'skiy (GK NII VVS) were the project test pilots. Interestingly, the tests at Bagerovo began with ground integration of the guidance system. Quite simply, the Tu-4K was parked at an elevated hardstand, with the *izdeliye* K positioned just over 1 km (0.62 miles) ahead of it as though it had been launched. The Tu-4's radar beam moved left/right and up/down a little, the way it would during an actual launch, and the test engineer beside the 'missile' recorded how the latter's control surfaces moved in concert with the beam, informing the Tu-4's radar operator thereof via field telephone! This allowed the guidance system to be tuned. Next, the Tu-4K made a number of flights without missiles, 'painting' and tracking the ships in the Black Sea with its radar.

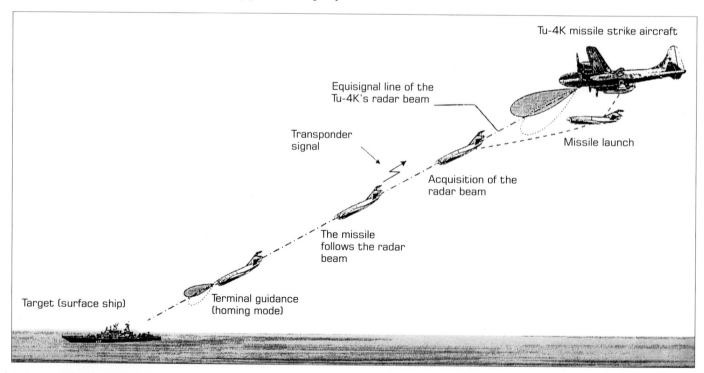

A drawing showing the KS-1's launch technique.

LII test pilots Sultan Amet-Khan, Fyodor I. Boortsev, Sergey N. Anokhin and Vasiliy P. Pavlov made numerous flights in the *izdeliye* K which was launched by the Tu-4K. The first such flight was performed by Amet-Khan in the first aircraft (K-1) in May 1951 (some sources say 4th January 1951). The Black Sea Fleet's *Svetlana* class light cruiser SNS **Krasnyy Kavkaz** (Red Caucasus; SNS = Soviet Navy ship) with a full displacement of 9,000 tons and a hull length of 167.9 m (550 ft 10 in) was used as the target; she had been decommissioned in 1949 to become a test vessel. During these initial tests the cruiser moved on its designated course 20-30 km (10.8-16.2 nm) from the coast off Cape Chauda. The demonstrator's pilots were requested to come as close to the ship as possible before breaking off the 'attack' – the guidance system had to be tuned in such a way as to make sure the real missile would hit below the waterline.

The Tu-4K always carried two 'manned missiles' but only one was launched in any test flight. The Tu-4K's propeller discs were uncomfortably close to the demonstrator's cockpit, and the pilots were reluctant to apply full power immediately after separation, fearing a collision. As a result, the demonstrator would drop 600-800 m (1,970-2,625 ft) below the Tu-4's flight path, which made it hard to acquire the radar beam. Another major problem was that, in spite of the flaps, *izdeliye* K had an extremely high landing speed – more than 300 km/h (186 mph) – and hence a long landing run. Test pilot Pyotr I. Kaz'min also participated in the program.

As in the case of the real missile, the engine of the *izdeliye* K was started by the missile launch operator sitting in the center pressure cabin of the Tu-4K 'mother ship'; however, the pilot of the thing could also start the engine if necessary – a feature which once saved his life. One day the Tu-4K prototype was flying a routine test mission with Nikol'skiy in the captain's seat and Sultan Amet-Khan in the cockpit of the demonstrator. Taking off from Bagerovo AB, the aircraft flew west almost as far as the city of Simferopol', then made a left turn and headed towards Cape Chauda. Nikol'skiy asked missile launch operator Aleksandr Shevelyov over the intercom if the *izdeliye* K was ready. Instead of giving a straight answer, Shevelyov started asking questions – 'how many minutes left to launch', 'aren't we rushing things' and the like. All at once there was a sharp jolt as the Tu-4K was flying at 3,000 m (9,840 ft) over the foothills of the Crimean Mountains. Shevelyov glanced out of the observation blister and saw that the *izdeliye* K was gone; he shouted in frightened tones over the intercom that they had lost the demonstrator. Immediately afterwards the navigator sitting in the extreme nose caught sight of the demonstrator which was rapidly losing altitude and turning towards the coast; Nikol'skiy had no radio communication with Amet-Khan and could not ask him what had happened. When asked if perhaps he had inadvertently pushed the launch button, Shevelyov was positive that he had not – the button had a hinged guard which had to be opened first. Nikol'skiy had no choice but to contact the tower at Bagerovo AB, reporting what had happened, and was told to observe the demonstrator insofar as possible and then return to base.

Sultan Amet-Khan's excellent airmanship and presence of mind saved the day; with the engine shut down, the *izdeliye* K was falling like a brick, and he had no more than two minutes to start the engine, accelerate the aircraft and bring it into level flight. He

Basic performance of the Tu-4K (Tu-4KS)	
Normal AUW with two missiles	52,000 kg (114,640 lb)
Maximum AUW	62,000 kg (136,680 lb)
Maximum speed at high altitude	485 km/h (301 mph)
Service ceiling	8,600 m (28,215 ft)
Normal range with two missiles	4,000 km (2,484 miles)

coped, levelling out a few dozen feet above the water and heading back to Bagerovo. This time the demonstrator came straight in and landed, skipping the usual pass over the runway ending in a zoom climb that signified a successful mission. When Amet-Khan climbed out and the ground crew ran up to him, asking what had happened, he just waved them aside, muttering something in disgust, and walked off to report to the approaching project leaders. Yet, it was not for nothing that Amet-Khan was known as an honest man. He frankly admitted that when the *izdeliye* K was still on the pylon, he, having nothing to occupy himself with until the moment of launch, started idly pushing the buttons on the control console, being certain that the electric power supply from the 'mother ship' was off. So it was – except for the release button, which was always power-on in case of an emergency, and when Amet-Khan pushed it the demonstrator fell off.

On another occasion the autopilot failed to disengage at the terminal guidance phase when Fyodor I. Boortsev was flying the *izdeliye* K. Boortsev had to exert considerable force in order to override the autopilot servos and avoid a collision with the target ship.

Incidentally, the high-priority and risky job of the test crew was well paid – as it should be. Legend has it that after a while the stingy commanders decided that the risk was not so high any more and that the test pilots' reimbursement should be reduced by a factor of ten (!). However, the original pay rates had been set out in a government directive endorsed by Iosif V. Stalin himself; thus, the new rates would have to be approved at the top level as well. Before the draft was sent to the top it was given for approval to Sultan Amet-Khan, who wrote a sarcastic comment on the document: *'My widow disagrees'.* Stalin's verdict was in the same macabre humorous vein: *'I agree with Amet-Khan's widow. (signed) Stalin'.*

Even as the tests continued, on 4th August 1951 the Council of Ministers issued a directive requiring 12 Tu-4s to be converted to Tu-4K standard and delivered to the Naval Aviation by the end of the year; a further 26 aircraft were to be converted in 1952.

After 150 manned flights had been made (including 78 by Sergey N. Anokhin), the appropriately modified K-4 demonstrator commenced a series of unmanned launches. As Russian folk wisdom goes, 'the first pancake comes out wrong'; the first unmanned launch of a KS-1 performed by Maj. N. P. Kazakov's crew over the Arabatskaya Strelka sand spit in the Sea of Azov in May 1952 was unsuccessful – the missile failed to acquire the radar beam because of an incorrectly set autopilot and fell into the sea.

That year the KS-1 cruise missile entered low-rate initial production at plant No.256 in Ivan'kovo township in the Kalinin Region. (Ivan'kovo was transferred to the Moscow Region in 1958 when the administrative boundary was changed; it was later absorbed by the nearby town of Doobna.) The missile had the in-house product code *izdeliye* B; the NATO codename was AS-1 *Kennel*.

Incidents continued when real KS-1 missiles commenced trials. On one occasion an inert KS-1 launched from the starboard pylon failed to 'fall through' before accelerating, as it normally would. Moving to the left, the missile struck the propeller of the No.3 engine and was thrown violently to the left – straight into the propeller disc of the No.2 engine! Finally falling clear, the missile entered a dive and fell into the Black Sea. The collision inflicted considerable damage; not only had the Tu-4K lost both inboard engines but a blade fragment had struck the No.1 engine, severing the control cable so that the engine stopped responding to throttle inputs. Capt. Nikol'skiy ordered the crew to prepare for bailing out; just then it transpired that, for whatever reasons, one of the test crew members had no parachute. Nikol'skiy had no choice but to land at Bagerovo, managing a safe emergency landing with the crash rescue teams standing by. When the aircraft came to a standstill, a mechanic put up a stepladder, climbed up on the port wing, fighting against the prop wash, opened an inspection hatch and shut down the engine manually.

A long investigation ensued as the designers tried – with scant success – to find the reason of the missile's abnormal behaviour. Nothing of the sort had happened in any of the previous launches, and all attempts of the Tupolev OKB, TsAGI and LII specialists to replicate the situation ended in failure. Basically, the real missile differed outwardly from the *izdeliye* K demonstrator only in lacking the cockpit canopy, and calculations showed that it could not have any significant effect on the missile's aerodynamics and provoke an abnormal separation. Yet, to be on the safe side, several initial-production KS-1s were fitted with a teardrop-shaped metal fairing replicating the canopy; this was eventually removed as unnecessary after many test launches.

State acceptance trials of the Kometa weapons system began in July 1952, continuing until January 1953 and involving 12 launches. Again, SNS *Krasnyy Kavkaz* sailing in the Black Sea about 100 km (54 nm) from Feodosiya served as the target – this time literally. After the ship put to sea, the controls were set so that she moved on a circular course at her maximum speed of 18 kts (33 km/h or 20.5 mph); then the crew was evacuated by motor torpedo boats coming up alongside, watching from a safe distance. Next, the Tu-4K began its missile attack. Instrumented test rounds having no warhead were used at first; as a rule, they impacted amidships and went clean through the hull, losing their wings in the process and punching huge holes in the ship's sides (the breach on the exit side was more than 10 m²/107.5 sq ft). Then the crew returned on the double, boarded the ship, making hasty repairs to stop her from sinking, and brought her back to base where she was patched up within a short time and readied for the next experiment. Eight of the 12 missiles launched in the course of the state acceptance trials scored direct hits.

When asked if a single KS-1 could sink a ship of this size, Navy specialists said this was impossible. Hence the 12th and final launch was done using a live missile to check this claim. On 21st November 1952 SNS *Krasnyy Kavkaz* put to sea for the last time. Everything (that is, the setting-up and evacuation procedure and

the Tu-4K's target run) was pretty much as usual until the missile hit. The explosion tore the cruiser apart – she broke in two and sank within three minutes. Afterwards, some critics maintained that the ship had been rigged with an additional explosive charge and such devastation would have been impossible without it. Interestingly, it was not until 1955 that SNS *Krasnyy Kavkaz* was officially struck off charge, so she was a ghost ship for two years; in 1964 her name passed to a Type 61 (*Komsomolets Ookrainy* class; NATO *Kashin* class) guided missile destroyer of the Black Sea Fleet.

The Tu-4K was not built as such; about 50 bombers were converted to this standard by plant No.22 in Kazan' and plant No.23 in Moscow. At the Kazan' plant the Tu-4K conversion job was known under the codename 'order 162'; the plant made four such conversions in 1952 and another 21 in 1953. The BD-KS missile pylons of production examples differed from those of the prototype, having a truncated rear end in order to avoid hampering the flaps' operation – and, of course, no recess for the canopy; some sources call the production model BD-Tu4.

The Kometa weapons system was officially accepted for service in early 1953 as the first of its kind in the Soviet Union. It deserves mention that Sergey N. Anokhin and Vasiliy P. Pavlov received the HSU title for their part in the creation of the system. Sultan Amet-Khan, a wartime fighter ace with 30 'kills' to his credit, was already Twice HSU by then; should he be awarded a third HSU title, a plausible explanation would have to be given in the press – which was impossible, considering the top secret status of the program. Therefore, Amet-Khan received 'only' the Order of Lenin and became a Stalin Prize laureate.

The Tu-4K saw service with two Naval Aviation regiments (in the Black Sea Fleet and the North Fleet) and two Air Force/DA regiments until the mid-1950s. After that, the obsolescent Tu-4K was replaced by the Tu-16KS *Badger-B* armed with the same missiles. Most of the Tu-4Ks were converted to Tu-4D transports.

Tu-4 with AAMs

Perhaps the most unusual weapon carried by the *Bull* in service was the RS-2-U AAM (alias K-5M or *izdeliye* I; NATO codename AA-1 *Alkali*) developed by OKB-2 under Chief Designer Pyotr D. Grooshin. The installation was meant to enhance the bomber's defensive capability in the rear hemisphere. The four missiles were carried back to front on APU-3 launch rails (*aviatsionnoye pooskovoye oostroystvo* – aircraft-mounted launcher) mounted under the rear fuselage and fired by the radar operator; guidance was by means of the Kobal't radar having a suitably modified directional pattern.

A few Tu-4s modified in this fashion saw service with the 25th NBAP (*nochnoy bombardirovochnyy aviapolk* – Night Bomber Regiment). Generally, however, the system proved unsatisfactory and did not gain wide use. Target lock-on was unstable, the launch range was rather short and the missiles were expensive, not to mention the fact that they were intended for the Air Defence Force, not the strategic bomber arm.

Tu-4R (B-4R) long-range reconnaissance/ECM aircraft

Concurrently with the effort to copy the B-29 and put the basic B-4 (Tu-4) into production the Tupolev OKB worked on a long-range photo reconnaissance version – the Soviet counterpart of the Boeing F-13 (later redesignated RB-29). In basic daytime reconnaissance configuration the mission equipment of the B-4R, known from 1948 as the Tu-4R (*[samolyot-] razvedchik* – reconnaissance aircraft), consisted of AFA-33/75 cameras with a 750-mm focal length and AFA-33/100 or AFA-33M aerial cameras with a 1,000-mm focal length. Apart from the cameras, the aircraft differed from the bomber version in having extra fuel tanks in both bomb bays, which held 10,500 kg (23,150 lb) of additional fuel to extend range. The all-up weight was 65,300 kg (143,960 lb), including 24,500 kg (54,010 lb) of fuel.

For night missions the aircraft was configured with NAFA-3S/50 or NAFA-5S/100 cameras; in this case one of the extra fuel tanks was removed so that the aircraft could carry FotAB flash bombs. Otherwise the Tu-4R was identical to the baseline *Bull*. The Tu-4's capacious bomb bays permitted the installation of extremely large film cassettes accommodating enough film for up to 195 300 x 300 mm (11.8 x 11.8 in) frames.

The Tu-4R was not built as such, but a number of bombers were converted into 'Photobulls' in service units as per necessity. These aircraft differed from the original Tupolev OKB project in that only three additional fuel tanks were carried (all in the forward bomb bay) while the rear bomb bay housed the cameras.

Later in their service career some Tu-4Rs were equipped with a PR-1 electronic intelligence (ELINT) and electronic countermeasures (ECM) system designed under the guidance of Chief Designer Ye. Ye. Freyberg. The system was capable of detecting enemy radars with a pulse rate frequency (PRF) of 150-3,100 MHz; it could also emit D-band (2,600-3,100 MHz) jamming signals, which is why it was sometimes called PR-1D or simply PR-D. From 1954 onwards the Tu-4R was equipped with SPS-1 wideband jammers (*stahntsiya pomekhovykh signahlov* – interference signal emitter), and some aircraft which survived into the late 1950s were equipped with the more capable SPS-2. Both models were borrowed from the Tu-16 (they were fitted to the Tu-16RM-1/Tu-16RM-2 *Badger-D* maritime reconnaissance version and the Tu-16SPS ECM version).

Additionally, a DOS chaff dispenser could be installed in one of the bomb bays, ejecting lengths of aluminium foil strip or short glass 'needles' to generate false radar returns. The ELINT/ECM suite was operated by two additional crewmen – an electronic warfare officer (EWO) and a technician. It has to be said the suite was not very popular with Tu-4 crews, especially since the bulky PR-1 deprived the crew of the much-needed rest area. By virtue of its additional ECM role the Tu-4R was fielded not only with specialised reconnaissance air regiments but also with squadrons of ordinary bomber regiments to support the operations of the basic bomber version; such squadrons were known as 'noise' squadrons.

Tu-4 ELINT and ECM versions
In the 1950s the Tu-4 served as the basis for a number of ELINT and passive/active ECM versions designed to support the operations of the basic bomber version. Since Soviet electronic equipment of the time was rather bulky and consumed a lot of power, there was no way the entire mission suite could be crammed into a single aircraft; hence a family of five aircraft working in concert

had to be created. Each aircraft carried equipment covering this or that waveband; together they took care of the entire range of the enemy air defence's electronic assets. The equipment was installed in the bomb bays.

No separate designations are known for the five variants; however, each set of equipment had its own code letter in Russian alphabetical sequence (A, B, V, G and D). The ciphers A and G denoted ELINT-configured aircraft; the ciphers B and V were aircraft with a mixed ELINT/ECM equipment suite, while the cipher D was for a pure ECM version. Only one set of equipment with this or that cipher could be installed at a time; hence at least two or three differently configured aircraft had to escort a bomber armada. This was obviously inefficient, hence no more than ten 'Electric *Bulls*' were in service and they saw only very limited use.

In 1956 the NII-108 research institute within the Ministry of Electronics Industry framework – later known as the Central Radio Technology Research Institute (TsNIRTI – *Tsentrahl'nyy naoochno-issledovatel'skiy rahdiotekhnicheskiy institoot*) – brought out an experimental ECM suite called *Zavesa* (Pall, or Curtain). The suite was too bulky to fit into a single aircraft and was therefore installed in two Tu-4s working as a pair. No further details are known.

Tu-4 radiation reconnaissance aircraft ('order 20')
The scant available information on this one-off aircraft is somewhat contradictory. One source states that in 1950 a single Tu-4 was outfitted with radiation reconnaissance equipment developed by the Physical Institute of the Soviet Academy of Sciences (FIAN – *Fizicheskiy institoot Akademii naook*); the conversion job was done by OKB-30 headed by Aleksandr P. Golubkov – the design office of the MMZ No.30 '*Znamya trooda*' (Banner of Labour) aircraft factory at Moscow-Khodynka. According to another source, in 1951-52 the Tupolev OKB and the Nuclear Physics Research Institute (NIIYaF – *Naoochno-issledovatel'skiy institoot yahdernoy fiziki*) headed by Igor' V. Kurchatov converted a production Tu-4 into a radiation reconnaissance aircraft equipped with Geiger counters in the bomb bay and a magnetic anomaly detector (MAD) in the rear fuselage. However, both sources concur that the aircraft was based in the Far East and used for gathering intelligence on US nuclear tests in the Pacific. Some reports say the converted Tu-4 operated from airbases in south-eastern China, monitoring the area where the tests were conducted. No separate service designation is known, although the aircraft was referred to in papers as 'order 20'.

Tu-4 escort fighter 'mother ship' (Project *Burlaki*)
When the Tu-4 entered service in 1948, it was clear that large bomber formations could not rely on their defensive armament alone and needed fighter escort. The potential adversary had jet fighters capable of flying nearly twice as fast as the Tu-4, and the bomber's prospects of reaching its targets unescorted were nebulous. This was clearly demonstrated by the Korean War where quasi-North Korean MiG-15s inflicted heavy losses on USAF B-29s attacking targets in North Korea.

The main problem was that the Soviet Union had no escort fighters compatible with the Tu-4. The Tu-83 long-range escort fighter based on the Tu-82 experimental twin-turbojet tactical bomber of 1949 remained a 'paper aeroplane'. The production

Above: Tu-4 '46 Black' (c/n 221001) was probably the first one modified under the Burlaki towed fighter concept.

Left: The rear end of '46 Black', showing the stowed drogue on the port side of the fuselage into which the towed fighter locked. The stripe above the rudder is red.

Below left: The drogue holder and the long conduit for the towing cable.

Below: In this view the conduit has been removed, showing that the cable went forward almost as far as the APU exhaust.

MiG-15 was clearly unsuitable; even the MiG-15S*bis* (*izdeliye* SD-UPB) with 600-litre (132 Imp gal) drop tanks had a range of only 2,520 km (1,565 miles). This was adequate for escorting IL-28 tactical bombers with a range of 2,400 km (1,490 miles) but definitely not enough for the Tu-4 with its 5,400-km (3,354-mile) range.

One way to tackle the problem was for the bombers to carry captive or 'parasite' fighters with them. This approach had been pursued both in the USSR, where Vladimir S. Vakhmistrov's team developed seven fighter/bomber combinations called *zveno* (flight, as a tactical unit) in 1931-39, and in the USA, where experiments were carried out after the Second World War with B-29s carrying

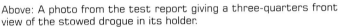

Above: A photo from the test report giving a three-quarters front view of the stowed drogue in its holder.

Above right: The drogue and holder of another Tu-4 similarly converted (note that the holder is of a darker colour).

Right: In an emergency (if normal uncoupling was impossible) the fighter's 'harpoon' could be jettisoned, the movable part remaining locked into the drogue as shown here. This aircraft is an early-production Tu-4 with three B-20E cannons in the tail barbette.

Below right: Tu-4 '46 Black' in flight with the drogue stowed.

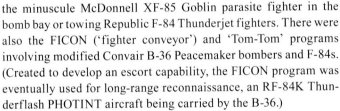

the minuscule McDonnell XF-85 Goblin parasite fighter in the bomb bay or towing Republic F-84 Thunderjet fighters. There were also the FICON ('fighter conveyor') and 'Tom-Tom' programs involving modified Convair B-36 Peacemaker bombers and F-84s. (Created to develop an escort capability, the FICON program was eventually used for long-range reconnaissance, an RF-84K Thunderflash PHOTINT aircraft being carried by the B-36.)

A similar program was conducted in the USSR. In 1950, responding to a Long-Range Aviation headquarters proposal, Aleksandr S. Yakovlev's OKB-115 together with OKB-30 began investigating ways of increasing the range and endurance of escort fighters without resorting to drop tanks. The argument was that a fighter weighed down by drop tanks becomes slow and sluggish, which spoils its chances in a dogfight with enemy fighters.

The solution was a system enabling the Tu-4 to tow MiG-15*bis* fighters, with automatic coupling and uncoupling. In theory, the bomber would still be able to carry a full payload and the fighter would be 'travelling light', its performance uncompromised by extra fuel. The system was codenamed Burlaki (pronounced *boorlakee*). In Russia, from the 16th century to the early 20th century, the burlaki (the plural of *boorlak*) were teams of strongmen whose job was to haul barges up rivers by means of ropes; the analogy with the towed fighter concept was obvious.

The Yakovlev/OKB-30 system utilised a drogue deployed by the Tu-4, the towing cable running through an external conduit on the port side to a winch in the rear fuselage operated by the tail gunner. A pneumatically-operated telescopic probe with a barbed tip was installed atop the fighter's nose on the fuselage centerline; it was promptly dubbed 'harpoon' and the appellation found its way into official documents as well. The *modus operandi* was as follows. The bomber paid out 80-100 m (260-330 ft) of cable, the fighter approached the drogue and 'fired' the 'harpoon' into it; then the fighter pilot shut down his engine and the fighter was towed by the Tu-4 like a glider. If enemy fighters attacked, the fighter pilot

Tu-4 '41 Red' (c/n 1840848) was further modified to feature a combined towing/in-flight refuelling system. Note the fatter conduit for the hose and the extra drogue using the towing cable as a guide.

started his engine, broke contact with the bomber and engaged the enemy, hooking up to the bomber again for the journey home.

(Incidentally, the Lockheed Company tested a similar system in mid-1947 on a modified P-80A-1-LO, 44-84995 (buzz number PN-995). However, unlike the B-29/Tu-4 story, this was probably one of the cases when engineers working on the same problem in different countries arrive at the same solution independently.)

For starters, the Yakovlev/OKB-30 system was tested on the first prototype of the Yak-25 single-engined straight-wing jet fighter of 1947 ('15 Yellow', c/n 115001) and a modified North American B-25J Mitchell bomber fitted with a BLK-1 winch deploying a drogue on a 150-m (490-ft) cable. Stage 1 lasted from 1st June 1949 to 30th September 1950; nine successful contacts were made, with Sergey N. Anokhin flying the bomber and Valentin Chapov flying the fighter.

Stage 2 involved 'the real thing'. The 46th Kazan'-built Tu-4 (aptly serialled '46 Black', c/n 221001) was equipped with a BLI-50E winch and drogue holder 'basket', and a Gor'kiy-built

MiG-15*bis* serialled '408 Red' (c/n N53210408) was fitted with a 'harpoon' identical to that of the Yak-25. The bomber was converted by OKB-30, using Yakovlev OKB drawings, and the fighter by Yakovlev's prototype construction plant, MMZ No.115 (MMZ = *Moskovskiy mashinostroitel'nyy zavod* – Moscow machinery plant).

Tests began at LII on 2nd February 1951 and were completed on 26th March. The results were encouraging; the conversion had virtually no adverse effect on either aircraft's performance, and reliable and safe contact could be made in daytime and at night without any trouble. The MiG-15's engine could be easily restarted at up to 6,000 m (19,685 ft). Contact was so smooth that the bomber's crew hardly felt anything at all, and the bomber's speed was reduced by only 10-12 km/h (6.2-7.4 mph) if engine rpm remained constant. Anokhin reported that the MiG-15*bis* handled well under tow and the procedure could be mastered by the average pilot in two or three training sorties.

Between 28th July and 24th August 1951 the Burlaki system passed its state acceptance trials – again with good results. The bomber and the MiG-15 were flown by project test pilots Anatoliy D. Alekseyev and Vasiliy G. Ivanov respectively; M. I. Panyushkin and Ol'ga N. Yamshchikova were the engineers in charge of the trials. According to the GK NII VVS report, connection and disconnection was possible in level flight at 300-360 km/h (186-223 mph) IAS and 200-9,000 m (660-29,530 ft), during turns with 15-20° bank and climb/descent at up to 10 m/sec (1,970 ft/min). In clear weather the 'air train' could briefly cruise at its service ceiling of 9,650 m (31,660 ft), which was 1,550 m (5,085 ft) less than the standard bomber's. Its top speed was 392 km/h (243 mph) at sea level and 490 km/h (304 mph) at 9,000 m versus 524 km/h (325 mph) for the standard Tu-4.

Yet the trials also revealed that the Burlaki had serious shortcomings. Firstly, the MiG-15's cockpit heating and pressurisation system did not work with the engine inoperative, and sitting for hours in a cockpit which became bitterly cold at 7,000-10,000 m (22,965-32,810 ft), wearing an oxygen mask, was a sore trial for the pilot. Secondly, the drag generated by the towed fighter slowed

Tu-4 '41 Red' with a combined towing/IFR system. The scrap views show Tu-4 '46 Black' in the towing-only version with the drogue stowed and deployed.

Right: Tu-4 '41 Red' with the towing drogue deployed. The refuelling drogue is still stowed.

Below right: Tu-4 '46 Black' towing MiG-15*bis* '408 Red'.

Bottom right: The Tu-4's towing drogue as seen by the pilot of a MiG-15*bis* about to make contact.

the Tu-4, and a slow bomber in the formation would inevitably slow down the entire formation, which was unacceptable. Maximum range at 6,000 m with the Tu-4's TOW of 63,320 kg (139,600 lb), including 2,000 kg (4,410 lb) of bombs, was only 3,920 km (2,430 miles) versus 4,740 km (2,944 miles) for the standard bomber, which was poor performance by the day's standards. Worse, the fighter had no chance of reaching its home base if it became separated from the bombers during a dogfight with enemy fighters – the pilot would have to eject.

The trials report contained many suggestions, such as providing a secure telephone link allowing the fighter pilot and the bomber crew to communicate while maintaining radio silence and adapting the system for the new and faster bombers then under development (the Tu-16 and Tu-95). The main proposal, however, was to change the ideology of the system completely; the probe and drogue were to be used for in-flight refuelling rather than towing. This led to the next phase of development (see page 94).

Despite the system's shortcomings, on 30th October 1951 the Council of Ministers issued a directive ordering the conversion of five more Tu-4s and five more *Fagot-Bs* to Burlaki standard for service trials. The bombers (c/ns 1840848, 2805003, 2805005, 2805110 and 2805203) were indeed modified by plant No.18. In contrast, the fighters were custom-built with the 'harpoons' by plant No.153 in Novosibirsk; they were serialled '2170 Red', '2175 Red', '2176 Red', '2190 Red' and '2204 Red' (c/ns 2115370, 2115375, 2115376, 2115390 and 2215304 respectively).

The trials were held in the 50th VA (*vozdooshnaya armiya* – Air Army, ≈ air force) at Zyabrovka AB (Gomel' Region, Belorussia) between 9th July and 8th September 1952. The Tu-4s were flown by five crews of the 57th *Smolenskaya* TBAD/171st *Smolensko-Berlinskiy* GvTBAP/3rd AE, while the MiG-15s were operated by ten crews (that is, pilot and technicians) of the 144th IAD/439th IAP/1st AE. (TBAD = *tyazholaya bombardirovochnaya aviadiveeziya* – Heavy Bomber Division (≈ Bomb Group (Heavy)); GvTBAP = *Gvardeyskiy tyazholyy bombardirovochnyy aviapolk* – Guards Heavy Bomber Regiment (≈ Bomb Wing (Heavy)); AE = *aviaeskadril'ya* – Air Squadron; IAD = *istrebitel'naya aviadiveeziya* – Fighter Division (≈ Fighter Group); GvIAP = *Gvardeyskiy istrebitel'nyy aviapolk* – Guards Fighter Regiment. The Guards units are the elite of the Soviet/Russian armed forces; this title was given for gallantry in combat, thus indicating this is a

World War Two-vintage unit. The honorary appellation *Smolensko-Berlinskiy* was given for the division's part in liberating Smolensk and taking Berlin. The 57th TBAD and the 171st GvTBAP had also been awarded the Order of the Red Banner of Combat.)

The objective was to evaluate the system's reliability and 'user-friendliness', fighter/bomber rendezvous techniques and formation flying techniques. The trials involved 142 hook-ups

Left: Tu-4s modified for fighter towing seen at Zyabrovka AB in the summer of 1952.

Below left: The captain's instrument panel of a Tu-4 equipped with the Burlaki system; the circle marks the control panel pertaining to the system.

Bottom: This drawing shows the fighter's position relative to the Tu-4 during towed flight/refuelling. Horizontal separation was 20 m (65 ft) and vertical separation was 5-8 m (16-26 ft).

(17 of them at night) and went without incident. The manoeuvring envelope was slightly narrower than during the state acceptance trials, with bank angles up to 15° and rates of climb/descent up to 7 m/sec (1,380 ft/min). The longest towed flight lasted 2 hours 30 minutes, including 2 hours 27 minutes with the engine shut down.

Before making contact the fighters zeroed in on the bombers, using their ARK-5 automatic direction finders which worked with the bomber's 1RSB-70 radio used as a short-range navigation system. During landing approach the fighters stayed connected right down to 300 m (980 ft). After extending the landing gear and setting the flaps 20° the fighter pilot waited for the signal from the bomber crew or the tower to break contact; receiving the go-ahead, he disengaged at 2-3 km (1.24-1.86 miles) from the runway threshold and landed normally.

The system was ultimately put to the test in two sessions of mock combat on 5th August 1952. A flight of fighter-towing Tu-4s was 'attacked' by four MiG-15s representing enemy fighters. The attackers were guided to their target by a ground controlled intercept (GCI) station using target information from an air defence radar.

On the first occasion, the towed fighters lost; paraphrasing the system's 'strongman' name, the Burlaki turned out to be strong in the arm but weak in the head. The bombers' flight leader spotted the 'enemy' at 12-15 km (7.45-9.3 miles) range as the fighters were making their first attack and gave the order to start the engines, disengage and repel the attackers. However, as the towed fighters did so the attackers managed to make a second 'firing pass'. If this had been for real, the bombers would have been shot down – probably taking their captive protectors with them!

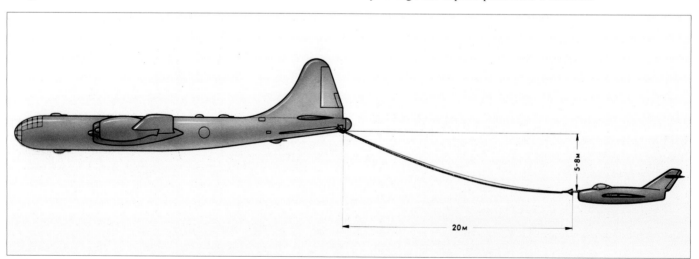

The second try was more successful; two pairs of MiGs took turns patrolling (flying top cover) and resting (that is, being towed). This time one pair of escort fighters was ready to repel an incoming attack; yet again the 'bad guys' were discovered a little too late and the protective pair just couldn't cope with them. Still, the 'enemy' fighters did not manage to repeat the attack before they found themselves counterattacked. The conclusion was that incoming enemy fighters needed to be spotted at least 4 minutes before they got within firing range so that the towed escort fighters could get ready. This could be done by fitting the bomber with a search radar, enabling the crew to spot enemy fighters at 60-80 km (37-50 miles) range.

Technology quickly made the Burlaki system obsolete and it never entered service. Firstly, the Tu-4 was replaced by the Tu-16 jet bomber whose cruising speed of 1,000 km/h (621 mph) matched the speed of many fighters. The *Badger*'s heavy defensive armament and ECM equipment gave it a good chance of reaching its target. Secondly, experiments began with in-flight refuelling (IFR) systems for fighters; some of them are described below. Work on towed captive fighters was discontinued in 1954.

Tu-4 in-flight refuelling tanker/receiver versions

As noted earlier, the Tu-4 belonged to the long-range bomber class. With a normal bomb load it could reach targets in Europe, northern Africa, the Middle East and Japan from bases on the vast territory of the USSR. Yet the USA brought out strategic bombers with far longer range, most notably the Convair B-36 Peacemaker. The Soviet military leaders wanted an aircraft with comparable range that could hit targets in North America. One of the steps towards this goal was an attempt to give the Tu-4 longer legs by giving it IFR capability.

The first Soviet experiments in this area dated back to the 1930s when attempts were made to extend the range of the Tupolev TB-3 heavy bomber. A sister aircraft equipped with an additional fuel tank acted as the tanker. Since the TB-3's airspeed was low, a hose could be paid out by means of a hand-driven winch as the tanker flew directly above the receiver aircraft; a crewman would then catch it and connect it manually to the fuel filler, whereupon transfer of fuel by gravity could begin. This method was most inconvenient, and the work did not progress beyond the experimental stage.

a) 'system of crossing cables'

IFR technique development in the Soviet Union resumed after the Second World War, now that the designers were armed with new technologies and facing new requirements. The Tu-4 bomber was the prime candidate for receiving an efficient IFR system, but tactical fighters were not left out either – at this stage anyway.

On 12th April 1949 Air Force C-in-C Pavel F. Zhigarev wrote thus to Minister of Aircraft Industry Mikhail V. Khroonichev: '*Our preliminary calculations for a production Tu-4 being refuelled by another Tu-4 show that there are no technical obstacles to increasing the range of a solitary Tu-4 carrying a 2,000-kg [4,410-lb] bomb load to 9,700 km [6,024 miles] with two fuel top-ups. [...] With further modifications, the Tu-4 can have a range of 15,000-16,000 km [9,316-9,937 miles] with two or three fuel top-ups, which*

Above right: The in-flight refuelling receptacle of a Tu-4 fitted experimentally with an early version of the Vakhmistrov IFR system. The tanker's hose is locked into position.

Right: A later version of the receptacle of the Vakhmistrov IFR system.

will increase the combat radius of a solitary heavy bomber to 6,000-7,000 km [3,726-4,347 miles]. Using bases on Rudolph Island, near Murmansk and in the east of the Chukotka Peninsula will put Canada and most of the USA within this radius. Thus, having Tu-4s refuelled by sister ships allows us to take aerial operations to North America; three-quarters of the route will be over the Polar Regions where strong air defence assets cannot be put up.'

A while earlier, in 1948, the design team headed by Vladimir S. Vakhmistrov developed the so-called 'system of crossing cables' based on a system developed by the British company Flight Refuelling Ltd. Before making contact both aircraft would deploy cables; the receiver's cable terminated in a stabilising drogue, stretching out almost horizontally, while the tanker's cable was provided with a weight to keep it in a near-vertical position, this weight incorpo-

Top left and above left: The trials of the Vakhmistrov system Unserialled Tu-4 c/n 222405 converted into a tanker begins to deploy the hose from the cutout where the rear ventral cannon barbette used to be.

Left: Seen from the starboard gunner's station, the tanker with the partly-deployed hose changes formation to engage the receiver aircraft's cable and grapple.

Top: Unserialled Tu-4 c/n 222401 was fitted with the refuelling receptacle of the Vakhmistrov system on the starboard side of the tail gunner's station.

Above: The hose is hauled into position by the receiver aircraft's grapple

Above: Tu-4 c/n 222401 leads the Tu-4 tanker (c/n 222405) in refuelling formation. Fuel can be seen streaming from the receptacle, suggesting this is the moment of disengagement.

Right: The two aircraft in refuelling formation, seen from astern.

Below right: The business end of the tanker's hose interlocked with the receiver aircraft's cable.

rating a snap lock. The tanker would fall into echelon port formation with the receiver aircraft, keeping 8-10 m (26-32 ft) behind and 3-4 m (10-13 ft) above it, and move into echelon starboard formation so that the cables crossed and the lock engaged. Then the receiver aircraft rewound the cables, the tanker paying out a hose attached to its own cable until it locked automatically into a receptacle on the receiver's rear fuselage, forming a loop. During fuel transfer the tanker stayed 12-15 m (40-50 ft) above the receiver and slightly behind it, with a distance of about 50 m (164 ft) between the fuselages. Fuel was transferred by pumps delivering up to 700 litres (154 Imp gal) per minute. When refuelling was completed the hose would be unlocked and rewound; the tanker then peeled off to starboard, the cables went taut and a special weak link in one of them gave way, disengaging the aircraft.

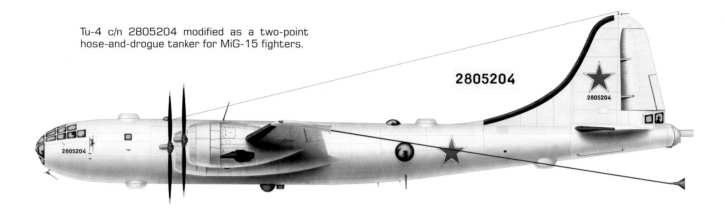

Tu-4 c/n 2805204 modified as a two-point hose-and-drogue tanker for MiG-15 fighters.

The system was flight-tested first on the Second World War-vintage Tu-2 bomber and then on the Tu-4; no pumps were fitted initially and fuel was transferred by gravity on the first attempts. The use of the rather inefficient and outdated 'system of crossing cables' can only be explained by the need to meet stringent deadlines. An operable IFR system had to be demonstrated to the government at all costs – just to 'give the dog a bone', and a better solution could be devised later.

Two Kazan'-built Batch 24 Tu-4s with no serials were modified for testing the system. The tanker (c/n 222405) had a hose drum unit (HDU) installed in the unpressurized rear fuselage; the rear ventral cannon barbette was removed to create an outlet for the hose. The receiver aircraft (c/n 222401) had the receptacle on the starboard side of the tail gunner's station; two versions of this receptacle were tested. Presumably the tail gunner acted as the refuelling system operator (RSO) in both cases.

After a deal of testing and refining the Vakhmistrov system underwent state acceptance trials and was rejected as inefficient and obsolescent.

b) adaptation of the Burlaki towed fighter concept
As mentioned earlier, the VVS considered using a probe-and-drogue system based on the Burlaki towed fighter concept.

Initially OKB-30, OKB-134 and the Yakovlev OKB modified the existing Burlaki system by adding new elements. The fighter's 'harpoon' incorporated a valve and was connected to the fuel system. The bomber was equipped with three kerosene tanks, a pump and an inert gas pressurisation system to reduce the risk of fire and explosion if hit by enemy fire. After the fighter had made contact with the tanker's towing drogue, a hose terminating in a smaller drogue was paid out, the drogue sliding along the towing cable, and the fighter accelerated, locking the two drogues together. (The original drogue was modified so as to allow fuel to pass through it into the probe.) 1,210 litres (266.2 Imp gal) of fuel could be transferred in six minutes. When refuelling was completed the smaller drogue was automatically disengaged and the hose rewound.

Two of the aircraft used for service trials of the Burlaki system – Tu-4 '41 Red' (c/n 1840848) and MiG-15*bis* '2204 Red' (c/n 2215304) – were converted for flight refuelling trials which took place at LII between 24th September 1954 and 2nd March 1955. The tanker was captained by Aleksandr A. Yefimov, with A. I. Vershinin as the RSO; the fighter was flown by Sergey N. Anokhin and Fyodor I. Boortsev. The program involved ten flights in the MiG-15*bis*, including five contacts at 2,000 m (6,560 ft) and 4,000 m (13,120 ft); on three occasions fuel was actually transferred. However, an attempt to repeat the perfor-

Right: Tu-4 c/n 2805204 two-point converted into an experimental hose-and-drogue tanker refuels MiG-15*bis* '342 Blue' and MiG-15 '074 Black'. The latter fighter sports a large Komsomol (Young Communist League) badge on the nose which may look like the numerals '17 Red' at a distance; the actual serial is applied in small digits and illegible in this photo. Note that the tanker retains the cannon barbettes but not the actual cannons, and a large sideways-firing cine camera pack is installed in lieu of the tail cannons to record the contact with the fighters.

Another view of the same refuelling formation flying high over a winter landscape.

Opposite page, bottom: The Tu-4 two-point tanker parked at Zhukovskiy. The stowed drogues can be seen protruding from the wingtips.

mance at 8,500 m (27,890 ft) failed because the system's rubber components froze up, losing their elasticity.

c) two-point probe and drogue system
Generally the 'wet Burlaki' system was considered excessively complex, and as early as August 1951 the Council of Ministers issued a directive which required a system allowing two MiG-15s to be refuelled at a time to be developed. OKB-918 led by Semyon M. Alekseyev took on the task. This design bureau, which absorbed the entire Vakhmistrov team, later became the Zvezda (Star) company best known for the K-36 ejection seat fitted to almost all current Russian combat aircraft.

An unserialled Kuibyshev-built Tu-4 (c/n 2805204) was converted into the two-point tanker at plant No.18. The arrangement proposed by OKB-918 was much simpler than the OKB-30/ Yakov-

lev system, utilising only one drogue and hose for each receiver aircraft. Two HDUs were installed in the forward bomb bay, the hoses running on supporting rollers inside the wings and exiting from specially-modified wingtips. The RSO sat in the tail gunner's station. The hoses were extended by the drag created by the drogues when the HDUs were set to 'unwind'. For night refuelling sessions the drogues had 'cats' eyes' reflectors around the rim that would shine when illuminated by the fighter's landing light. Since the tanker was to work with jet aircraft, part of the fuel tankage was isolated and set aside for kerosene. The cannon barbettes were there but no armament was fitted; the tail guns were replaced by a cine camera unit to record the refuelling sequence.

A Moscow-built MiG-15 *Fagot-A* serialled '074 Black' (c/n 3810704) and three *Fagot-Bs*, including Kuibyshev-built '342 Blue' (c/n 123042 or 133042) and '618 Blue', were fitted with

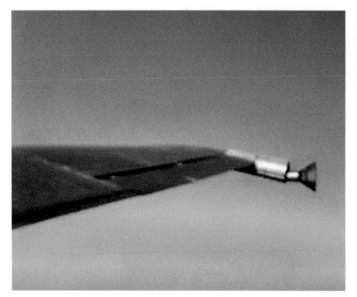

fixed telescopic refuelling probes offset to port on the intake upper lip; the conversion work was done by the Novosibirsk factory in May 1952. Test flights began in 1953, with a considerable delay due to late equipment deliveries for the tanker. The Tu-4 was flown by Pyotr I. Kaz'min, Stepan F. Mashkovskiy, Leonid V. Chistyakov and other LII test pilots, while Sergey N. Anokhin and Vladimir N. Pronyakin flew the fighters; V. Ya. Molochayev and S. N. Rybakov were the engineers in charge. When the fighter's tanks were full (this took four or five minutes), the transfer pumps were shut down and the contact was broken automatically.

At first, Mikoyan OKB engineers were apprehensive about having the probe near the intake, fearing the drogue would generate excessive turbulence at the air intake lip and provoke a compressor stall. These fears were possibly caused by knowing that in the USA, an F-84 fitted experimentally with the probe-and-drogue IFR system had the probe mounted on the starboard wing, well clear of the air intake. However, trials showed these fears to be unfounded.

Several versions of the hose were tried before the system was satisfactory. The original hose incorporating a reinforcing wire spiral was not durable enough. On the other hand, a 'soft' hose with no reinforcing wire flexed excessively and fighter pilots found that just a little turbulence made 'hitting the tanker' very difficult. Another problem was the considerable amount of residual fuel in the hose after the transfer pumps were shut down; as it broke contact with the tanker, the fighter was liberally doused with fuel, some of which even entered the cockpit. Still, the system was simple, reliable and offered a high fuel transfer rate.

The combination of the Tu-4 tanker and two MiG-15*bis* receivers was presented twice for state acceptance trials (the first time was in 1953). Vasiliy G. Ivanov was project test pilot, Aleksey G. Solodovnikov, Viktor S. Kipelkin and Mikhail S. Tvelenyov also flew the fighters; again, Ol'ga N. Yamshchikova was engineer in charge. The system failed the trials both times because of problems with the supporting rollers inside the wings which caused hose oscillation and failure of the fighters' refuelling probes due to the whiplash effect of the hose. Also, unlike the USAF, the Soviet Air Force had no need to fly its fighters over long distances.

d) single-point probe and drogue system

A further 12 Kuibyshev-built Tu-4s were converted into single-point hose-and-drogue tankers in 1952. These aircraft participated in the tests of the Myasishchev M-4 *Bison-A* bomber when the latter was equipped with an IFR probe. Again, part of the fuel system was set aside for kerosene which was transferred to the M-4. The tankers were flown by Stepan F. Mashkovskiy, Pyotr I. Kaz'min, Leonid V. Chistyakov and other LII test pilots. Later, the probe-and-drogue IFR system was used successfully on the Myasishchev 3MN/3MS *Bison-B* and 3MD *Bison-C* heavy bombers, most aircraft of the Tu-95/Tu-142 *Bear* family, the Tu-126 *Moss* airborne warning and control system (AWACS), the Tu-22KD/RD/PD/UD *Blinder* and Tu-22M0/M1/M2 *Backfire-A/B* supersonic long-range bombers. In the mid-1970s the system was finally adapted for tactical aircraft (the Sukhoi Su-24M *Fencer-C* bomber) as well.

Opposite page:
Far left: The starboard drogue of Tu-4 c/n 2805204 is shown here just as it begins to deploy, aligning itself with the slipstream.

Left: The port drogue and hose as filmed by the cine camera mounted on the port wing of one of the MiG-15s. The offset IFR probe and the front end of the outer boundary layer fence are just visible.

Below left: The nose of the tanker as seen from a chase plane. Note the stripes on the hose showing how much of it has been paid out.

Above right: This unserialled Tu-4 (c/n 2805703) is one of the three tankers involved in the state acceptance trials of the 'wingtip-to-wingtip' IFR system.

Right: This (likewise unserialled) Tu-4 is one of the three receiver aircraft participating in the trials. The c/n looks like 2805120, which is impossible, suggesting a retouched photo.

e) wing-to-wing system

Another wing-to-wing refuelling system was developed by LII test pilots Igor' I. Shelest and Viktor S. Vasyanin. They witnessed the testing of the Vakhmistrov system and were quick to perceive its weaknesses. Since both men had received an engineering education, they had no trouble preparing a technical proposal, with drawings and all, which they showed to the institute's top brass. The Shelest/Vasyanin system clearly had major advantages over the 'system of crossing cables': the refuelling process was completely automated, the tanker and receiver aircraft flew on parallel courses and neither had to enter the other aircraft's wake vortex.

Shelest and Vasyanin were allowed to go ahead with the project, working in parallel with Vakhmistrov's team. Initially a pair of Tu-2 bombers were equipped with the new system by LII. On 16th July 1949 LII test pilot Sultan Amet-Khan and one of the system's inventors successfully performed a fully automatic in-flight refuelling in these aircraft. The process was duly documented on cine film and this was demonstrated to Andrey N. Tupolev, who immediately recognised the soundness of the idea and approached the government, lobbying for support. He got his way; a few days later LII received an official government assignment to install and test the new IFR system on the Tu-4 bomber. To this end the institute received two brand-new *Bulls* for conversion as the tanker and receiver aircraft.

Since Shelest and Vasyanin had developed the system at their own initiative, they had to draw up the system's specifications for themselves, relying on their experience as test pilots, engineering

intuition and common sense. Recorded in the official test report, the requirements sounded as follows:

'1. Making contact and keeping formation during refuelling shall be safe and relatively simple, presenting no problem for first-line service pilots after a few conversion training sessions;

2. Contact shall be established quickly in daytime and at night, given adequate visibility, and in moderate turbulence;

3. The means of contact (the hose and whatever goes with it – Auth.) shall remain stable in flight and present no danger of damage to the aircraft's vital components in the event of an incident;

4. The tanker, not the receiver aircraft, shall perform all the necessary manoeuvres during contact and fuel transfer;

5. The bomber's IFR equipment shall be as lightweight and compact as possible and the bomber crew shall have the smallest possible additional workload during refuelling;

6. The refuelling process shall be automated and the IFR equipment shall be remotely controlled from the pressurized cabins;

7. The overall time required, including engagement and disengagement, shall not exceed 20 minutes;

8. The volume of transferable fuel shall be 35-40% of the total fuel capacity if the tanker and receiver aircraft are sister ships;

9. Installation of IFR equipment in the tanker shall not render it unusable in the bomber role and shall not cause deterioration of flight performance.'

Items 4 and 5 of the above are especially noteworthy. Since the bomber is out on a combat mission, it will have to penetrate enemy air defences; then thing will really get ugly and all available crew

A Tu-4 tanker (c/n 222202) deploys the hose, which is engaged by the receiver's parachute-stabilised grapple visible on the left.

resources will be needed. Therefore it is perfectly natural that the bomber crew need not waste adrenalin on the refuelling procedure – they should engage the autopilot and the crew of the tanker, which would not be penetrating the air defences, should do the rest. A few years later, however, when the Shelest/Vasyanin system was adapted to the Tu-16, the Air Force command – not active pilots but chairborne commanders – demanded that the crew of the receiver aircraft should play the active part in the refuelling procedure. The argument in support of this requirement was positively staggering: *'Suppose the bomber crew decides not to fulfil*

the mission and returns to base, claiming they had not been refuelled. Who will take the blame then?'

Actually Shelest and Vasyanin proposed three versions of their IFR system. Version 1 (described below) was implemented on the Tu-4 on a small scale; Version 2 designed for fighters underwent preliminary tests on two modified MiG-19 *Farmer-A* day fighters designated SM-10/1 and SM-10/2 but was not adopted for Air Force service. Version 3 was used successfully for many years on the Tu-16, the tanker version of the *Badger-A* being designated Tu-16Z (*zapravhshchik* – refuelling tanker).

Refuelling under the wing-to-wing system is in progress; as in the case of the Vakhmistrov system, the receiver aircraft leads the formation.

In the version used by the *Bull* the hose deployed from the tanker's port wingtip was pulled over to the starboard wingtip of the receiver aircraft by a special cable, forming a loop. This left both aircraft enough room for manoeuvring and allowed refuelling to take place in turbulent conditions. All the principal manoeuvres and actions involved in formating with the receiver aircraft and making contact were performed by the tanker, while the bomber kept its intended course.

The tanker's equipment fit included the fuel transfer hose 30 m (98 ft) long positioned along the port wing's rear spar, a winch and cable for rewinding the hose; this arrangement incurred a smaller weight penalty than an HDU in the bomb bay. The equipment further included a cluster of three booms tipped with spring-loaded ring clips for capturing the bomber's contact cable, a fuel transfer pump, a sealing chamber for connecting the hose to the fuel delivery pipeline, a hose ejector and a control and indication system. The receiver aircraft featured a winch and cable terminating in a weight equipped with a snap lock and a stabilising drogue, a hose receptacle and fuel transfer pipeline, shutoff valves and a control and indication system. The engagement/disengagement process and fuel transfer were remote-controlled and automated. The entire system was housed inside the airframes of both aircraft, creating no extra drag and having virtually no adverse effect on their performance and handling.

The refuelling procedure was as follows. Maintaining speed, heading and altitude, the receiver aircraft deployed a contact cable 100-120 m (330-390 ft) long, tipped by a drogue-stabilised weight known as ***grooz-zamok*** (locking weight). The tanker closed in from behind, falling into echelon starboard formation with the bomber and placing its port wingtip over the cable. Then the tanker moved away and forward so that the cable slid along the wing undersurface onto a retractable boom at the wingtip, engaging a ring clip connected by springs to the fitting at the tip of the hose. The ring slipped off the boom and the bomber's RSO started rewinding the cable; at the same time the hose ejector forced the outer end of the hose from the tanker's wingtip. Next, the weight at the end of the cable locked into the ring, connecting the cable with the hose; as the cable was rewound it extracted the hose from the tanker's wing, pulling it up to the receiver aircraft. When the hose was fully extended a fitting at the inner end locked into the sealing chamber; meanwhile the fitting at the other end automatically engaged the receptacle under the bomber's starboard wingtip and contact was established, whereupon the transfer pump was switched on.

When the bomber's tanks were full the shutoff valves automatically stopped further fuel delivery. After that the fuel lines were scavenged with compressed nitrogen forcing the remaining fuel back into the tanker's fuel tanks and the hose was released. The tanker rewound the hose, the bomber paying out the contact cable at the same time. As the end of the hose disappeared inside the tanker's wingtip the connection was broken, the ring parting company with the hose and remaining on the cable, which the bomber then rewound.

The system allowed the tanker to make up to three contacts with a bomber during a sortie. Each time a fresh boom and ring clip would be extended from the tanker's wingtip and a special mechanism would attach the ring to the fitting at the end of the hose.

As a matter of fact, the Tu-4 equipped as the receiver aircraft had a duplicated set of IFR equipment for maximum reliability – two winches with contact cables and two refuelling receptacles. Each cable could be used three times, as there was room for only three contact rings on the weight at the end of the cable; thus the bomber could 'hit the tanker' up to six times in a single sortie, which made for very long range indeed.

Refuelling was done at indicated airspeeds of 320-350 km/h (198-217 mph) with the aircraft cruising at the service ceiling. The fuel transfer rate was 800 litres (176 Imp gal) per minute and a maximum of 10,400 litres (2,288 Imp gal) could be transferred at a time. (Some sources state the Tu-4's fuel transfer rate with the 'wing-to-wing' system was 2,000 litres (440 Imp gal) per minute.)

Being test pilots, the inventors were concerned about developing an acceptable piloting technique for the tanker crew right from the start. Igor' Shelest took great pains to make sure the system worked OK, testing it over and over again; he was well aware that a single fatal accident could cost both him and his 'co-author' their lives. Reprisal came swiftly in those days, and a death sentence would be more than likely.

The tanker crew ran into a few problems during early test flights. As the aircraft approached the contact cable deployed by the bomber, the tanker pilots tended to look at their port wingtip over the shoulder, which was inadmissible during formation flying as it created a risk of mid-air collision. The problem was fixed by installing a forward-facing light on the tanker's wingtip for night operations; this emitted a narrow beam of light which the tanker pilot could see clearly. Also, the pilots were now assisted by the senior gunner who sat in the center pressure cabin, controlling the dorsal and ventral cannon barbettes. Using the lateral sighting blister, the senior gunner gave directions over the intercom to the pilot who could now look ahead and left, focusing his attention on the receiver aircraft. He smoothly manoeuvred the tanker into position without bank or sideslip, keeping the required interval by checking the bearing on the bomber and using the bomber's tail to gauge the distance. As a result, the tanker now 'hooked up' to the bomber flawlessly on the first try.

Subsequent test flights performed by Sultan Amet-Khan and Aleksey P. Yakimov gave interesting results: after making a few successful contacts the tanker pilots no longer needed assistance from the gunner, learning to judge the distance and position relative to the bomber correctly just by watching it. For instance, Mark L. Gallai made contact with the bomber successfully on his first night refuelling mission, even though he had no prior experience of night tanker operations. Later, when the Shelest/Vasyanin IFR system found its way to service units flying the Tu-4, service pilots mastered the knack after making ten training flights.

State acceptance trials of the 'wing-to-wing' IFR system were to commence in April 1950. In fact, they began in August 1950, continuing until November, and were completed with unsatisfactory results. Shelest and Vasyanin had to rework the system. A year later the Tu-4 tanker and receiver aircraft with the revised system were submitted to GK NII VVS again for check-up trials; a month later, in December 1951, the military test pilots started practicing night IFR operations. This time the system was recommended for service – providing that the remaining deficiencies were corrected.

Since the new system had shown good results during tests, in 1952 the Air Force suggested building a small batch of Tu-4 tankers and IFR-equipped Tu-4 bombers. The government picked up the ball: on 28th March 1952 the Council of Ministers issued directive No.1523-530 ordering the Tupolev OKB, plant No.18 and a number of other enterprises in the aviation industry to convert three Tu-4s into tankers and equip a further three *Bulls* with refuelling receptacles. The document specified the amount of transferable fuel as 10,000 litres (2,200 Imp gal) and the refuelling time as 20 minutes; one tanker was to be capable of topping up at least three bombers in a single sortie. The aircraft were to be submitted for state acceptance trials in August 1952.

The three tankers converted in keeping with the said directive (c/ns 2805701, 2805702 and 5805703) and the three receiver aircraft (c/ns 2805606, 2805608 and 2805610) commenced check-up trials at GK NII VVS on 27th May 1953. The trials lasted nearly two months, with Aleksandr V. Sarygin and S. K. Musatov as project test pilots. Still, even though the 'wing-to-wing' system met the specifications, it was not recommended for service in the then-current form. Night operations were extremely complicated because there were no signal lights facilitating formation keeping. Furthermore, the IFR equipment turned out to be unreliable, and the hose extractor installed in the front part of the forward bomb bay rendered the tanker unable to carry FAB-250M-46 and FAB-500M-46 bombs whose fuses were obstructed by the hose.

More modifications ensued, and in August 1954 the DA finally took delivery of the first three tankers and three IFR-capable bombers. In April 1955 DA crews put the system to good use, mak-

The one-off Tu-4D (*izdeliye* 76) transport/troopship, c/n 2303201, with two P-90 paradroppable cargo pods fitted under the wings.

ing three sorties in which the bombers covered a distance of 8,200 km (5,100 miles) with two fuel top-ups en route. In July 1956, worn-out IFR equipment forced a pause in such flights until October when the aircraft had been repaired. Ten more tankers and ten bombers with IFR receptacles were delivered in late 1955, followed by a further 13 tankers and 13 receivers in 1956. By the end of 1957 the DA had 23 qualified tanker crews and 21 bomber crews qualified for IFR. Not many *Bulls* were thus upgraded for the simple reason that the far more capable jet-powered Tu-16 featuring the same system was about to enter large-scale production.

Tu-4 fuel carrier

The Tu-4's large fuel tankage rendered it suitable for the 'fuel carrier' role. A number of *Bulls* were adapted for delivering fuel to fighter and tactical bomber bases and refuelling tactical aircraft on the ground. To this end some of the fuel cells were isolated from the bomber's fuel system and used for carrying jet fuel (kerosene); simple refuelling equipment with hoses, pumps and a control panel was fitted.

Tu-4TRZhK liquid oxygen tanker

Similarly, the need to supply aviation and missile units with liquid oxygen (LOX) spawned a version designated Tu-4TRZhK (*trahnsportnyy rezervouar zhidkovo kisloroda* – LOX transportation reservoir). Heat-insulated LOX tanks and associated equipment were installed in the bomb bays of the otherwise standard aircraft. Normally the equipment was only fitted during military exercises when units scattered far and wide had to be supported.

Two more views of the same aircraft, showing all defensive armament has been retained.

This page, top to bottom:
A P-90 cargo pod (the early version with stabilising fins at the rear, as shown on the preceding page) with ground handling wheels attached, loaded with an 85-mm D-44 field gun.

The production version of the P-90 (possibly designated P-90MS) lacking the fins but featuring bumpers under the front end to prevent a nose-over on touchdown. A 57-mm ZiS-2 anti-tank gun is loaded in this case.

The less-than-successful P-98M pod designed with larger SP guns in mind, with three ground handling wheels attached.

The P-110K cargo pod designed to work with a parachute/retro-rocket system, which is displayed alongside. Note the hinge lines allowing the sides to swing open after touchdown for unloading.

Opposite page, top: Here the Tu-4D (*izdeliye* 76) is seen at a later date, loaded with production-standard P-90 pods.

Right: P-90, P-98 and P-110K (left to right) cargo pods at the premises of plant No.468 during tests.

Below right: The aftermath of an unsuccessful drop of a P-90 pod: the GAZ-69 jeep inside is visibly damaged in the hard landing.

Far right, above: A P-90 pod is ready for hooking up to the Tu-4D (*izdeliye* 76), whose engines and flight deck are under wraps.

Far right: A P-110K pod with an 85-mm gun inside suspended under the starboard wing of a Tu-4D.

Tu-4D military transport/troopship aircraft (first use of designation, *izdeliye* 76)

As early as 1947 the Tupolev OKB considered using the Tu-4 as a military transport carrying various combat vehicles on underwing racks (with or without streamlined cargo pods or fairings). Three payload options were proposed. Option 1 was two self-propelled howitzers weighing 3,800 kg (8,380 lb) each. At an all-up weight of 60,000 kg (132,275 lb), including 14,100 kg (31,085 lb) of fuel, the Tu-4 loaded with these howitzers would have performance as detailed in the first table on page 104.

Option 2 reportedly consisted of two 76.2-mm (3-in) OSU-76 SP howitzers. (This appears very unlikely, as the OSU-76 was an

experimental lightweight version of the production SU-76, featuring the same ZiS-3 gun on a different chassis from a light tank and a weight reduced from 14 tons (30,860 lb) to 4 tons (8,820 lb), hence OSU for *obleg**chon**naya samo**khod**naya [artille**reey**skaya] oosta**nov**ka* (lightened SP howitzer). It had been built in only three

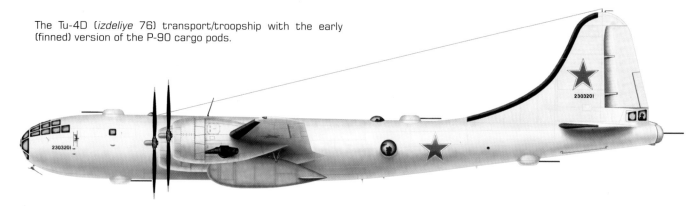

The Tu-4D (*izdeliye* 76) transport/troopship with the early (finned) version of the P-90 cargo pods.

Estimated performance of the Tu-4 loaded with two 3,800-kg SP howitzers (Option 1)		
	without fairings	with fairings
Top speed at sea level, km/h (mph)	340 (211)	405 (251.5)
Rate of climb at sea level, m/sec (ft/min)	1.7 (334)	2.9 (570)
Climb time to 1,000 m (3,280 ft), minutes	11	7
Range at 1,000 m, km (miles)	3,400 (2,111)	4,400 (2,732)
Service ceiling, m (ft)	2,500-3,000 (8,200-9,840)	5,000 (16,400)
Take-off run, m (ft)	2,500 (8,200)	2,200 (7,220)

Estimated performance of the Tu-4 loaded with two OSU-76 SP howitzers (Option 2)		
	at sea level	at 4,000 m (13,120 ft)
Top speed, km/h (mph)	389 (241)	418 (260)
Rate of climb, m/sec (ft/min)	2.4 (472)	1.1 (216)
Climb time to 1,000 m (3,280 ft), minutes	7	37*
Range at 1,000 m, km (miles)	2,600 (1,614)	–
Service ceiling, m (ft)	5,000 (16,400)	–
Take-off run, m (ft)	2,400 (7,870)	–

* To 4,000 m (13,120 ft)

copies back in 1944 and had shown disappointing results, which is why it did not enter production.) This time the fairings were a must; calculations showed that without fairings the aircraft would be unable to take off even with a reduced weight of 54,500 kg (120,150 lb) because of the extremely high drag. The Tu-4's estimated performance in this configuration with a 61,000-kg (134,480-lb) TOW is shown in the second table on this page.

Option 3 was one OSU-76 enclosed by a fairing. With a 54,000-kg (119,050-lb) TOW, estimated range in this configuration was 3,350 km (2,080 miles).

In the early 1950s the Air Force proposed converting existing Tu-4 bombers to transport/troopship aircraft designated Tu-4D (*desahntnyy* – for paradropping). This involved outfitting the bomb bays with detachable floors and tip-up seats to accommo-

date 28 paratroopers in extremely austere conditions. Additionally, artillery pieces and combat vehicles, such as the BTR-40 light armoured personnel carrier (a 4x4 wheeled APC similar in size and role to the White M3A1), could be carried under the wings in two P-90 or P-98 paradroppable cargo pods suspended on specially designed BD5-4D pylons between the inner and outer engines. Small items were to be paradropped in P-85 cylindrical pods dimensionally similar to the FAB-3000M-46 bomb, up to four of which were to be carried internally. The Tu-4D could be reconverted to bomber configuration in case of need.

Design work on the Tu-4D began at Aleksandr P. Golubkov's OKB-30 in September 1951; the aircraft bore the in-house designation *izdeliye* 76. The P-90, P-98 and P-85 pods were developed and manufactured by plant No.468 in Moscow, which specialised in

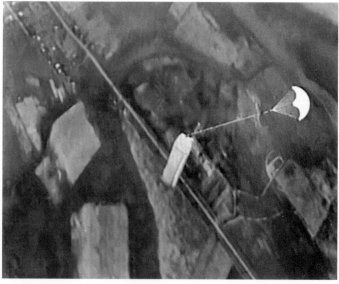

Top: Lower view of the Tu-4D (*izdeliye* 76). In contrast to the preceding photos, the radar and the surrounding skin panel appear to have been removed.

Top right: The aircraft opens the forward bomb bay doors for paradropping personnel.

Above: A parachutist leaves the front bomb bay of the Tu-4D (*izdeliye* 76).

Above right: Another view of the same machine during a paradrop. One trooper's parachute seems to have opened prematurely.

Far left: The Tu-4D (*izdeliye* 76) drops the starboard P-90MS pod.

Left: A P-90 pod is seen falling earthwards a few moments after release. The stabilising parachute is open, and the four-canopy main parachute system will deploy shortly.

paradrop equipment, while the pods' parachute systems were developed by the Paradropping Systems Research & Experimental Institute (NIEI PDS – *Naoochno-issledovatel'skiy eksperimentahl'nyy institoot parashootno-desahntnykh sistem*). The program also involved other R&D establishments. Thus, a branch of the Tupolev OKB manufactured a scale model of the Tu-4D with spinning propellers, which underwent wind tunnel tests at the Siberian Aviation Research Institute (SibNIA – *Sibeerskiy naoochno-issledovatel'skiy institoot aviahtsii*) in Novosibirsk with scaled-strength models of the P-90 pods supplied by plant No.468.

A single Tu-4 from the last Moscow-built batch (no serial, c/n 2303201) was converted as the Tu-4D (*izdeliye* 76) prototype and tested at LII and GK NII VVS in various configurations (with P-85, P-90 and P-98 pods and paratroopers). This aircraft retained the standard defensive armament. The trials confirmed the viability of the underwing cargo pods and of the idea in general. The tests performed by LII jointly with plant No.468 and NIEI PDS showed that, with a take-off weight of 54,500 kg (120,150 lb), the Tu-4D fitted with P-90 pods had a combat radius of 950 km (590 miles) when operating singly and 800 km (496 miles) when operating in squadron strength.

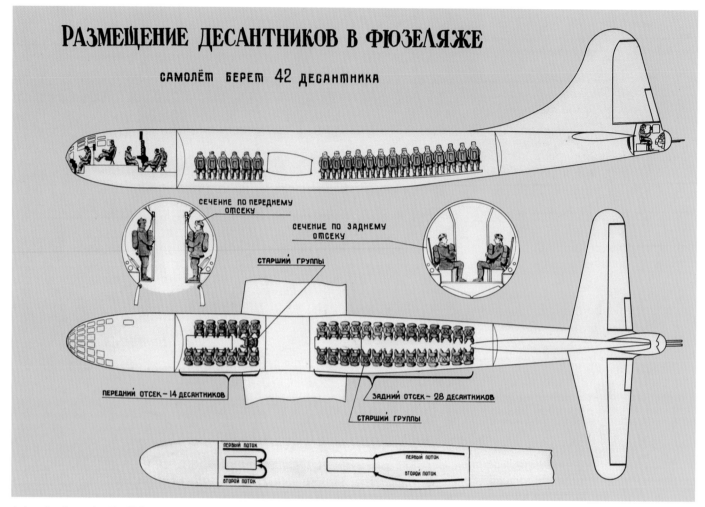

РАЗМЕЩЕНИЕ ДЕСАНТНИКОВ В ФЮЗЕЛЯЖЕ

САМОЛЁТ БЕРЕТ 42 ДЕСАНТНИКА

СЕЧЕНИЕ ПО ПЕРЕДНЕМУ ОТСЕКУ

СЕЧЕНИЕ ПО ЗАДНЕМУ ОТСЕКУ

СТАРШИЙ ГРУППЫ

ПЕРЕДНИЙ ОТСЕК – 14 ДЕСАНТНИКОВ

ЗАДНИЙ ОТСЕК – 28 ДЕСАНТНИКОВ

СТАРШИЙ ГРУППЫ

ПЕРВЫЙ ПОТОК

ВТОРОЙ ПОТОК

ПЕРВЫЙ ПОТОК

ВТОРОЙ ПОТОК

A drawing from the Tu-4T (*izdeliye* 4T) project documents, showing the 42-seat troopship configuration with 14 troopers in the front bomb bay and 28 in the rear cabin; one trooper in each bay is the jumpmaster.

On 10th July 1954 the Council of Ministers issued directive No.1417-637 ordering the conversion of Tu-4s for transport/ paradrop duties, followed two weeks later by MAP order No.454 to the same effect. The plan was to modify 300 *Bulls* in 1955 to the same standard as the aircraft tested at GK NII VVS. Conversion was to take place in service conditions; the schedule was set forth explicitly – 30 aircraft in the first quarter of the year, 100 aircraft in the second quarter, 120 more in the third quarter and the final 50 in the fourth quarter. The Tupolev OKB was required to turn over the complete set of working drawings and documents to the Air Force not later than 1st September 1954. The demands were met, but the actual Tu-4Ds were rather different aircraft (see below).

Tu-4T military transport/troopship aircraft (*izdeliye* 4T)
The same CofM directive No.1417-637 and MAP order No.454 tasked the Tupolev OKB with developing another transport/troopship derivative of the *Bull* designated Tu-4T (*trahnsportnyy* – transport, used attributively). This time the specification was rather more stringent; the aircraft was to deliver and paradrop 40-46 troopers, or two 57-mm (2.24-in) ASU-57 SP guns, or a conventional artillery piece with a tractor and a supply of ammunition, or 5,000-6,000 kg (11,020-13,230 lb) of other materiel. The first

aircraft thus converted, complete with two sets of underwing cargo pods, was to commence state acceptance trials in December 1955.

The ADP of the Tu-4T (in-house designation *izdeliye* 4T) was completed in 1955. In preparing it the designers relied heavily on their experience with the Tu-4D, incorporating the same design features, namely troop seats and other appropriate changes to the bomb bays, reinforced wings to enable carriage of cargo pods (the thickness of the spar webs was doubled) and appropriate hoists for attaching the pods. The following payload options were possible:

• 42 paratroopers with full kit seated in the bomb bays and the center cabin;

• a 57-mm anti-tank gun or 76.2-mm field gun (such as the ZiS-2 and ZiS-3 respectively) with a supply of ammunition and a GAZ-69 jeep in two P-90MS paradroppable pods under the wings (the gun crew of eight was accommodated in the rear bomb bay); each pod could hold up to 3,200 kg (7,050 lb) of cargo;

• two ASU-57 (ASU-57P) SP guns, or 85-mm (3.34-in) field guns (such as the D-44), in P-98 paradroppable pods under the wings (again, the crews were accommodated in the rear bomb bay);

• 3,000 kg (6,610 lb) of cargo (ammunition, engineering troops materiel, foodstuffs etc.) in four P-85 paradroppable pods carried internally; the total weight of the cargo was 4,800 kg (10,580 lb);

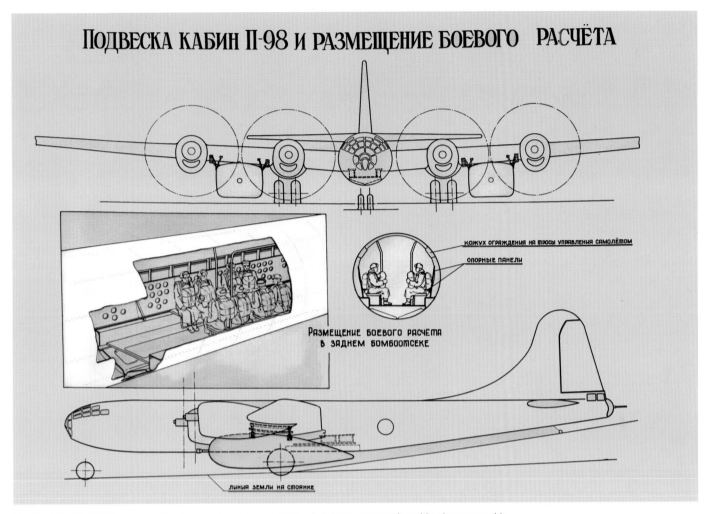

ПОДВЕСКА КАБИН П-98 И РАЗМЕЩЕНИЕ БОЕВОГО РАСЧЁТА

КОЖУХ ОГРАЖДЕНИЯ НА ТРОСЫ УПРАВЛЕНИЯ САМОЛЁТОМ

ОПОРНЫЕ ПАНЕЛИ

РАЗМЕЩЕНИЕ БОЕВОГО РАСЧЁТА
В ЗАДНЕМ БОМБООТСЕКЕ

ЛИНИЯ ЗЕМЛИ НА СТОЯНКЕ

The Tu-4T with P-98 cargo pods attached and a gun crew of eight accommodated in the rear cabin.

• 2,500-2,600 kg (5,510-5,730 lb) of cargo in PDMM-47 bags (*parashootno-desahntnyy myakhkiy meshok* – soft paradropping bag) wrapped in PDUR-47 straps (*parashootno-desahntnyy ooniversahl'nyy remen'* – versatile paradropping strap) and carried in special cassettes in the bomb bays.

In order to cut empty weight and maximise the payload the Tu-4T's defensive armament was reduced to the tail gun barbette; the remote control and sighting system for the dorsal and ventral barbettes, the standard bomb racks and release mechanisms (with one exception) and the Kobal't radar were deleted. Accordingly the apertures for the barbettes and the radome were faired over. The center pressure cabin was eliminated, its forward pressure dome at fuselage frame 37 and the pressurized crawlway being deleted to combine the cabin and the rear bomb bay into a large troop cabin, and the aperture in the forward cabin's rear pressure dome at frame 13 was sealed off. The sighting blisters in the center cabin were deleted; the dorsal aperture was faired over, while the lateral ones became emergency exits enabling evacuation in the event of a belly landing and were closed by flush covers incorporating small circular windows offset upward to admit some light into the cabin. The auxiliary oil tank in the center fuselage catering for the engines, the armour plate at frame 40, the armoured box for the defensive fire control computer under the center cabin floor and the other equipment in the cabin (control panels, cameras, bunks etc.) were also removed. Some of the avionics and equipment (the RV-2 radio altimeter, the Bariy IFF transponder, the SD-1 DME, the R-800 radio, the propeller de-icing system alcohol tank etc.) were relocated. The wing structure was revised to incorporate hardpoints for the cargo pods – the front spar, ribs 7L/7R and 9L/9R were reinforced, and auxiliary ribs 6A were added in both wings. This required the No.4 fuel tanks to be shortened, reducing overall fuel capacity by 470 litres (103.4 Imp gal). The Tu-4T had a crew of seven – captain, co-pilot, flight engineer, navigator, bombardier (he was responsible for accurate paradropping of the cargo), radio operator and tail gunner.

In troopship configuration the front bomb bay (in effect, the front troop cabin) and the rear troop cabin featured easily removable floors which incorporated entry/paradropping hatches closed by doors in flight. The rear cabin floor was situated some 450 mm (1 ft 5$\frac{23}{32}$ in) lower than the center pressure cabin floor of the standard bomber. Boarding was via stepladders which were stowed above the wing center section carry-through box. The troop cabins featured lightweight collapsible seats with rigid backrests and overhead guide tubes for hooking up the paratroopers' static lines.

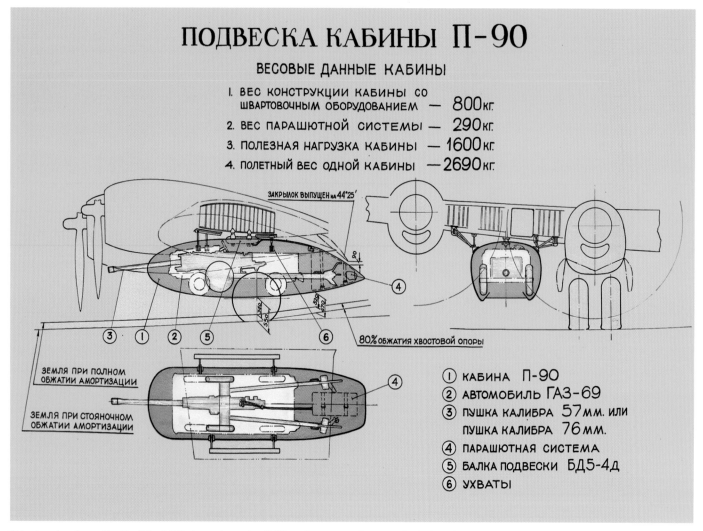

ПОДВЕСКА КАБИНЫ П-90

ВЕСОВЫЕ ДАННЫЕ КАБИНЫ

1. ВЕС КОНСТРУКЦИИ КАБИНЫ СО ШВАРТОВОЧНЫМ ОБОРУДОВАНИЕМ — 800 кг.
2. ВЕС ПАРАШЮТНОЙ СИСТЕМЫ — 290 кг.
3. ПОЛЕЗНАЯ НАГРУЗКА КАБИНЫ — 1600 кг.
4. ПОЛЕТНЫЙ ВЕС ОДНОЙ КАБИНЫ — 2690 кг.

ЗАКРЫЛОК ВЫПУЩЕН на 44°25′

80% ОБЖАТИЯ ХВОСТОВОЙ ОПОРЫ

ЗЕМЛЯ ПРИ ПОЛНОМ ОБЖАТИИ АМОРТИЗАЦИИ

ЗЕМЛЯ ПРИ СТОЯНОЧНОМ ОБЖАТИИ АМОРТИЗАЦИИ

① КАБИНА П-90
② АВТОМОБИЛЬ ГАЗ-69
③ ПУШКА КАЛИБРА 57 мм. ИЛИ ПУШКА КАЛИБРА 76 мм.
④ ПАРАШЮТНАЯ СИСТЕМА
⑤ БАЛКА ПОДВЕСКИ БД5-4д
⑥ УХВАТЫ

This drawing from the Tu-4T's ADP shows how the P-90 pods were to be attached, detailing payload options and weight data.

The front cabin accommodated 14 paratroopers – seven to port, six to starboard and the forward-facing jumpmaster at the rear; the rear cabin had 14 seats on each side, the jumpmaster occupying the fifth seat to port (immediately aft of the hatch).

In both cabins the troopers sitting on the starboard side jumped first. The cabins were equipped with red/yellow/green signal lights and signal bells telling the paratroopers when to hit the silk; an SPU-14 intercom was provided, allowing the navigator to communicate with the jumpmasters. Normally it was the navigator who opened the bomb bay doors, but the cabins featured emergency switches allowing the jumpmasters to open the doors.

Since the troop cabins were unpressurized, the paratroopers were equipped with oxygen masks and individual KP-22 breathing apparatus (*kislorodnyy pribor* – oxygen apparatus) installed over their heads. Each cabin had an oxygen supply control panel, the oxygen being stored in 18 cylindrical bottles inside and above the wing centre section carry-through box and five more in the rear fuselage. A cabin heating and ventilation system was provided, featuring a BO-40 petrol-fuelled heater (*benzo'obogrevahtel'*) installed under the cabin roof above the wing centre section carry-through box; the heater had a dorsal 'elephant's ear' air intake

(with hot-air de-icing) and a dorsal exhaust. The warm air was distributed in the cabins via a system of ducts and manifolds running along the sidewalls at floor level; each cabin had two ventilation air outlets. Each cabin featured overhead lights, a first aid kit (on the port side), a Thermos flask with a hot beverage (two in the rear cabin) and a toilet bucket (two in the rear cabin).

The control system cables running through the troop cabins, and certain equipment items located in the cabins, were protected by special covers preventing accidental damage by the paratroopers. Changes were made to the pressurisation system, which now catered for the flight deck and the tail gunner's station only.

For carrying vehicles and field guns in P-90MS and P-98 cargo pods, BD5-4D pylons with release mechanisms could be installed under the wings between the inner and outer engine nacelles (at ribs 6A). When not fitted they were to be stored inside the wing centre section carry-through box. The pods were secured by sway braces to fittings at ribs 7 and 9.

The P-90MS had an airfoil section and a hole for the gun barrel in the nose fairing which was detachable for loading/unloading; the PG-3715-54 four-canopy parachute system (*parashoot groozovoy* – cargo parachute) with a total area of 330 m² (3,552 sq ft) was

ПОДВЕСКА КАБИНЫ П-98

ВЕСОВЫЕ ДАННЫЕ КАБИНЫ

1. ВЕС КОНСТРУКЦИИ КАБИНЫ С ТРОСОВОЙ СИСТЕМОЙ
   И ШВАРТОВОЧНЫМ ОБОРУДОВАНИЕМ . . . . - 1300 КГ *1500*
2. ВЕС ПАРАШЮТНОЙ СИСТЕМЫ . . . . - 1000 КГ *420*
3. ПОЛЕЗНАЯ НАГРУЗКА КАБИНЫ . . . . - 3200 КГ *2500*
4. ПОЛЕТНЫЙ ВЕС ОДНОЙ КАБИНЫ . . . . - 5500 КГ *4500*

45 (ЗАКРЫЛОК ВЫПУЩЕН
НА 44° 25')

80% ОБЖАТИЯ ХВОСТОВОЙ ОПОРЫ

СХЕМА ПОДЪЕМА

ВИД ПО СТР. А
(НА РОЛИКИ ПОДЪЕМА)

ЗЕМЛЯ ПРИ СТОЯНОЧНОМ
ОБЖАТИИ АМОРТИЗАЦИИ

ЗЕМЛЯ ПРИ ПОЛНОМ
ОБЖАТИИ АМОРТИЗАЦИИ

1. КАБИНА П-98
2. АРТИЛЛЕРИЙСКАЯ САМОХОДНАЯ УСТАНОВКА АСУ-57
3. АРТИЛЛЕРИЙСКАЯ САМОХОДНАЯ УСТАНОВКА АСУ-57п
4. АВТОМОБИЛЬ ГАЗ-69
5. ПУШКА КАЛИБРА 85мм
6. ПАРАШЮТНАЯ СИСТЕМА
7. БАЛКА ПОДВЕСКИ БД5-4д
8. УХВАТЫ
9. УСИЛЕНИЕ НЕРВЮРЫ №7
10. НЕРВЮРА №6А
11. УСИЛЕНИЕ 1-го ЛОНЖЕРОНА
12. ДОРАБОТКА БЕНЗОБАКА №4
13. ШТАНГА ПОДЪЕМА
14. ЛЕБЕДКА БЛ-47
15. КАТУШКА ДЛЯ ТРОСА

A similar drawing for the P-98 cargo pod. The inset in the lower right-hand corner shows how the pod was hoisted into position.

stowed in the flattened rear end. The pod itself, together with cargo tie-down fittings, weighed 800 kg (1,760 lb), the parachute system added another 290 kg (640 lb) and the payload was 1,600 kg (3,530 lb); the overall weight of the loaded pod was thus 2,690 kg (5,930 lb). The pod's ground clearance was 550 mm (1 ft $9^{21}/_{32}$ in) with the aircraft at rest and 340 mm (1 ft $1^{25}/_{64}$ in) at maximum main gear oleo compression; the minimum clearance under the pod's rear end during rotation on take-off was 320 mm (1 ft $0^{39}/_{64}$ in).

The P-98 was deeper and wider, with a blunter front end, but was similar in design. Curiously, the ASU-57 paradroppable SP gun (*aviadesanteeruyemaya samokhodnaya [artillereeyskaya] oostanovka*) was loaded into the pod back to front and fitted entirely within the pod, whereas the ASU-57P amphibious version (*plavayushchaya* – swimming) was loaded facing forward, so that the gun barrel protruded. The pod itself and the MKS-8 parachute system weighed 1,300 kg (2,870 lb) and 1,000 kg (2,205 lb) respectively, the maximum payload was 3,200 kg (7,050 lb) and the overall weight of the loaded pod was 5,500 kg (12,130 lb). ground clearance was 280 mm (approx. $11^{1}/_{32}$ in) with the aircraft at rest and 202 mm ($7^{61}/_{64}$ in) at maximum oleo compression; the minimum clearance under the rear end during rotation was 17 mm ($0^{43}/_{64}$ in).

The pods were lifted into position by means of BL-47 bomb hoists with special extenders. With both types of pod, maximum flap deflection was 44°25' (versus the normal landing setting of 45°); the clearance between the flaps and the pods was 90 mm ($3^{35}/_{64}$ in) for the P-90 and 45 mm ($1^{49}/_{64}$ in) for the P-98. The latter pod was designed with roadability in mind – it could be towed tail-first by a jeep, using special wheels, or transported by a flatbed lorry (in that case the nose fairing was detached and placed inside the pod). Transportation by rail (on a flatcar) was also possible.

When the Tu-4T was configured for carrying 1,200-kg (2,645-lb) P-85 cargo pods with MKS-2K parachute systems, the troop cabin floors and other associated equipment was removed and KD4-248 bomb cassettes were fitted (between frames 17-18 in the front bomb bay and between frames 31-32 in the rear bomb bay). Being too bulky to fit inside the bays side by side, the P-85s were staggered vertically so that the upper pod was carried on the port side in the front bay and on the starboard side in the rear bay; the parachutes' static lines were hooked up to the bomb cassettes. If PDMM-47 bags and PDUR-47 straps were used, the aircraft was equipped with special containers 1.52 m (4 ft $11^{27}/_{32}$ in) wide with ventral clamshell doors; the bags were placed vertically in rows of three, the static lines being attached to DP-1 locks on the bay roof. The front bay held five such rows and the rear bay held six, giving a total of 33 bags; the drop occurred when the bombardier released the D4-48 locks securing the clamshell doors. The aircraft could be easily reconfigured in service conditions to suit a specific mission.

The P-90 pod underwent state acceptance trials between 20th November 1954 and 15th February 1955; these involved ten test drops from the Tu-4T. The GK NII VVS trials report released on 22nd March 1955 stated that *'the P-90 cargo pod carried by the Tu-4 aircraft modified for paradropping duties enables transportation and delivery by paradropping of a GAZ-69 gun tractor or a 57-mm gun with ammunition (25 rounds), or a 76-mm gun with ammunition (25 rounds). When carried in non-drop configuration it enables transportation of single spare parts and equipment*

Tu-4T weights (as per ADP)					
	Option 1*	Option 2	Option 3	Option 4	Option 5
Standard Tu-4 empty weight, kg (lb):	37,000 (81,569)	37,000 (81,569)	37,000 (81,569)	37,000 (81,569)	37,000 (81,569)
minus (deleted items)	3,601 (7,938)	3,601 (7,938)	3,601 (7,938)	3,601 (7,938)	3,601 (7,938)
plus (added mission equipment)	1,764 (3,888)	1,530 (3,373)	1,753 (3,864)	2,310 (5,092)	1,530 (3,373)
Resulting Tu-4T empty weight, kg (lb)	35,163 (77,519)	34,929 (77,004)	35,152 (77,495)	35,709 (78,723)	34,929 (77,004)
Take-off weight, kg (lb):	55,921 (123,282)	55,500 (122,354)	55,837 (123,097)	57,594 (126,970)	55,921 (123,282)
crew	630 (1,388)	630 (1,388)	630 (1,388)	630 (1,388)	630 (1,388)
cannon ammunition	240 (529) †	407 (897)	407 (897)	407 (897)	407 (897)
engine oil	1,000 (2,204)	1,000 (2,204)	1,000 (2,204)	1,000 (2,204)	800 (1,763)
fuel	13,848 (30,529)	12,376 (27,284)	13,848 (30,529)	13,848 (30,529)	6,774 (14,933)
paradroppable load	5,040 (11,111)	6,158 (13,575)	4,800 (10,582)	6,000 (13,227)	11,960 (26,366)
Weight over the drop zone, kg (lb):	48,130 (106,106)	47,310 (104,300)	48,050 (105,930)	49,840 (109,876)	51,080 (112,610)
including fuel	6,427 (14,168)	4,576 (10,088)	6,427 (14,168)	6,462 (14,246)	2,564 (5,652)
fuel remaining	46.5%	37%	46.5%	46.7%	37.8%
Normal landing weight, kg (lb) ‡	37,753 (83,229)	37,519 (82,713)	37,742 (83,205)	38,299 (84,433)	37,519 (82,713)
Maximum landing weight, kg (lb) §	50,000 (110,230)	50,000 (110,230)	50,000 (110,230)	50,000 (110,230)	50,000 (110,230)
including fuel	8,427 (18,578)	7,376 (16,261)	8,511 (18,763)	6,754 (14,889)	1,674 (3,690)
fuel remaining	60.8%	59.6%	61.5%	48.7%	24.7%
CG position, % mean aerodynamic chord:					
take-off (gear down/gear up)	31.1/30.5	29.4/28.8	26.3/25.7	26.2/25.6	29.5/28.9
landing (gear down/gear up)	20.4/21.2	21.1/21.9	20.6/21.4	20.7/21.6	21.1/21.9

* Option 1: 42-seat troopship configuration. Option 2: Two P-90MS pods and jeep/gun crew of eight in the rear cabin. Option 3: Four P-85 pods (internally). Option 4: 6,000 kg (13,230 lb) of cargo in PDMM-47/PDUR-47 packaging (internally). Option 5: Two P-98 pods and gun crews (eight) in the rear cabin.

† In troopship configuration the cannon ammunition complement is reduced for CG reasons.

‡ No load, ammunition expended, 1,500 kg (3,310 lb) of fuel and 460 kg (1,014 lb) of oil remaining.

§ With load and ammunition.

Tu-4T performance (as per ADP)					
	Option 1*	Option 2	Option 3	Option 4	Option 5
Take-off weight, kg (lb)	55,921 (123,282)	55,500 (122,354)	55,837 (123,097)	57,594 (126,970)	55,500/– (122,354)†
Maximum speed at sea level, km/h (mph) ‡	410 (254)	360 (223)	410 (254)	408 (253)	340/340-350 (211/211-217)
Maximum rate of climb at sea level, m/sec (ft/min) ‡	3.2 (630)	2.47 (486)	3.2 (630)	2.9 (570)	1.89/2.0-2.5 (372/393-492)
Service ceiling, m (ft) ‡	9,800 (32,150)	8,200 (26,900)	9,800 (32,150)	9,500 (31,170)	6,850/6,000-7,000 (22,470/19,685-22,965)
Take-off run, m (ft)	1,290 (4,230)	1,240 (4,070)	1,290 (4,230)	1,440 (4,270)	1,350 (4,430)/–
Take-off field length, m (ft)	2,580 (8,460)	2,930 (9,610)	2,580 (8,460)	2,880 (9,450)	3,190 (10,465)/–
Landing run, m (ft)	1,070 (3,510)	1,100 (3,610)			
Landing field length, m (ft)	1,750 (5,740)	2,215 (7,270)	1,750 (5,740)	1,750 (5,740)	2,215 (7,270)/–
at landing weight, kg (lb)	47,850 (	50,000 (110,230)	47,850 (	47,850 (	50,000 (110,230)
Combat radius until					
fuel burnout at 1,000 m (3,280 ft), km (miles)	n.a.	1,590 (987)	n.a.	n.a.	720/700-800 (447/435-496)
Maximum range at 3,000 m (9,840 ft), km (miles)	4,500 (2,795)	n.a.	4,500 (2,795)	4,345 (2,698)	n.a.

* Options as above

† The data for Option 5 are OKB estimates (first figure)/Air Force operational requirement (second figure)

‡ At maximum continuous power. The service ceiling is given with the weight reduction due to fuel burnoff taken into account.

Top: Head-on view of Tu-4 '23 Red' (c/n 2806202) in 'production-standard' Tu-4D configuration during trials at GK NII VVS, with P-90 pods attached.

Above: Unlike the Tu-4D (*izdeliye* 76), the 'production-standard' Tu-4D had no armament, though the radar was still there.

Right: Another view of the same aircraft.

for the Tu-4 and Tu-16, [...] ground support equipment and other cargoes.'

The Tu-4T prototype passed its state acceptance trials in 1956. These included verification of the paradropping equipment; the aircraft's performance and handling with underwing cargo pods and in 'clean' configuration were also checked. With P-90MS pods the aircraft handled almost like an ordinary Tu-4, except for the higher control wheel forces from the ailerons in turbulent conditions. The control forces arising when the pods were dropped

were quite small and easily corrected by trimming the aircraft. With cargo pods, the Tu-4T could maintain level flight in the event of a single-engine failure at up to 1,000 m (3,280 ft). In that case, maximum speed was 327 km/h (203 mph) with P-90MS pods and 305 km/h (189 mph) with P-98 pods, the all-up weight being 55,500 kg (122,360 lb) and 53,500 kg (117,950 lb) respectively.

Still, the Tu-4T remained a one-off because the twin-turboprop Antonov An-8 *Camp* and the four-turboprop An-12 *Cub* had appeared by then; these purpose-built transports suited the needs

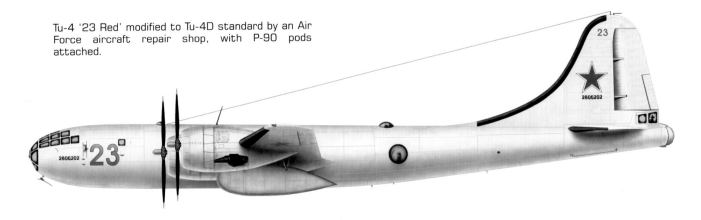

Tu-4 '23 Red' modified to Tu-4D standard by an Air Force aircraft repair shop, with P-90 pods attached.

of the Air Force's transport command (VTA – *Voyenno-trahnsportnaya aviahtsiya*, Military Transport Aviation) much better.

Tu-4D military transport/troopship aircraft
(second use of designation)

Despite having the same designation, the aircraft that eventually entered service as the Tu-4D was different from the *izdeliye* 76 prototype – and from the Tu-4T, for that matter. Most notably, it lacked defensive armament altogether (some sources, though, claim the tail barbette was retained). Unlike the Tu-4T, the center cabin was left unchanged; thus, the troop capacity was limited to 28 paratroopers, just like on the original Tu-4D.

Reports on how many Tu-4s were modified to 'Tu-4D Mk II' standard vary. Some sources claim all 300 aircraft stipulated by the aforementioned CofM directive No.1417-637 were modified and transferred from the DA to the VTA by the end of 1955. Other sources state no more than 100 Tu-4s were actually converted by plant No.23 in 1955. Two further *Bulls* modified to Tu-4D standard in Air Force workshops (c/ns 230121 and 2806202) passed check-up trials at GK NII VVS in the spring of 1956. The tests showed that in 'clean' configuration (without external stores) the Tu-4D had almost identical flight performance to the standard bomber; with P-90 pods it was 64 km/h (39 mph) slower at sea level and 79 km/h (49 mph) slower at 5,000 m (16,400 ft). Rate

Above left: A Tu-4D drops a test load during trials of the PN-81 retro-rocket system developed by plant No.468 for paradropping heavy items.

Left: A display model depicting the airborne command post conversion of the Tu-4; note the extra windows and the lack of armament.

Right: '29 Red' (c/n 2207510), the prototype of the Tu-4NM drone launcher conversion.

of climb was reduced by 1.1 m/sec (216 ft/min); the take-off run and the landing run increased by 190 m (620 ft) and 80 m (260 ft) respectively.

The P-90 pod was approved for operational use, even though it did give rise to a number of complaints. In particular, the pods were prone to rolling over on landing if there was a crosswind at ground level; the shock absorption was rather inadequate, with occasional damage to the contents as a result. Also, preparing the P-90 for a mission was a time-consuming process; a mission crew of eight manning the vehicles/guns being paradropped needed 8 hours 40 minutes to prepare two P-90s, more than half the time (4 hours 40 minutes) being required to pack the PG-3517-54 parachute systems alone! Also, VTA Commander Air Marshal Nikolay S. Skripko and Airborne Troops (VDV – *Vozdooshno-desahnt-nyye voyska*) Commander Lt.-Gen. Vasiliy F. Marghelov pointed out that the P-90 had been dropped onto snow (which cushioned the impact to a certain degree) during trials and that it was advisable to hold additional tests in the summer.

The larger P-98M pods designed for paradropping 2,500 kg (5,510 lb) of cargo turned out to be a lemon, as they impaired the Tu-4D's flight performance to an unacceptable degree. A third model of cargo pod developed by plant No.468 – the P-110 – was tested with good results; it was about the same size as the P-98 and had the same payload but was lighter, grossing at 1,000 kg (2,205 lb). With a 54,500-kg (120,150-lb) TOW, the Tu-4D's speed was reduced from 417 to 364 km/h (from 259 to 226 mph) at sea level and from 477 to 422 km/h (from 296 to 262 mph) at 7,000 m (22,965 ft), and rate of climb dropped from 3.3 to 2.6 m/sec (from 649 to 511 ft/min), but the Air Force was happy with these figures. A version of the same pod tested on the Tu-4D, the P-110-RP (P-110K), differed in having a single parachute assisted by a retro-rocket instead of a four-canopy parachute.

The Tu-4D served on with the VTA until the early 1960s when it was replaced by the An-12 medium transport.

Tu-4USh (Tu-4UShS) navigator trainer

In the late 1950s several dozen Tu-4 bombers were converted to navigator trainers to fill the needs of the log-range bomber arm (DA). The Soviet equivalent of the TB-29 was designated Tu-4USh (*oochebno-shtoormanskiy* – for navigator training); occasionally the designation was rendered as Tu-4UShS (*oochebno-shtoor-manskiy samolyot* – navigator trainer aircraft). The cannon armament was deleted while the Kobal't radar and OPB-5SN bombsight were replaced by an RBP-4 Rubidiy-MM2 radar (Rubidium; NATO codename *Short Horn*) and an OPB-11R bombsight as fitted to the Tu-16 and a limited bomber capability was retained; thus the Tu-4USh could be used for training not only navigators but also bombardiers (who doubled as navigators on some Soviet bomber types). The trainees sat in one of the bomb bays, taking turns to man the navigator/bomb-aimer's station.

The aircraft entered service with the Soviet Air Force's Chelyabinsk Military Navigator College (ChVVAUSh – *Chelyabinskoye vyssheye voyennoye aviatsionnoye oochilishche shtoormanov*) in early 1957, replacing the IL-28 bomber; two of the college's squadrons had nine aircraft each. The Tu-4USh remained in service until the end of 1960 when it was replaced by the Tu-124Sh-1, one of two navigator trainer versions of the Tu-124V *Cookpot* twin-turbofan short-haul airliner.

Tu-4 communications relay aircraft

In the 1950s at least one Tu-4 was converted into a communications relay aircraft for maintaining radio communication between Soviet Navy ships and submarines and the Navy's shore-based command and control centers – a role known as TACAMO (TAke

Left: Close-up of the La-17 ramjet-powered target drone on the Tu-4NM's starboard wing rack. Note the ram-air turbine in the nose driving a generator.

Below left: A side view of the same drone.

Below: This view shows how the Tu-4NM's ejector racks worked – the La-17 swung down on parallel arms during launch to make sure that the drone's fin cleared the bomber's wing.

Bottom: Tu-4NM '29 Red' taxies out for a sortie.

Charge And Move Out). Its main distinguishing feature was a long trailing wire aerial unwound from a drum in the rear fuselage.

Tu-4 airborne command post conversion

A few Tu-4s stripped of all armament were converted into 'flying headquarters' or airborne command posts (ABCPs). The 'war room' was located in the center pressure cabin which was suitably outfitted and provided with extra windows to admit some light.

Tu-4NM drone launcher aircraft

Six late-production Kazan'-built Tu-4s, including '29 Red' (c/n 2207510), were adapted in 1952 to carry a pair of Lavochkin La-17 (*izdeliye* 201) jet-propelled target drones under the wings for training air defence missile crews; they were designated Tu-4NM (*nositel' misheney* – drone launcher). The aircraft were lightened by removing all armament and some equipment items; nevertheless, they climbed awfully slowly, requiring two hours to reach

Tu-4NM '29 Red' with two La-17 target drones.

8,000 m (26,250 ft) and guzzling fuel at an awful rate. Hence the air launch idea was dropped; in service the La-17 used a ground launcher, taking off with the help of two solid-fuel rocket boosters.

Tu-4M target drone conversion
Also in 1952, the decision was taken to convert obsolete or time-expired aircraft, including Tu-4s, into target drones for training

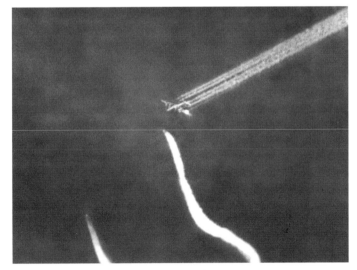

Above: The final moments of a Tu-4M drone contrailing across the sky as two surface-to-air missiles streak towards it.

Top right: Many Tu-4s ended up as ground targets at practice ranges. Two of the four bombers seen here are already destroyed.

Above right: Kaboom! A decommissioned Tu-4 takes a direct hit from an air-to-surface missile.

Right: The aftermath of a successful practice strike. This is an early Kazan'-built Tu-4 (note the three B-20E cannons in the tail turret).

Above: CCCP H-1139 is the best-known of the four demilitarised Tu-4s operated by the Polar Aviation. Note that the forward dorsal cannon barbette has been replaced with an astrodome, not simply faired over like the other four.

Left: Soviet Polar researchers pose with a Polar Aviation Tu-4 at an Arctic airfield. The nose titles read *Polyarnaya aviahtsiya*. Note the cargo glider – a Tsybin Ts-25 or a Yakovlev Yak-14 – in the background.

Left: Another improvised photoshoot beside CCCP H-1139. This time the crew is depicted (that is, all except the man on the left, who is apparently not an airman but a researcher).

air defence crews. After taking off and climbing to the required altitude the bomber's crew would bail out and the Tu-4 flew on towards destruction, controlled by an on-board computer or by a ground control station via a radio channel. The drone version was designated Tu-4M (*mishen'* – target) and was the Soviet counterpart of the QB-29.

Tu-4 long-range ice reconnaissance aircraft conversion

In the spring of 1954, when the Long-Range Aviation began phasing out the *Bull*, the Polar Aviation (or, to be precise, the Main Directorate of the North Sea Route responsible for shipping in the polar regions and support of Soviet Arctic research) took delivery of its first two demilitarised Tu-4s. These were placed on the Soviet civil register as CCCP H-1138 (that is, SSSR N-1138 in Cyrillic characters; c/n 2302801) and CCCP H-1139 (c/n 2805710); a further two aircraft, CCCP-H1155 (c/n 2208009) and CCCP-H1156 (c/n 2208407), followed a year later. (In 1922-1958 the CCCP- country prefix was followed by a code letter denoting the aircraft's owner plus up to four digits. N stood for the Main Directorate of the North Sea Route. Strictly speaking, the first two of the abovementioned registrations should have been painted on as CCCP-H1138 and CCCP-H1139.)

The Tu-4s were adapted for operations along the North Sea Route. All armament was removed, and the forward dorsal barbette was replaced by an astrodome; the center pressure cabin was converted into a researchers' bay and also served as a canteen. The

Above: The pilot of CCCP H-1139 poses with his mount. Note the Polar Aviation pennant painted on the aircraft's tail.

Right: CCCP H-1139 is unloaded after bringing supplies to an Arctic airfield. A Mil' Mi-4 helicopter is parked nearby.

Here, Tu-4 CCCP H-1139 disgorges a number of 200-litre (55-gallon) fuel drums. The 'putt-putt' is running to keep the aircraft's interior warm, as evidenced by the open escape hatch and the smoke coming out of the exhaust near the digit '3'.

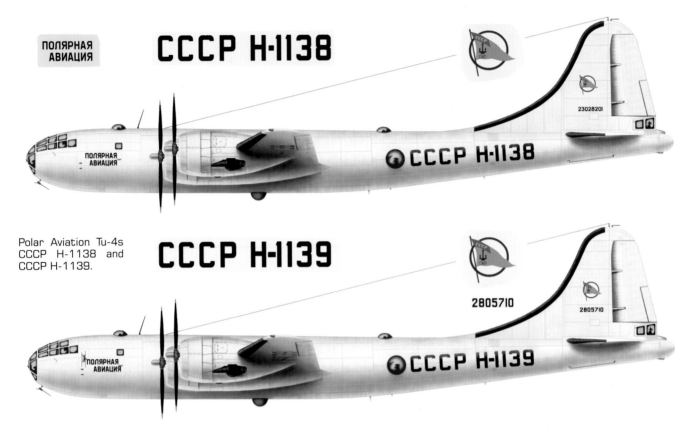

ПОЛЯРНАЯ АВИАЦИЯ

CCCP H-1138

CCCP H-1139

Polar Aviation Tu-4s
CCCP H-1138 and
CCCP H-1139.

Polar Aviation Tu-4s were based near Murmansk and occasionally operated from Tiksi and Amderma, flying long-range ice reconnaissance missions over the Greenland Sea between the coast of Greenland and Spitsbergen (Svalbard) Island, over Greenland itself, the Franz-Joseph Land archipelago and the North Pole in the direction of Canada. Such missions typically lasted 20-25 hours, so the aircraft carried as much fuel as it could hold and the take-off weight usually exceeded 60,000 kg (132,275 lb). These aircraft were also used as fuel tankers, landing on ice fields to deliver precious fuel and other supplies to Soviet drifting ice stations in the Arctic Ocean.

At least one of the four Tu-4s remained in service in the early 1960s, being reregistered CCCP-92648 under the new five-digit Soviet civil aircraft registration system introduced in 1958. Under this system, which remains in use in Russia to this day, the first two digits are a code denoting the aircraft type for flight safety reasons, allowing quick identification by air traffic controllers, though some registration blocks correspond to no specific type and are a 'mixed bag' of various aircraft.

Tu-4 modified for 360º movie shooting

Believe it or not, the Tu-4 also had altogether civil applications. In the late 1950s the Tupolev OKB received a request from the Soviet Union's Central Documentary Film Studios (TsSDF) to develop a system enabling 360º film shooting from the air by means of several remote-controlled cine cameras installed under an aircraft's belly on an extendable mount. A special term, *tsirkorama* ('circo-rama' – a portmanteau of 'circle' and 'panorama'), was coined for such 360º movies.

Basically the customer's demands were that the camera mount was to be extended pretty far into the slipstream in order to create a total 'eye in the sky' effect (that is, to prevent parts of the aircraft from getting into the picture). Furthermore, the surface of the mount was to remain parallel to the fuselage waterline at all times and the whole structure was required to be adequately stiff. The Tu-4 turned out to be the best camera platform for the job.

The OKB's branch office in Tomilino (Moscow Region) considered several possible locations for the camera mount, alternative ways of attaching it to the fuselage and alternative actuation system designs. When the optimum design had been chosen this unusual installation was manufactured by the OKB's experimental facility in Moscow (MMZ No.156 *'Opyt'*; the name translates as either 'experiment' or 'experience'). The installation was a long externally mounted truss whose rear end was hinged to the fuselage underside. The camera mount was hinged to the front end and maintained in horizontal position by mechanical linkages as the truss tilted; extension and retraction was by means of cables and pulleys. The cameraman operated the cine cameras remotely from the cabin.

After completing its flight test program the modified aircraft (identity unknown) was turned over to the Central Documentary Film Studios. The 'circorama' movies shot with the help of this aircraft were shown in a purpose-built movie theatre at the Soviet Union's prime showground, the National Economy Achievements Exhibition (VDNKh – *Vystavka dostizheniy narodnovo khoziaistva*) in Moscow, and Tupolev OKB employees were among the first spectators. (As a point of interest, two helicopter types – the Mil' Mi-4 *Hound* and the Yakovlev Yak-24 *Horse* – were also used for making such movies.)

Boeing B-29-5-BW '256 Black' following conversion as a 'mother ship' for the DFS 346 V1.

Tu-4 c/n 230503 was similarly converted into a 'mother ship' for the DFS 346 V3.

Previous page: B-29-5-BW '256 Black' at Tyoplyy Stan airfield following modification for the 'mother ship' role. It carried the unpowered first prototype of the DFS 346 high-speed research aircraft (the rear view shows that the nozzles of the twin-chamber rocket motor are blanked off). Note the cine camera under the wingtip to capture the moment of release.

Above: This side view of '256 Black' shows that a second cine camera pointing obliquely forward was fitted under the tail gunner's station.

Left: Front view of the 346 V1 under the wing of the B-29 on a special twin rack; the limited ground clearance is noticeable. The pilot of the DFS 346 flew the aircraft in a prone position.

Below left: Another view of the first DFS 346 carried by B-29 '256 Black'.

Top right: Tu-4 c/n 230503 was converted in the same way for the same mission. Unlike '256 Black', it had the cannon barbettes (but no cannons). Here the 346 V1 or 346 V2 is suspended under the wing.

Above right: Here, in contrast, Tu-4 c/n 230503 carried the ill-fated third prototype (346 V3) which made two powered flights before crashing in the third flight. Note the dark colour of the rocket aircraft.

Right: The Tu-4 becomes airborne with the light-coloured 346 V1 or 346 V2 attached.

Boeing B-29 and Tu-4 'mother ships' for experimental aircraft

Upon completion of the ASh-73TK engine's flight tests (see below) B-29-5-BW '256 Black' was ferried to the Kazan' aircraft factory and converted into a 'mother ship' for rocket-powered experimental aircraft in accordance with MAP order No.210 dated 16th April 1948. A special rack was installed under the starboard wing between the Nos. 3 and 4 engines for carrying the B-5 aircraft designed by Matus R. Bisnovat or the German '346' (DFS 346, aka Siebel Si 346). The latter had been designed by Hans Rössing who completed development of this aircraft while working in captivity in the USSR as head of OKB-2.

The converted bomber was flown to the airfield in Tyoplyy Stan on the south-western outskirts of Moscow where trials of the '346' had been going on since 1948. (Now the airfield is long since gone, and Tyoplyy Stan is a residential area.) In 1949 and 1950 '256 Black' made a number of test flights with three prototypes of the '346'; the experimental aircraft was released at altitudes up to 9,700 m (31,820 ft), making both gliding and powered flights. After the '346' program B-29 '256 Black' remained in use at LII as a testbed for some time yet until finally struck off charge and scrapped.

A Moscow-built Tu-4 (no serial, c/n 230503) was similarly converted for carrying the '346' rocket-powered aircraft – probably in accordance with the same MAP order No.210. As was the case with B-29 '256 Black', the '346' was carried aloft on a special rack mounted between the Nos. 3 and 4 engines. Tu-4 c/n 230503 served in the 'mother ship' role until the termination of the program in 1951 when the third prototype (346 V3) crashed during its third powered flight.

ShR-1/ShR-2 bicycle landing gear testbed

When the newly resurrected design bureau led by Vladimir M. Myasishchev (OKB-23) started work on the M-4 (*izdeliye* M) *Bison-A* strategic bomber in March 1951, the designers opted for a bicycle landing gear. The twin-wheel nose unit envisaged originally and the main unit equipped with a four-wheel bogie absorbed

30% and 70% of the aircraft's weight respectively. This arrangement was not totally new to Soviet aircraft designers by then, as the I-215D proof-of-concept aircraft designed by Semyon M. Alekseyev had successfully flown in October 1949. However, there was as yet no experience of using a bicycle landing gear on a heavy aircraft (the 'aircraft 150' twin-turbojet medium bomber designed by the captive German designer Brunolf W. Baade did not make its first flight until 5th September 1952), and the need arose to verify the novel arrangement on a suitably converted bomber. Hence OKB-23 developed a testbed version of the Tu-4 for studying the behaviour of a heavy aircraft with a bicycle landing gear during take-off, landing and taxying, as well as for aircrew training. The aircraft was designated ShR-1, the Sh denoting *shassee* (landing gear) and the R being the Tu-4's product code (*izdeliye* R).

Above: The ShR-1 testbed jacked up in the hangar of plant No.23 for installation of the non-retractable experimental bicycle landing gear. The massive nosewheels (actually stock Tu-4 mainwheels), the four-wheel main bogie and the crude outrigger struts with stock Tu-4 nosewheels are clearly evident.

Left: Close-up of the ShR-1's main gear unit, showing the tandem oleos and the mounting truss allowing installation in three different positions.

Opposite page, top: The crew of the ShR-2 testbed in the flight deck.

Below: The ShR-2 (the same Tu-4 c/n 230322) on the apron at Zhukovskiy, showing the nose gear bogie with small wheels and the photo calibration crosses.

The ShR-1 bicycle landing gear testbed.

A brand-new *Bull* manufactured by the co-located production plant No.23 (no serial, c/n 230322) was delivered to the Myasishchev OKB for conversion. The experimental landing gear was fixed. The standard nose unit gave way to a new levered-suspension strut with larger 1,450 x 520 mm (57 x 20.4 in) twin wheels – that is, stock Tu-4 mainwheels. The scratchbuilt main unit consisted of two stock Tu-4 main gear oleos mounted in tandem and rigidly connected; to these was hinged a four-wheel bogie with 1,450 x 520 mm wheels. The entire assembly was attached to a hefty truss made of welded steel tubes which was installed near the rear bomb bay; the main gear unit could be fitted in three different positions, absorbing 72%, 85% or 90% of the total weight. This unusual design feature was introduced to see how the changing weight distribution and wheelbase affected the aircraft's field performance and manoeuvrability on the ground (the reason was that different versions of the M-4's preliminary design (PD) project featured an aft-retracting or forward-retracting nose gear unit, with attendant changes in wheelbase and weight distribution). The nosewheels and mainwheels were equipped with hydraulic brakes.

The outrigger struts installed a short distance outboard of the Nos. 1 and 4 engines consisted of stock Tu-4 nose gear units mated to special truss-type mountings which allowed the length of the outriggers to be adjusted. Interestingly, the experimental landing gear could be removed and the normal gear reinstalled for positioning flights. The armament and the radar were not fitted.

The conversion job was completed in January 1952; Stage 1 of the test program began in April, lasting until June. The ShR-1 made 50 high-speed runs and 34 flights to check the aircraft's stability and controllability, the operation of the nose gear steering mechanism and the optimum position of the main gear and outrigger struts. As the main gear was moved aft the load on the nose gear gradually increased from 10.6% to 20.8% and then to 28.6%. Test data were recorded automatically.

Meanwhile, the M-4's twin-wheel nose unit had been rejected in favour of a four-wheel bogie with electrically steerable front wheels. Hence the Tu-4 testbed was modified accordingly and redesignated ShR-2; the nose bogie had four 950 x 350 mm (37.40 x 13.77 in) stock Tu-4 nosewheels. In this guise the aircraft underwent further tests in 1953, making 24 taxi runs and 17 flights; the nose gear bogie absorbed 20% of the weight versus 40% on the real *Bison*. The tests allowed the take-off technique to be mastered

well in advance of the M-4's maiden flight – the aircraft required no rotation to become airborne, taking off almost of its own will (and thus created a psychological problem: the pilot had to resist the urge to haul back on the control yoke). They also showed that the bicycle landing gear offered excellent ground handling.

UR-1/UR-2 control system testbed

The bicycle landing gear was not the only novel feature of the M-4. For the first time in Soviet aircraft design practice the *Bison* incorporated fully powered controls with reversible and irreversible hydraulic actuators in the aileron, rudder and elevator control circuits, plus an artificial feel mechanism.

Testing these features on a ground rig ('iron bird') was not enough – the actuators had to be put through their paces in flight. Hence a Tu-4 (identity unknown) was converted into the UR control system testbed by OKB-23; U stood for *[sistema] oopravleniya* – control system and the R was a reference to the Tu-4.

The first version designated UR-1 featured reversible actuators; tests in this configuration were completed in March 1952, involving 12 flights totalling about 20 hours. The test flights showed that the powered controls worked acceptably, even though a few failures did occur; incidentally, the powered controls improved the *Bull*'s handling dramatically. In April 1952 the aircraft was refitted with irreversible actuators and redesignated UR-2. The test results obtained on this aircraft enabled OKB-23 engineers to simplify the M-4's control system, utilising a simple spring-loaded artificial feel mechanism instead of a complex automatic device. Still, further

The same aircraft after modification as the ShR-2.

230322

Top: Tu-4LL '9 Black' with VD-3TK development engines. The AV-28 contraprops had smaller diameter than the standard V3-A3 props.

Above and below: This side view of the same aircraft shows the longer and slimmer outer engine nacelles. Note that the cannons have been retained.

tests with a pneumatically-operated automatic artificial feel system were conducted on the UR-2 in September 1952.

KR crew rescue system testbed

Later, the UR-2 control system testbed was further modified for testing the M-4's crew rescue system and redesignated KR, the K

denoting *katapool'tee*ruyemoye *kres*lo (ejection seat) and the R referring to the Tu-4. Its *raison d'être* was the *Bison*'s unconventional crew rescue system with movable ejection seats, the captain and co-pilot ejecting consecutively through a common hatch.

The KR was equipped with ejection seats for the pilots, navigator, flight engineer and dorsal gunner. Between October 1952 and

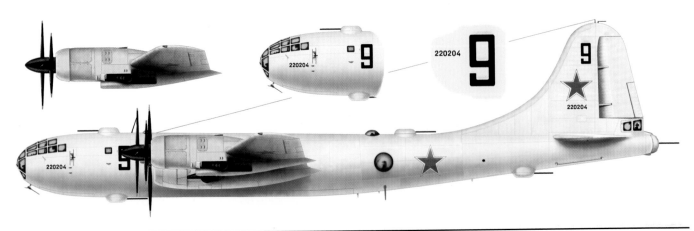

January 1953 it performed 12 test flights in which 20 ejections were made, including seven with live parachutists. As a result, some changes had to be introduced into the ejection seat's design.

Boeing B-29 engine testbed
As part of the effort to reverse-engineer the B-29, one of the Superfortresses interned in the Far East – B-29-5-BW '256 Black' (ex-42-6256, c/n 3390) – was handed over to LII, becoming a testbed for the 2,400-hp Shvetsov ASh-73TK 18-cylinder radial selected to power the Tu-4. The development engine was fitted instead of the No.3 Wright R-3350-23 Duplex Cyclone; this did not lead to any changes in the aircraft's appearance. In keeping with the aircraft's testbed role all defensive armament was removed and the apertures for the dorsal and ventral barbettes were faired over.

Flight tests of the ASh-73TK began in September 1946; the B-29 was flown by test pilots Nikolay S. Rybko (LII) and I. I. Shooneyko. After passing a rigorous test program the ASh-73TK was used successfully for many years on the Tu-4 and the Beriyev Be-6 *Madge* flying boat.

Tu-4LL engine testbeds with piston and turboprop engines (including 'aircraft 94/1' and 'aircraft 94/2')
The Tu-4 was also used for testing new piston, turboprop and turbojet engines. Several *Bulls* were modified for this role, bearing the designation Tu-4LL (*letayushchaya laboratoriya* – lit. 'flying laboratory'). (Note: In Russian the term *letayushchaya laboratoriya* is used indiscriminately to denote any kind of flying testbed or research/survey aircraft.)

In January 1950 the ninth Kazan'-built Tu-4 (c/n 220204), appropriately serialled '9 Black', was modified for testing the 2,000-hp Dobrynin/Skoobachevskiy M-251TK (alias VD-3TK) liquid-cooled engine created by OKB-36 in Rybinsk, Yaroslavl'

Region. This supercharged 24-cylinder radial-block engine (with six blocks of four cylinders arranged at 60° to each other) was intended for the Alekseyev Sh-218 attack aircraft, which in the event never materialised; 36 such engines were manufactured. Two VD-3TKs were fitted in place of the Nos. 1 and 4 ASh-73TKs, driving AV-28 four-blade contra-rotating propellers of 3.6 m (11 ft 9^{47}⁄₆₄ in) diameter developed by OKB-120, aka the Stoopino Machinery Design Bureau (SKBM – *Stoopinskoye konstrooktorskoye byuro mashinostroyeniya*). The water radiators were chin-mounted, while the oil coolers were buried in the wing leading edge between the engines. The Tu-4LL with VD-3TK engines underwent tests until July 1950.

In the second half of the 1950s the Tupolev OKB converted two *Bulls* into engine testbeds. Initially these two aircraft were used to test the engines intended for the Tupolev 'aircraft 80' and 'aircraft 85' long-range bombers – the 4,000-hp Shvetsov ASh-2TK and 4,700-hp ASh-2K 28-cylinder four-row air-cooled radials (in the latter case, with an SKBM AV-55 propeller) and the 4,300-hp Dobrynin M-253K (VD-4K) 24-cylinder water-cooled radial-block engine with an SKBM AV-48 four-blade propeller. On both aircraft the development engine was installed in the No.3 position. (The ASh-2TK never flew on anything except a testbed, but the other two engines subsequently powered the '85' prototypes.)

When turboprop engines came on the scene, these two Tu-4LLs were further modified for testing early Soviet turboprops – the 5,163-ehp Kuznetsov TV-2 (TV stands for *toorbovintovoy [dvigatel']* – turboprop), the 6,250-ehp TV-2F (*forseerovannyy* – uprated), the 7,650-ehp TV-2M (*modifitseerovannyy* – modified) and the 12,500-ehp 2TV-2F coupled engine. Actually one of these two aircraft – the fifth Moscow-built Tu-4 (no serial, c/n 230501) – was modified twice, the two configurations having separate designations. The second version, known as **'aircraft 94/1'**, had the

Tu-4LL c/n 230501 in late guise as the 'aircraft 94/2' with the 2TV-2F engine. Note the constant-chord blades of the contraprops.

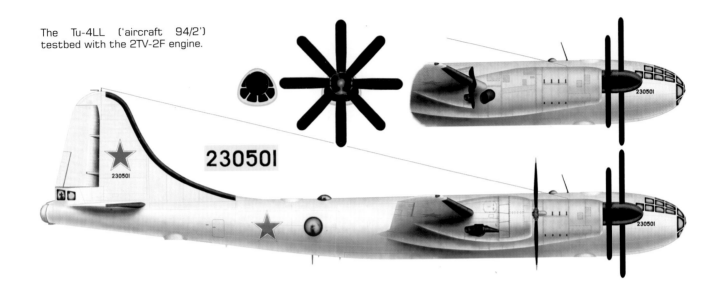

The Tu-4LL ('aircraft 94/2') testbed with the 2TV-2F engine.

230501

No.3 ASh-73TK replaced by the entire forward/center fuselage of Tupolev's 'aircraft 91' naval strike aircraft, or Tu-91 *Bychok* (Goby fish; NATO reporting name *Boot*). This housed a single TV-2M direct-drive turboprop aft of the cockpit driving SKBM AV-44 three-blade contra-rotating propellers of 4.4 m (14 ft 5^{15}⁄$_{64}$ in) diameter via an extension shaft; as on the actual Tu-91, the engine exhaust pipe was bifurcated, exiting on both sides aft of the wing trailing edge. An air data boom looking like a tripod was installed on the fuselage nose. The aircraft was flown in this guise by a test crew captained by Yuriy T. Alasheyev in 1954.

Version 3 of Tu-4LL c/n 230501 (**'aircraft 94/2'**) had a 2TV-2F engine in the same position – effectively two TV-2Fs side by side driving four-blade contra-rotating propellers via a common reduction gearbox. The engine nacelle was of the same type as orig-

inally developed for the Tupolev '95-1' strategic bomber (the first prototype of the Tu-95 *Bear-A*), with a wide air intake below the propeller spinner. The propellers had wide constant-chord blades.

After completing their trials programs on the Tu-4LL the TV-2M and the 2TV-2F went on to power the actual Tu-91 and '95-1' prototypes respectively. Both aircraft had an unhappy fate, the Tu-91 falling victim to a whimsical head of state (Nikita S. Khrushchov's predilection towards missiles was the bane of the Soviet aircraft industry's existence) while the '95-1' was lost in a crash caused by an uncontained failure of the reduction gearbox and uncontrollable engine fire, which led the Tupolev OKB to select a different engine type for the Tu-95.

In 1954 one more *Bull*, a late-production Moscow-built example (no serial, c/n 2303001), was urgently converted into a Tu-4LL

Left: Tu-4LL c/n 230501 in early configuration as 'aircraft 94/1' with 'half a Tu-91' instead of the No.3 engine. The steel cable running from the hub of the AV-44 contraprops is for a vibration sensor. Note the bulletproof glass windows flanking the engine air intake; these were meant to facilitate weapons aiming for the Tu-91's pilots.

Top right: With ground power connected and a test engineer watching, the 'aircraft 94/1' ground-runs the TV-2M turboprop in the 'half a Tu-91' at Zhukovskiy. Note the oil cooler air intakes in the wing leading edge and the shape of the fairing for the starboard main gear unit.

Above right: The same scene from a different angle. The turboprop's starboard exhaust pipe is particularly evident here.

in the wake of the '95-1' bomber's crash. This aircraft served as a testbed for the 12,500-ehp Kuznetsov TV-12 direct-drive turbo-prop driving SKBM AV-60 four-blade contraprops of 5.6 m (18 ft 4^{15}/$_{32}$ in) diameter. Again, the engine was installed in the No.3 position, the nacelle protruding far ahead of the wing leading edge so that the propeller's rotation plane was almost in line with the flight deck. The pitots were mounted on a tripod-like structure ahead of the flight deck glazing to keep them out of the turbulent airflow from the prop; a cine camera pointing at the development engine was installed in a fairing aft of the nosewheel well.

A crew captained by Mikhail A. Nyukhtikov performed the flight tests. On completion of the test program TV-12 engines were fitted to the second prototype of the *Bear* (the '95-2'), the engine subsequently entering production as the NK-12. (In passing, to this day it remains the world's most powerful turboprop – the NK-12M and subsequent versions deliver 15,000 ehp for take-off. The NK-12 has done sterling service for nearly 60 years now, powering such assorted types as the Tu-95, Tu-142 *Bear-F* anti-submarine warfare aircraft, Tu-114 *Cleat* long-haul airliner, Tu-126 *Moss* AWACS, An-22 *Cock* heavy transport and Alekseyev A-90 *Orlyonok* (Eaglet) transport/assault wing-in-ground effect craft.)

The next Tu-4LL converted by LII for testing turboprop engines in the mid-1950s ('22 Red', c/n 221203) had two development engines – initially 4,000-ehp Kuznetsov NK-4s, later identically rated Ivchenko AI-20s – in the Nos. 1 and 4 positions driving SKBM AV-68I four-blade propellers of 4.5 m (14 ft 9^{11}/$_{64}$ in) diameter. Both engines had been developed for the An-10 *Ookraïna* (the Ukraine; NATO reporting name *Cat*) and the Il'yushin IL-18 *Moskva* (Moscow, the 'second-generation' IL-18 of 1957; NATO reporting name *Coot*) and the AI-20 was eventually selected to power both types. Unusually, at one time the port turboprop was installed above the wing as on the IL-18, while the starboard turboprop was mounted below the wing as on the An-10! This 'lopsided' installation served to check the behaviour of the engine in different operating conditions.

In most cases the development engine was again installed in the No.3 position. Tu-4LL c/n 225402 (serial unknown) was an exception, with a pair of TV-2 engines (c/ns 16 and 17) driving

AV-41V four-blade contra-rotating propellers instead of the Nos. 1 and 4 ASh-73TKs. The aircraft was custom-built as a testbed by plant No.22 in 1951 and was known as 'order 165'. In this configuration it made 27 flights totalling 72 hours 51 minutes. Unfortunately the aircraft was lost on 8th October 1951 during a test flight. At an altitude of 5,000 m (16,400 ft) the No.1 turboprop started up normally but the No.4 engine would not – and caught fire when the

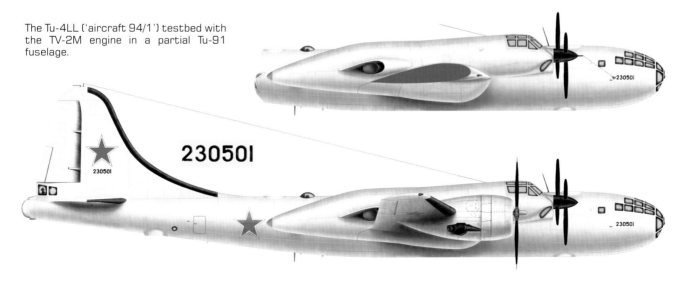

The Tu-4LL ('aircraft 94/1') testbed with the TV-2M engine in a partial Tu-91 fuselage.

230501

230501

Above: Tu-4LL c/n 2303001 with a TV-12 (NK-12) development engine and AV-60 propeller. Note the tripod-like air data probe on the nose. The c/n has been retouched away by the military censor.

Left: The same aircraft taxies out for a test mission. Note the faired cine camera aft of the nosewheel well pointing at the development engine.

attempt was repeated. The fire was caused by a leak of fuel into the turboprop's nacelle through the joint between the engine and the extension jetpipe. The crew managed a forced landing, escaping from the burning aircraft which was destroyed by the fire.

One peculiarity of the turboprop testbeds was that the power of the development engine exceeded that of the original ASh-73TK by a factor of 2 to 5. This required major structural changes; in particular, the engine bearer(s) had to be designed anew to convey the much higher forces. Besides, the development engine(s) ran on kerosene instead of aviation petrol, which meant a separate fuel system had to be provided for the turboprop(s). Tu-4LL testbeds with experimental turboprop engines were operated by the Tupolev

The Tu-4LL with the TV-12 development engine.

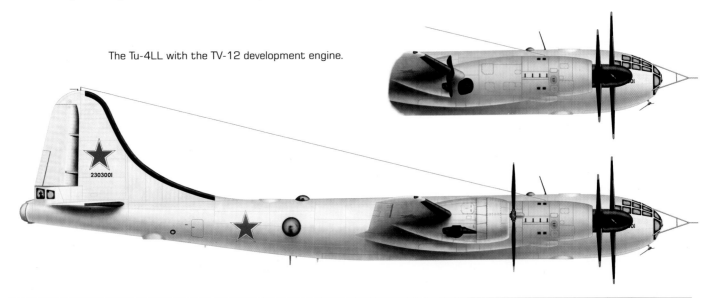

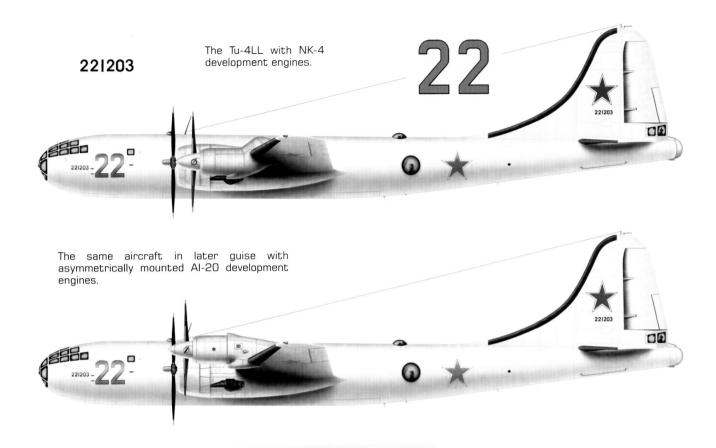

221203

The Tu-4LL with NK-4 development engines.

The same aircraft in later guise with asymmetrically mounted AI-20 development engines.

Right: Tu-4LL '22 Red' (c/n 221203) featuring two NK-4 turboprops with AV-681 propellers in the Nos. 1 and 4 nacelles in an An-10/An-12 style position

Below: The same aircraft in later guise, with AI-20 turboprops installed asymmetrically (the No. 1 engine is in an IL-18 style nacelle).

OKB and LII in 1951-60. Interestingly, LII sources state that three, not four, examples were equipped with turboprops.

Tu-4LL engine testbeds with jet engines (including DR-1/DR-2)

Three other Tu-4LLs were used to test new turbojet and turbofan engines. These aircraft were configured differently – the development engine was installed in the front bomb bay in a special nacelle which could be raised or lowered hydraulically on a system of levers. The nacelle was semi-recessed for take-off and landing to give adequate ground clearance, extending clear of the fuselage into the slipstream before start-up. It was also possible to extend the engine on the ground when the aircraft was parked over a special trench with concrete-lined walls and a jet blast deflector for ground runs. When the nacelle was retracted the air intake was blanked off by a movable shutter to prevent windmilling and foreign object damage. In an emergency (for instance, if the hydraulic retraction mechanism failed) the development engine could be jettisoned to permit a safe landing, special pyrotechnical guillotines cutting the fuel and electric lines. The test engineer and his assistant sat in the center pressure cabin; the cabin floor incorporated a glazed window for inspecting the nozzle of the development

engine. (A similar testbed based on the Superfortress existed in the USA – the XB-29G converted from B-29B-55-BA 44-84043.)

Specifically, in the course of the Myasishchev M-4 bomber's development OKB-23, jointly with LII, converted a Moscow-built Tu-4 (no serial, c/n 230113) into a testbed for the 5,000-kgp (11,020-lbst) Lyul'ka AL-5 axial-flow turbojet which was considered as a possible powerplant for the M-4 at an early development stage. In this guise the aircraft was known at the Myasishchev OKB as the

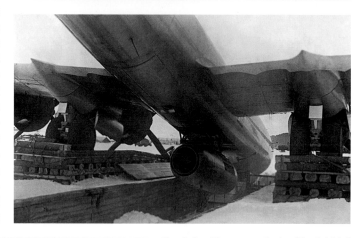

Top left: Close-up of the No. 4 NK-4 engine and its curious dog-leg nacelle on Tu-4LL '22 Red'.

Top and above: The Tu-4LL (DR-2) engine testbed sits over a concrete-lined trench for ground-running the AM-3 development engine. Here the nosewheels are actually in the trench and the mainwheels rest on specially constructed makeshift ramps to give the aircraft a strong nose-down attitude, simulating a steep dive.

Left: Another view of the DR-2 parked over the same trench, this time in a more normal attitude. The pylon of the lowered AM-3 turbojet and the faired air intake cover ahead of the front bomb bay can be seen here.

The Tu-4LL (DR-2) engine testbed with the AM-3 development engine.

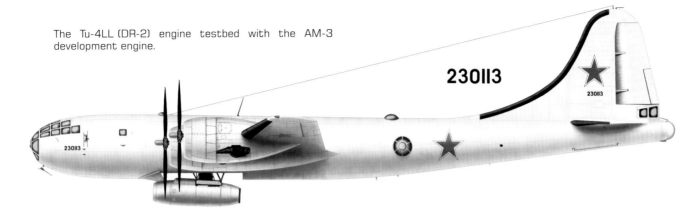

Top: One more Tu-4LL with a turbojet engine (c/n 230314). Note the heat shield aft of the development engine.

Center and above: Two more views of the same aircraft in flight with the turbojet running.

Top: Yet another Tu-4LL with a turbojet (c/n 2302501) parked over the trench. The development engine is in take-off/landing position.

Above: Close-up of the engine pod on Tu-4LL c/n 2302501. Note the retractable cover closing the air intake to prevent windmilling.

Left, center left and below left: Here, Tu-4LL c/n 2302501 taxies out for a test mission at Zhukovskiy; the small ground clearance of the semi-retractable development engine pod is readily visible. Note the metal strips on the lateral observation blisters. The other Tu-4 in the background is devoid of armament and thus is apparently also a testbed of some sort.

Bottom left: Tu-4 '80 Red' which served as a testbed for the RBP-4 Rubin radar; note the development radar's radome aft of the usual 'thimble' for the Kobal't radar.

Above right: A drawing from OKB-51 project documents showing how the 16KhA Priboy air-to-surface missiles were to be carried by the Tu-4. The missile racks are mounted on truss-type supports directly under the outer engine nacelles.

Right: On the actual Tu-4 weapons testbed used for testing the 16KhA missile, the missiles were suspended on more solid pylons between the inner and outer engines, Tu-4K style. The checkerboard colour scheme indicates this missile is an inert test round.

DR-1; D stood for *dvigatel'* (engine, a reference to the engine testbed role) while *izdeliye* R was again the *Bull*'s product code.

Later, when the much more powerful Mikulin AM-3 axial-flow turbojet rated at 8,700 kgp (19,180 lbst) for take-off was selected for the M-4, this replaced the AL-5 development engine on Tu-4LL c/n 230113 in January 1952 and the in-house designation at OKB-23 was changed to **DR-2**. The testing of the AM-3 was an arduous process, but it enabled the M-4 (and the Tu-16) to complete their test programs quickly and enter production.

Later two further unserialled Moscow-built Tu-4s (c/ns 230314 and 2302501) were similarly converted into Tu-4LLs. Apart from the two engine types mentioned above, the three aircraft were used to test a wide range of axial-flow jet engines up to 1962 – the 2,700-kgp (5,950-lbst) AM-5F, the 3,300-kgp (7,275-lbst) Mikulin AM-9B (RD-9B) afterburning turbojet, the 5,110-kgp (11,265-lbst) Tumanskiy R11-300 afterburning turbojet, the 900-kgp (1,980-lbst) Tumanskiy RU19-300 turbojet, the 6,830-kgp (15,060-lbst) Lyul'ka AL-7 non-afterburning turbojet, the 7,260-kgp (16,000-lbst) AL-7P and the 9,200-kgp (20,280-lbst) afterburning AL-7F, the 8,440-kgp (18,610-lbst) Klimov VK-3 afterburning turbofan, the 6,270-kgp (13,820-lbst) VK-7 turbojet, the Dobrynin VD-5, the 11,000-kgp (24,250-lbst) VD-7 afterburning turbojet and the 5,400-kgp (11,900-lbst) Solov'yov D-20P commercial turbofan. The method of installation and the testing techniques were basically the same as on the DR-1/DR-2 testbed, except in the case of the VD-5 and VD-7; these bulky turbojets were positioned too low even when semi-recessed, necessitating installation of a special fixed landing gear with longer struts.

In addition to the M-4 and Tu-16, the jet engines tested on the Tu-4LL powered a wide selection of aircraft types – the '98' (Tu-98) *Backfin* supersonic medium bomber prototype, the Tu-22 *Blinder* supersonic medium bomber, the Tu-110 medium-haul airliner, the Tu-124 *Cookpot* short-haul airliner, the Mikoyan/Gurevich SM-1, SM-2, SM-9, MiG-19, Ye-5, Ye-6, Ye-7, I-1, I-7 and MiG-21F *Fishbed-C* experimental and production fighters, the Sukhoi Su-7 *Fitter-A* fighter-bomber and its Su-7U *Moujik* trainer version, the Sukhoi T-3, S-1, Su-9 *Fishpot-B* and Su-11 *Fishpot-C* interceptors, the Yakovlev Yak-25 *Flashlight-A*, Yak-27 *Mangrove* and Yak-28 *Brewer/Firebar* tactical aircraft family, the Yak-30 advanced trainer, the Il'yushin IL-40 attack aircraft and IL-46 tactical bomber, the Myasishchev 3M/3MS/3MN/3MD which passed manufacturer's flight tests and state acceptance trials in the 1950s and early 1960s.

Tu-4 avionics testbeds

The Tu-4 also served for verifying various new avionics systems (mostly of a military nature). These included the Rym-S targeting system, a remote guidance/targeting system for torpedo boats and the PRS-1 Argon gun ranging radar (*pritsel rahdiolokatsionnyy strelkovyy* – radio gunsight) intended for the tail cannon barbettes of heavy aircraft. The latter program was especially intensive – a small batch of Tu-4s (including c/n 2805103?) was actually built with PRS-1 gun ranging radars; the radar later became a standard fit on such bombers as the Tu-16, Tu-95 and M-4.

The Tu-4 served as the testbed for the DISS-1 Doppler speed and drift sensor system (*doplerovskiy izmeritel' skorosti i snosa*), the ARK-5 automatic direction finder (*avtomaticheskiy rahdio-kompas*), the RSIU-2 and RSIU-3 communication radios, a blind landing system cryptically designated **Booivol-Kod** (Buffalo-Code), short-range radio navigation (SHORAN) slot antennas

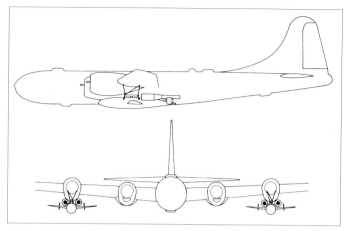

built into the fin, very low frequency trailing wire aerials (TWAs) and many other avionics items, some of which became a standard fit for Soviet aircraft.

One more Tu-4 coded '80 Red' (c/n 224001?) served as a testbed for the RBP-4 Rubin (Ruby, pronounced *roobin*; NATO codename *Short Horn*) navigation/bomb-aiming radar. This radar found use on the Tu-16 *Badger-A*, Tu-22 *Blinder-A* and Tu-95

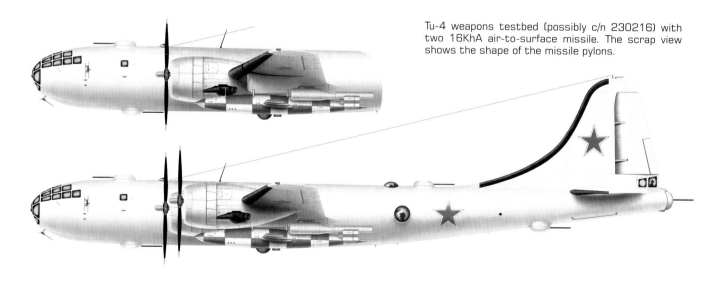

Tu-4 weapons testbed (possibly c/n 230216) with two 16KhA air-to-surface missile. The scrap view shows the shape of the missile pylons.

Top and above: The G-310 interceptor prototype makes a ballistic test launch of a G-300 air-to-air missile. Note the multiple nose radomes and the open bomb bay doors.

Left: This sequence of stills from on-board cine camera footage shows how the G-300 AAM is propelled clear of the aircraft by the ejector rack during launch.

Bear-A bombers and some of their variants, the An-8 and An-12 transports, and the Tu-104 *Camel* twin-turbojet medium-haul airliner derived from the Tu-16. The RBP-4 was housed in a rather flat-bottomed radome located a short way aft of the standard 'thimble' radome for the Kobal't radar.

Tu-4 weapons testbeds with guided weapons
Additionally, the *Bull* found use for testing a selection of new air-to-surface weapons.

a) Chelomey's missile: In 1948-49 the Soviet Air Force considered the possibility of equipping the Tu-4 with cruise missiles developed by Vladimir N. Chelomey's OKB-51. As mentioned earlier, this OKB had reverse-engineered the German Fieseler Fi 103 'buzz bomb' (better known as the V-1), a few examples of which had fallen into Soviet hands; the Soviet version was designated 10Kh. It served as the progenitor of a whole family of missiles built along the lines of the V-1 but having different wing shapes and different engines. The most advanced of these was

perhaps the 16Kh, named *Priboy* (Surf), which had two pulse-jets instead of one, tapered wings and a reworked tail unit with twin fins and rudders replacing the V-1 style engine pylon/rudder assembly of the earlier models.

The missile was 7.6 m (24 ft 11$^{7}/_{32}$ in) long, with a fuselage diameter of 0.84 m (2 ft 9$^{5}/_{64}$ in), a wing span of 4.68 m (15 ft 4$^{1}/_{4}$ in) and a wing area of 4.91 m² (52.85 sq ft). The original D12 engines turned out to be unsatisfactory and were replaced by 251-kgp (553-lbst) D14-4 pulse-jets on the definitive version; both engines were developed in house. The 16KhA live version had a launch weight of 2,557 kg (5,637 lb) and carried a 950-kg (2,094-lb) HE warhead; a target drone version designated 16KhM (*mishen'*) was also offered. The missile had a top speed of some 900 km/h (559 mph), a range of 190 km (118 miles) and a cruise altitude of 1,000 m (3,280 ft).

According to the original project, the Tu-4 converted into a missile platform for the 16Kh would have two beam-type racks attached on rather rickety truss-type structures directly under the outer engine nacelles. However, the actual aircraft used as a test-bed for the 16KhA was more like the Tu-4K, featuring more solidly built faired pylons installed between the inner and outer engines. Several such aircraft were reportedly in use with OKB-51 but only one of them is has been positively identified (c/n 230216); this aircraft was delivered to the OKB in November 1951.

Unfortunately, on 19th February 1952 Tu-4 c/n 230216 crashed when both starboard engines failed during the course of a training flight. During the ensuing right-hand turn the aircraft lost speed and stalled at about 80 m (260 ft); rolling sharply to the right, the Tu-4 impacted in a wings-vertical attitude. Only three members of the ten-man mixed OKB-51/LII crew parachuted to safety; the others – captain Mikhail A. Kotyrev, co-pilot P. I. Boordonov, flight engineer I. P. Alekseyev, navigator Konstantin I. Trekov, radio operator F. A. Sychov, flight mechanic V. V. Artamonov and engineer in charge of the tests B. N. Fadeyev – were killed in the crash. Investigation showed that the aircraft had been fuelled incorrectly and the small amount of fuel in the starboard wing tanks was quickly used up, the Nos. 3 and 4 engines quitting due to fuel starvation.

b) Bisnovat's missile: In the autumn of 1951 a Tu-4 was transferred from a first-line DA unit to OKB-293 headed by Matus R. Bisnovat which was based at plant No.293 in Khimki, a northern suburb of Moscow. In 1948 OKB-293 had started work on the Shtorm (Sea Storm) coastal defence missile system; the subsonic anti-shipping missile forming its core was based on the '5' rocket-propelled experimental aircraft designed by Viktor F. Bolkhovitinov. The Tu-4 was to be converted into a testbed for development of the Shtorm missile.

c) guided bombs: Soviet development work on guided bombs, initially based on World War Two German types such as the Fritz X, began in 1950. That year the OKB-2 weapons design bureau led by Aleksandr D. Nadiradze (renamed GosNII-642 in December 1951) developed the UB-2000F (*oopravlyayemaya bomba* – guided bomb), which was aimed by means of an OPB-2UP optical sight and guided via a radio command channel. During trials in 1953 a modified Tu-4 carried two such bombs on underwing pylons. However the results proved disappointing and the 2,240-kg (4,940-lb) UB-2000F with a 1,795-kg (3,960-lb) warhead had to undergo a redesign, eventually being included into the VVS inventory two years later as the UB-2F *Chaika* (Seagull, aka *izdeliye* 4A22).

The 5,100-kg (11,240-lb) UB-5000F with a 4,200-kg (9,260-lb) warhead was developed and tested by OKB-2 in 1954. It was of similar design but featured a TV command guidance system with a TV camera in the nose transmitting a 'bomb's eye view' to a display at the bomb-aimer's station; the bomb-aimer manually corrected the weapon's flight path. Pretty soon, however, the Soviet Union discontinued work on guided bombs and the weapons system was never introduced on the Tu-4.

d) cannon installations: One Tu-4 is known to have served as a testbed for a tail barbette mounting 37-mm (1.24 calibre) Nudelman N-37 cannons.

G-310 extra heavy interceptor

Perhaps the most unusual role fulfilled by the *Bull* was that of a heavy interceptor – in fact, The Heaviest Interceptor That Ever

Was. In the second half of 1950 the aforementioned KB-1, which had been responsible for the Kometa weapons system, took on an even more important task – development of the S-25 *Berkut* (Golden Eagle) air defence system protecting Moscow against a possible US nuclear air strike. According to the original plans, the S-25 system was to include not only V-300 surface-to-air missiles (SAMs) developed by Semyon A. Lavochkin's OKB-301 as part of the weapons system – the first Soviet SAM, known to the West as the SA-2 *Guideline* – but also heavy interceptors armed with G-300 AAMs. Since a single nuclear bomb would be enough for destroying any major target, bombers carrying nukes were expected to operate singly, not in large formations, which called for a long-range interceptor that would destroy whatever aircraft that managed to slip through two lines of SAMs before they came close enough to drop the bomb.

The interceptor was required to have long endurance allowing it to patrol the distant approaches to the objective being protected and was to operate independently from ground-based air defence radars, which called for a powerful on-board radar suite. In other words, it was to be an interceptor and an airborne early warning (AEW) system rolled into one. The radar suite was to comprise four D-500 *Taïfoon* (Typhoon) radars with a detection range of 80-100 km (49-62 miles) scanning the front, rear, upper and lower hemispheres; the D-500 was developed by NII-17 under Viktor V. Tikhomirov. The armament was to be four AAMs with a 'kill' range of 40-50 km (24.8-31 miles) carried on ejector racks. Initially the G-300 was to have semi-active radar homing (SARH), but eventually KB-1 opted for beam-riding missiles, learning from experience with the Kometa weapons system.

In spite of SARH, guidance accuracy was not expected to be particularly high – comparable to that of the V-300 SAM, requiring a warhead weighing at least 100 kg (220 lb) in order to guarantee destruction of the target. Another requirement was a maximum 'kill' altitude of at least 20,000 m (65,620 ft), which called for a powerful rocket motor. All this boiled down to a launch weight of at least 1,000 kg (2,205 lb). KB-1 issued the operational requirement for the G-300 missile to the Lavochkin OKB in early 1951. Without much ado, the Lavochkin OKB designed the AAM as a scaled-down version of the V-300 (*izdeliye* 205), retaining the tail-first aerodynamic layout but reducing the length from 11.3 to 8.3 m (from 37 ft 0⅞ in to 27 ft 2⁴⁹⁄₆₄ in) and the body diameter from 650 to 530 mm (from 2 ft 1¹⁹⁄₃₂ in to 1 ft 8⁵⁵⁄₆₄ in). Like the larger version, the G-300 had a liquid-propellant rocket motor running on triethylamine xylidine and nitric acid. The AAM had the in-house product code *izdeliye* 210.

The choice of a suitable weapons platform was a major problem. Considering the weight of the four missiles and the radar suite, the aircraft had to have a large payload. After analysing the available choices, KB-1 rejected the IL-28 and Tupolev Tu-14 *Bosun* twin-turbojet tactical bombers, as these lacked the space for the radar suite and could only carry two G-300 missiles. The only option was the Tu-4 which, though admittedly inferior in speed, service ceiling and rate of climb, offered the required payload and space.

The modified aircraft bore the designation G-310. According to the advanced development project, with a 54,000-kg (119,050-lb) take-off weight, including four missiles and enough fuel for an

Above: The rear fuselage and tail unit of a proposed glider tug version of the Tu-4 undergoing static tests

Below: The glider tug failed the tests, the fin skin buckling at 59% of the design load. The damage is evident in this view.

on-station loiter time of four to five hours, the aircraft would have a service ceiling of 8,000-9,000 m (26,250-29,530 ft); with a 48,000-kg (105,820-lb) take-off weight and an on-station loiter time of 1.8 hours the service ceiling increased to 10,000 m (32,810 ft).

On 3rd November 1951 the Council of Ministers issued a directive terminating the work on *izdeliye* 210 and initiating development of a new version of the G-300 AAM which received the product code *izdeliye* 211; development of this missile was completed at the end of the year. In 1952 the Tupolev OKB and the Kazan' aircraft factory converted a brand-new Tu-4 (c/n 226404) into the G-310 prototype. The standard armament and the Kobal't-M radar were removed. The aircraft had a rather bizarre appearance, featuring twin guidance system radomes low on the forward fuselage sides and two larger radomes in tandem on top of the forward fuselage; the missiles were carried on special underwing ejector racks.

It is known that in 1952 the Kazan' aircraft factory built a custom-made Tu-4 equipped to carry four missiles and known as 'order 168'. It seems likely that the aircraft in question was the G-310.

In May-June 1952 the G-310 underwent initial flight tests, making ten flights, some of them with dummy G-300 missiles. Three engineers in charge of the tests – P. Limar from LII, V. Vaïner from the Lavochkin OKB and V. Bogdanov from the Tupolev OKB were assigned to the aircraft. At this stage the G-310 attained a service ceiling of 9,900 m (32,480 ft) with four missiles and a 49,500-kg (109,130-lb) take-off weight; because of the draggy external stores and radomes the maximum speed of 525 km/h (326 mph) at 9,000 m was 30-35 km/h (18.6-21.75 mph) lower than the standard bomber's.

As part of the test program, LII built a whole trajectory measurement infrastructure which included optical (cine theodolite) and radar tracking posts, the RTS-2 data link system (*rahdiotelemetricheskaya sistema*), three Tu-2 bombers outfitted with cameras as chase planes, and test equipment installed on the G-310. All measurements were synchronised by a time-keeping system; the missile launches were also filmed by AKS-3 cine cameras on board the G-310.

In October-November 1952 the aircraft made five ballistic launches of the G-300 AAM (that is, without the guidance system) at the GK NII VVS facility in Akhtoobinsk, Astrakhan' Region (Vladimirovka AB). It turned out that the missile's speed and range met the specifications. The launch proceeded safely; the missile dropped 100 m (330 ft) below the aircraft's flight path and any initial oscillations were quickly negated by the APG-301R autopilot. In January-June 1953 the Lavochkin OKB made revisions to the missile, eliminating the deficiencies noted during the tests. On 4th-19th August 1953 a further series of seven ballistic launches of the G-300 with the revised APG-301S autopilot was made at altitudes of 5,000-9,000 m (16,400-29,530 ft), and preparations were in hand for testing the fully guided version of the missile in the third quarter of 1954.

In 1952 a group of military experts headed by Minister of Defence Marshal Aleksandr M. Vasilevskiy joined in the preparations for the anticipated deployment of the G-310/G-300 system (sometimes referred to as Berkut-2). According to the plan, a special mission air division would be formed, comprising two regiments, each with three squadrons of G-310s; it would operate from two main bases – one within the ring of SAM batteries around Moscow and one outside it – and have two reserve bases located in the same fashion. The problem was that only three airfields within a 45-km (27.9-miles) radius of Moscow were suitable for the Tu-4; these were Moscow-Vnukovo airport, Chkalovskaya AB (GK NII VVS) and the LII airfield in Zhukovskiy. In March 1952 Marshal Leonid A. Govorov suggested using Chkalovskaya AB and Migalovo AB in Kalinin (now Tver') as the main bases, with Zhukovskiy and Dyagilevo AB in Ryazan' as the reserve bases. However, having the special mission air division stationed at any of these airfields would entail serious complications in their normal operation because of the tighter security measures, if nothing else. As a result, it was decided to build a new base north of Moscow... and so began the story of Moscow-Sheremet'yevo airport.

Then, however, new complications arose. Firstly, KB-1 was nowhere near completing development of the G-300's guidance system and was holding up the works. Secondly, the obsolete Tu-4 was clearly not the best choice as an interceptor. Being slow, it could only engage a high-speed target in head-on mode – an attack in pursuit mode was impossible. For the same reason the G-310 could only loiter near its objective and required as much as 1.5 hours to reach its service ceiling, which meant a scramble was out of the question. Therefore, on 14th August 1953 MAP issued order No.507ss cancelling all further work on the program. (The *izdeliye* 211 missile was briefly revived in 1953-54 as the weapon for the Lavochkin OKB's La-250 Anaconda supersonic interceptor – but that's another story.)

Project versions

In addition to the versions described above, the Tu-4 had several versions that never came off the drawing board.

Tu-4 torpedo-bomber and minelayer (project)

Believe it or not, a minelayer and torpedo-bomber version of the Tu-4 for the Soviet Navy was under consideration. In his letter No.19029s to Minister of Aircraft Industry Mikhail V. Khroonichev on 24th January 1949, Deputy Chief of General Staff (Air Force) Lt.-Gen. Nikolay P. Dagayev asked the minister's opinion on the draft Council of Ministers directive titled 'On the conversion of a production Tu-4 aircraft into a [naval] minelayer and high-altitude torpedo-bomber variant'. He further informed Khroonichev that a special technical session at OKB-156 in November 1948 had looked into the possibility of using the Tu-4 as a minelayer and high-altitude torpedo-bomber. It had been established that the Tu-4 could carry 12 AMD-1000 naval mines (*aviatsionnaya meena donnaya* – air-dropped bottom mine) or three 450-mm (17³⁄₄ in) calibre 45-36AVA high-altitude torpedoes without requiring major structural changes. The mines would have to be carried on a purpose-built cassette-type rack; the torpedoes would require installation of special bearers and modifications to the bomb bay doors. The draft directive also envisaged the capability to carry RAT torpedoes (*raketnaya aviatsionnaya torpeda* – rocket-propelled air-droppable torpedo).

On 7th February Khroonichev wrote back. In letter M-34/304ss he informed Dagayev that he considered the conversion inexpedient, as it would entail major structural changes. He stated that the Tupolev OKB was working on a version of the Tu-4 featuring enhanced cannon armament and new bomb armament (that is, NR-23 cannons and standardised bomb armament – *Auth*.); this

aircraft would have the potential to be used as a minelayer and torpedo-bomber.

Tu-4 glider tug (project)

In the late 1940s, when heavy assault gliders were still on the agenda, someone came up with the idea of using the Tu-4 as a glider tug. At that time the Soviet Air Force's only glider tug was the Il'yushin IL-12D – a military transport/assault version of the IL-12B *Coach* airliner which, apart from transporting materiel and dropping paratroopers, could work with Yakovlev Yak-14 *Mare* and Tsybin Ts-25 *Mist* assault gliders.

The conversion of the Tu-4 for glider towing involved some structural changes to the rear fuselage. This is borne out by the fact that a suitably modified Tu-4 rear fuselage underwent additional static tests. The results were disappointing: skin buckling was detected at 59% of the design load. Eventually the idea was dropped because assault gliders were outdated and were being withdrawn by the Soviet Air Force.

Tu-4 ECM version with Gazon suite (project)

In addition to the ECM-configured Tu-4Rs and the experimental ECM version mentioned earlier, it deserves mention that around 1956 NII-108 began development of the *Gazon* (Lawn) multiple-jammer ECM suite which was intended for installation in the Tu-4. However, the program was terminated in 1958 when the DA phased out its Tu-4s replacing them with Tu-16s.

Tu-4 VIP version (project)

In 1951, in keeping with a ruling of the CofM Presidium's Commission on Defence Industry Matters (VPK – *Voyenno-promyshlennaya komissiya*), the Tupolev OKB undertook design studies on a VIP version of the Tu-4 which would have been the Soviet equivalent of the US Air Force's Boeing VB-29. The aircraft was stripped of all armament; the forward pressure cabin accommodated three or four passengers seated behind the crew, while the center pressure cabin was transformed into sleeping quarters with two bunks. Auxiliary fuel tanks could be installed in the bomb bay, extending range by 1,500 km (931 miles). The aircraft was never built.

Tu-4 television relay aircraft (project)

Apart from the TACAMO version described above, in the late 1940s a version of the Tu-4 equipped for TV broadcasting signals relay was proposed. This is how Leonid L. Kerber described it:

'Television started booming in those years. The government was bombarded with complaints that TV broadcast reception was difficult or downright impossible a mere 100 km [62 miles] from the capital. Then the Ministry of Communication suggested equipping the Tu-4 with a signal relay system featuring a rigid aerial 10 m [32 ft 9⁴³⁄₆₄ in] long that would be deployed in flight. Picking up the signal from the Shaboovka tower (the lattice-like tower of Moscow's first TV center in Shabolovka Street – Auth.), the relay system would amplify it and ensure onward broadcast with the aircraft flying at 10,000 m [32,810 ft]. The Ministry of Communication and our ministry, the Ministry of Aircraft Industry, failed to reach an agreement on the issue, so the proposal was forwarded to V. M. Molotov's commission (the Council of Ministers' Information Com-

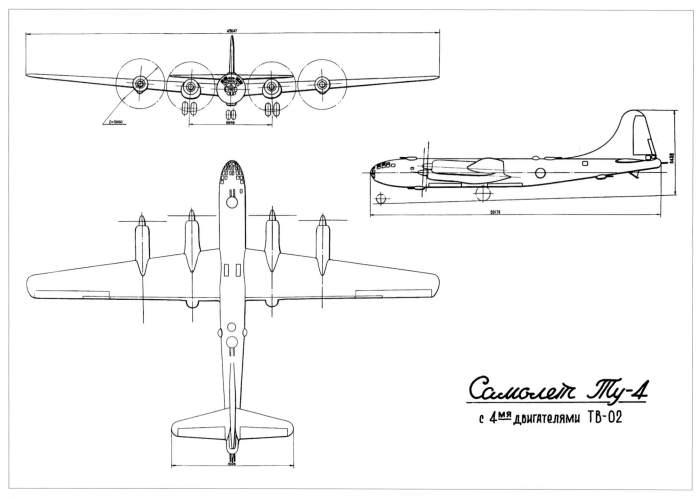

A three-view drawing from the project documents of a Tu-4 version re-engined with Kuznetsov TV-02 turboprops.

mittee, which was chaired by Vyacheslav M. Molotov until March 1949 – *Auth.*). *A meeting was arranged. Tupolev was away from Moscow at the time, so A[leksandr] A. Arkhangel'skiy and I went to attend the meeting. Molotov gave us an audience in the building formerly occupied* (before the October Revolution of 1917 – *Auth.*) *by [Pavel P.] Ryabushinskiy's bank, which is across from the former Stock Exchange building in Il'yinka Street. He started by asking why MAP was turning down such an enticing and honourable task.*

Arkhangel'skiy suggested that a specialist – that is, me – speak his mind. Molotov nodded assent. I stated that the relay system would require a lot of electric power, which the aircraft could not provide. Also, should the aerial fail to retract before landing, this would spell death to both the aircraft and the crew.

Molotov eyed me irritably – our big brass did not like being confronted by technical difficulties – and said: "Provide the power. As for the [possible] crash – firstly, make sure that the aerial will retract, and secondly, we should always have a back-up aircraft standing by." Then he dismissed us and never returned to the subject. The idea was dropped.'

'Aircraft 79' long-range bomber (project)
As early as 1947 the Tupolev OKB considered the possibility of fitting the Tu-4 with M-49TK liquid-cooled Vee-12 engines; very little information is available on this engine but it was almost certainly a product of the Mikulin OKB. Designated 'aircraft 79', the bomber was included in the work plan of the OKB's PD projects section; it would have looked rather like the Boeing XB-39 development aircraft (41-36954 *Spirit of Lincoln*) powered by Allison V-3420 engines. Eventually, however, 'aircraft 79' never got beyond the technical proposal stage due to the unavailability of the intended M-49TK engines.

Estimated performance of the 'aircraft 94' at maximum all-up weight	
Maximum all-up weight	63,300 kg (139,550 lb)
Fuel load	22,050 kg (48,610 lb)
Bomb load	2,000 kg (4,410 lb)
All-up weight over the target	47,800 kg (105,380 lb)
Top speed:	
at 6,000 m (19,685 ft)	676 km/h (420 mph)
at 10,000 m (32,810 ft)	650-680 km/h (404-422 mph)
Cruising speed	550-600 km/h (341-372 mph)
Range	5,400-6,300 km (3,354-3,913 miles)
Cruising altitude over the target	11,000-12,000 m (36,090-39,370 ft)
Take-off run	1,000-1,200 m (3,280-3,609 ft)

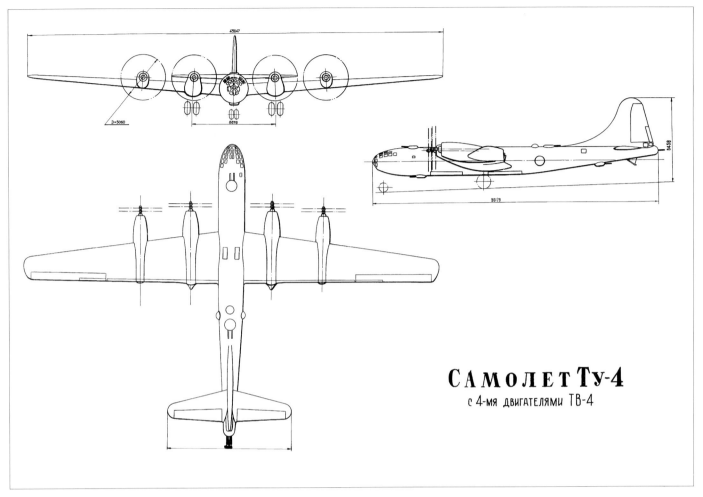

СAMOЛET Ту-4
с 4-мя двигателями ТВ-4

A three-view drawing from the project documents of a Tu-4 version re-engined with Kuznetsov TV-4 turboprops driving AV-41 propellers.

Tu-4 version with VK-2 turboprops (project)

In 1950 the Tupolev OKB began project studies aimed at re-engining the existing Tu-4 fleet with turboprops. One of the early proposals involved the VK-2 developed by Vladimir Ya. Klimov's OKB-117 – one of the first Soviet turboprop engines; it had a take-off rating of 4,200 ehp (some sources state 4,750 ehp) and a specific fuel consumption (SFC) of 326-355 g/hp·hr.

Versions with two and four VK-2s were considered in April 1950. Calculations showed that with a take-off weight of 62,000 kg (136,690 lb) and a 5,000-kg (11,020-lb) bomb load the twin-engined version would have a range of 7,155 km (4,446 miles) versus 5,030 km (3,125 miles) for the standard bomber; however, the top speed was reduced to 487 km/h (302 mph). The four-engined version would have a range of 4,445 km (2,762 miles) and a top speed of 636 km/h (395 mph). Either way, this was not good enough.

'Aircraft 94' long-range bomber (project)

On 22nd August 1950 the Council of Ministers issued directive No.3653-1519, ordering the Tupolev OKB to develop a version of the Tu-4 powered by four Kuznetsov TV-2 (sometimes called TV-02) or TV-022 turboprops. The bomber received the in-house designation 'aircraft 94'.

Developed by Nikolay D. Kuznetsov's OKB-276 from the TV-022 (a direct copy of the captured 5,000-ehp Junkers Jumo 022 turboprop), the basic TV-2 delivered 5,163 ehp for take-off during trials, with an SFC of 297 g/hp·hr versus the progenitor's 300 g/hp·hr. Maximum continuous power and SFC were 4,409 ehp and 313 g/hp·hr; cruise power and SFC were 3,740 ehp and 328 g/hp·hr (versus 3,000 ehp and 210 g/hp·hr for the TV-022). With accessories installed the engine had a dry weight of 1,700 kg (3,750 lb); it drove AV-41B four-bladed contra-rotating propellers of 4.2 m (13 ft 9 in) diameter. (The above SFC figures for the TV-2 are as per 'aircraft 94' ADP documents; the real engine showed a much better SFC of 257 g/hp·hr at take-off power and 198 g/hp·hr in cruise mode.)

A drawing from the project documents shows that the TV-02-powered Tu-4 was dimensionally unchanged; the engines were housed in slender cylindrical nacelles, the propeller rotation planes being ahead of the front bomb bay. Interestingly, the drawing shows the propellers' diameter as 5.06 m (16 ft 7^7/$_{32}$ in).

The normal bomb load was set at 1,500 kg (3,306 lb) and the maximum bomb load at 6,000-12,000 kg (13,227-26,455 lb).

As can be seen from the above figures, the performance of 'aircraft 94' with TV-2 engines was only some 15-20% higher than that of the standard Tu-4. This was definitely not good enough for the Soviet Air Force, and the project was shelved.

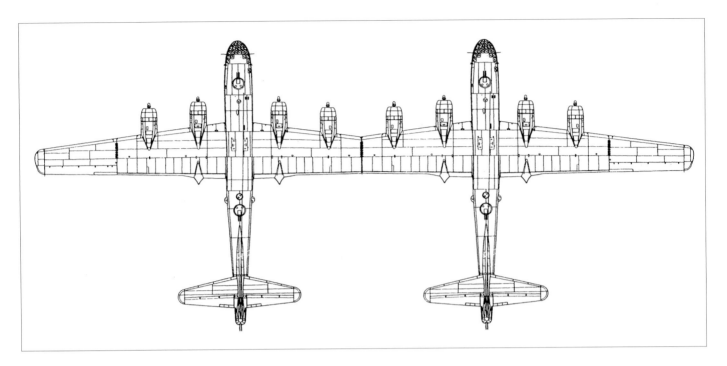

Tu-4 long-range bomber with TV-4 turboprops (project)
Later the Tupolev OKB dusted off the idea of re-engining the Tu-4, proposing that Kuznetsov TV-4 turboprops (sometimes reported as TV-04) rated at 6,300 ehp for take-off be installed. The engines were again housed in slender nacelles; however, the engines were positioned differently, with a high-set thrust line, their jetpipes passing above the wings, and drove contra-rotating propellers of 5.06 m diameter. The bomber's overall dimensions were likewise unchanged.

In March 1951 Minister of Aircraft Industry Mikhail V. Khroonichev wrote to Vice-Chairman of the Council of Ministers Nikolay A. Boolganin: '*Regarding [Chief Designer Tupolev's proposal concerning] the modification of the Tu-4 aircraft by replacing the four ASh-73TK engines with four TV-04 turboprop engines, I consider this proposal acceptable if the scope of the redesign is relatively small and if the aircraft is submitted for state acceptance trials in September-October 1951, not in May-June 1952, so that series production of turboprop-powered Tu-4s could begin in the second quarter of 1952 at the latest. It would be more expedient to modify the Tu-4 to take the 5,000-ehp TV-022 turboprop engines which have passed 100-hour bench tests; this would give the Tu-4 equipped with four TV-022 turboprops an aggregate power of 20,000 ehp versus 9,600 ehp with the current 2,400-hp ASh-73TK engines. Comrade Tupolev's proposal to equip the Tu-4 with TV-04 turboprops rated at 6,300 ehp is unacceptable because there is no such engine (sic) and it cannot be submitted for state acceptance trials this year. Designer Comrade Kuznetsov has indeed undertaken to uprate the TV-022 from 5,000 to 6,300 ehp, but this engine will be submitted for state acceptance trials in the first quarter of 1952, not this year, and for all practical purposes this means development of a new engine. Spending significant funds on developing such an engine is hardly expedient, and the best way of improving the Tu-4's flight performance is to replace it with twinjet fast bombers (! – sic) with a range of 5,000-6,000 km [3,105-3,730*

miles] and a speed of 900 km/h [559 mph]. As you know, the designers Comrades Tupolev and Il'yushin have been tasked with developing such bombers by a decision of the government; therefore it is, of course, more advisable to concentrate resources on the development of these jet bombers as a replacement for the Tu-4. Also, re-equipment of the Tu-4 with 6,300-ehp turboprops will give an aggregate power of 25,000 ehp and will inevitably require major structural reinforcement of the Tu-4 because its structural strength reserves even with the existing ASh-73TK engines are marginal, to say nothing of the fact that the time schedule proposed for the Tu-4's turboprop conversion means we will only be able to produce such aircraft in 1953. By then it is feasible to complete development of fast jet bombers...'

Hence the TV-4 powered version again remained a 'paper aeroplane'; there was no point in pursuing this project further because 'clean sheet of paper' designs of long-range bombers were being developed around new powerful Soviet turbojets and turboprops. These advanced aircraft featuring swept wings were expected to outperform the Tu-4 by far.

'Twin Tu-4' extra-heavy bomber
The Tu-4's most improbable project version was born in 1950. From time to time aircraft manufacturers in various parts of the world have developed twin versions of their aircraft by joining two stock fuselages/vertical tails with a new wing section. Thus, the North American P-51D Mustang fighter mutated into the P-82 Twin Mustang night fighter, the Messerschmitt Bf 109F was similarly duplicated as the Bf 109Z (*Zwilling* – Siamese twins) heavy fighter, which never flew, and the Heinkel He 111H-6 bomber served as the basis for the He 111Z. The latter was a specialised glider tug designed for towing ultra-heavy assault gliders, but have you ever heard of a twin-fuselage bomber?

On 1st July 1950 Maj. L. I. Martynenko, who was head of Section 2 of GK NII VVS's Scientific & Technical Department, and

Opposite page: A simplified rendition of how the 'Twin Tu-4' proposed by Martynenko and Rooter might have looked. The standard cannon barbettes are retained because full details of the proposed defensive armament are not known.

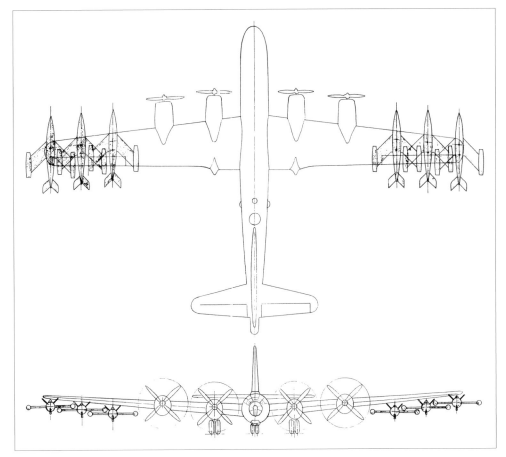

Right: A drawing from the Yakovlev OKB's ADP documents for the Yak-40 interceptor, showing one of the envisaged air launch options with the fighters carried three-abreast under the bomber's outer wings.

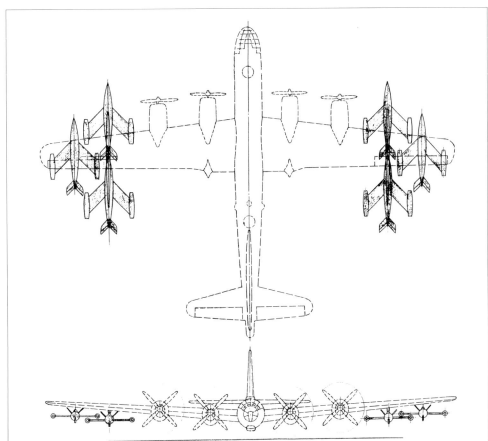

Another drawing from the Yak-40's project documents showing an alternative carriage arrangement, with a combined 'line astern/echelon formation'.

Capt. D. M. Rooter, an engineer in the same department, came up with the idea of 'duplicating' the Tu-4 in order to create a bomber with an all-up weight of 120-130 tons (264,550-286,600 lb) within a short time. Unlike the He 111Z, which featured a new constant-chord inter-fuselage wing section carrying an extra engine, the 'inventors' did not bother to suggest designing any new structural components. Quite simply, two standard Tu-4s were to have an outer wing panel detached (the port wing on one aircraft and the starboard wing on the other bomber) and then be conjoined at the wing center sections' outer ribs. The result was a rather bizarre inter-fuselage wing section whose chord was smallest at the centerline, with a kinked leading edge.

Martynenko and Rooter maintained that the 'Twin Tu-4' – we'll call it that, since no separate designation has been quoted for this project – would offer such benefits as a crew reduced from 22 (for two regular *Bulls*) to 13-14 and a weight saving achieved by deleting some redundant equipment items, such as one of the radars and some of the cannon barbettes. This, according to the project authors, would allow the aircraft to carry more fuel and a greater bomb load; the higher overall wing aspect ratio would give it a higher lift/drag ratio. Also, a more fuel-efficient engine operation mode could be used. All this was to give the bomber much longer range as compared to the basic Tu-4. According to Martynenko and Rooter's estimates, range with a 3,000-kg (6,610-lb) bomb load would increase from the standard bomber's 6,380 km (3,964 miles) to 11,000 km (6,835 miles) at an altitude of 3,000 m (9,840 ft) and from 5,050 to 9,450 km (from 3,138 to 5,872 miles) at 8,000 m (26,250 ft), which equals an improvement of 73% and 87% respectively. With in-flight refuelling (by then the wingtip-to-wingtip IFR system had been put through its paces on the Tu-4), range could be further extended to 13,000 km (8,077 miles) with a single top-up and 16,000 km (9,942 miles) with two top-ups.

Maximum speed over the target at 10,000 m (32,810 ft) was to increase from the standard Tu-4's 545 km/h (338 mph) to 610 km/h (378 mph) with the engines at maximum continuous power. The advertised increase in bomb load was particularly impressive; with a practical range of 5,000 km (3,105 miles) and a cruise altitude of 8,000 m, the 'Twin Tu-4' was to carry a whopping 23,500 kg (51,810 lb) of bombs versus 6,000 kg (13,230 lb) for two ordinary *Bulls*!

According to Martynenko and Rooter, the 'twin *Bull*' could be put into production within a few months – or, *'given heightened interest in the matter, [production can be organised] within a few weeks'*. In the proposal they wrote that a 'clean sheet of paper' bomber powered by all-new engines would take years to develop and field in sufficient numbers, whereas 'the twin aircraft will be the optimum solution until then'. They went on to conclude that *'according to the expert opinion of Test Pilot 1st Class [Mikhail A.] Nyukhtikov, the twin aircraft will have acceptable stability and handling'*.

The proposal was submitted for approval to Gheorgiy M. Malenkov, Secretary of the Communist Party Central Committee, and duly brought to the attention of the Air Force and MAP top brass. In turn, MAP requested the opinion of Chief Designer Andrey N. Tupolev. The latter gave a thumbs-down, rightly pointing out that joining two Tu-4 airframes in the manner proposed by Martynenko and Rooter was unacceptable from a structural strength standpoint – the narrow wing chord at the centerline joint could not ensure adequate stiffness. Tupolev also stated that the proposed straightforward approach would not work – the 'Twin Tu-4' would entail major structural changes, requiring a reinforced landing gear, larger ailerons, a new horizontal tail linking the vertical tails etc. Moreover, such an aircraft would be an operational nightmare, requiring centralised control of the eight engines, the control surfaces, various equipment etc – not to mention the fact that its landing gear track would be enormous, rendering it unable to use the majority of the existing runways and taxiways! Tupolev further stated that the advertised range, speed and payload figures were definitely exaggerated and, in fact, some of the aircraft's performance parameters were markedly inferior to those of existing bombers.

This verdict was furthered to Vice-Chairman of the Council of Ministers Nikolay A. Boolganin. Of course, the 'Twin Tu-4' idea was not pursued further.

Tu-4 'mother ship' with Yak-40 parasite fighters (project)
In the 1950s an Italian aviation magazine published an artist's impression showing a Tu-4 carrying two MiG-15 fighters on truss-type pylons under the outer wings. Nothing of the sort existed in reality; this was probably not so much a hoax but rather the result of information on the Burlaki program leaking to the West and being misinterpreted.

However, there was a time when the Tu-4 was actually intended to carry parasite fighters – and not just two but six! Between January and June 1948 OKB-115 headed by Aleksandr S. Yakovlev conducted project studies of a single-seat fighter designated Yak-40 (the first aircraft to bear this designation). The diminutive fighter was 7.5 m (24 ft 7^9/$_{32}$ in) long, with a wingspan of 5.05 m (16 ft 6^{13}/$_{16}$ in), and the all-up weight was only 1,800 kg (3,970 lb). It was to be powered by two 850-kgp (1,870-lbst) ramjet engines mounted at the tips of the mid-set wings; the latter and the butterfly tail (!) had 45° sweepback. The armament consisted of two 23-mm cannons flanking the cockpit, which was fitted with an ejection seat and enclosed by a frameless teardrop canopy. A generator with a ram-air turbine provided electric power.

The normal launch mode for point defence was by means of a wheeled dolly and solid-fuel rocket boosters, the fighter landing on a centerline skid and two skids under the engine nacelles. However, the Yakovlev OKB also suggested using the Yak-40 as a captive escort fighter for Tu-4 formations. Six such fighters were to be suspended under the bomber's outer wings. In one version the fighters were stacked vertically, their wings overlapping to enable the carriage of three Yak-40s within the limited space outside the outer engines' propeller discs; in another version two fighters on each side were carried in tandem and the third outboard of them.

The project did not reach the hardware stage. Later, in the mid-1960s, the Yak-40 designation was reused for a successful three-turbofan short-haul airliner (NATO reporting name *Codling*).

* * *

In addition, several versions of the Tu-4 were developed in China in the 1960s and 1970s; these are dealt with in Chapter 8.

Chapter 5

The Tu-4 in detail

The following structural description applies to the standard production Tu-4 conventional bomber.

Type: Four-engined long-range heavy bomber designed for day and night operation in visual meteorological conditions (VMC) and instrument meteorological conditions (IMC). The airframe was of riveted all-metal construction, with flush riveting on all exterior surfaces.

Fuselage: Semi-monocoque stressed-skin structure of circular cross-section; maximum diameter 2.9 m (9 ft 6³/₁₆ in), maximum cross-section area 6.6 m² (71.04 sq ft). The fuselage structure included 61 stamped frames and 40 stringers (stringer 0 located dorsally on the centerline, stringer 20 located dorsally on the centerline, and stringers 1L/19L and 1R/19R). The fuselage tapered towards the ends, only a small portion between frames 13 and 29 being cylindrical. Some of the frames were I-section mainframes absorbing the main structural loads; the regular frames were of either channel section or Z-section. Near cutouts the structure was reinforced by additional structural members (the nosewheel well and the bomb bay apertures were flanked by beams). The fuselage skin was made of D16AT duralumin sheet, mostly 0.8-1.8 mm (0¹/₃₂ to 0⁵/₆₄ in) thick, the

A full frontal of a production Tu-4 bomber. Note the instrument landing system aerial under the nose and the aerials on the sides of the flight deck section.

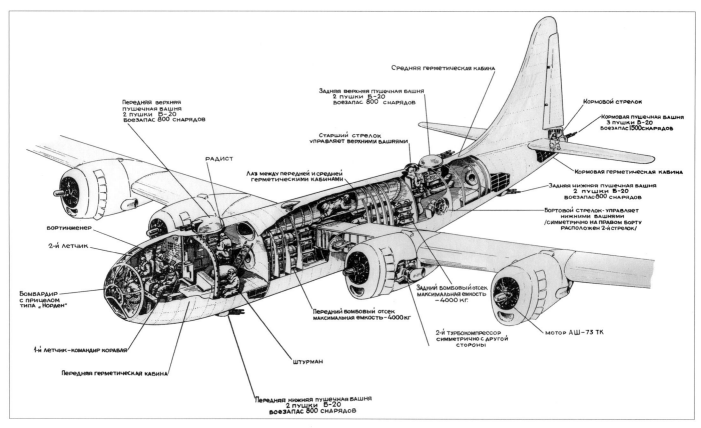

Передняя верхняя пушечная башня 2 пушки Б-20 боезапас 800 снарядов

РАДИСТ

Старший стрелок управляет верхними башнями

Лаз между передней и средней герметическими кабинами

Средняя герметическая кабина

Задняя верхняя пушечная башня 2 пушки Б-20 боезапас 800 снарядов

Кормовой стрелок

Кормовая пушечная башня 3 пушки Б-20 боезапас 1500 снарядов

Кормовая герметическая кабина

Задняя нижняя пушечная башня 2 пушки Б-20 боезапас 800 снарядов

Бортовой стрелок-управляет нижними башнями /симметрично на правом борту расположен 2-й стрелок/

БОРТИНЖЕНЕР

2-й ЛЕТЧИК

БОМБАРДИР с прицелом типа „НОРДЕН"

1-й летчик-командир корабля

Передняя герметическая кабина

ШТУРМАН

Передний бомбовый отсек максимальная емкость-4000 кг

Задний бомбовый отсек максимальная емкость -4000 кг.

2-й турбокомпрессор симметрично с другой стороны

мотор АШ-73 ТК

Передняя нижняя пушечная башня 2 пушки Б-20 боезапас 800 снарядов

A cutaway drawing of the Tu-4 representing the early production version armed with eleven B-20 cannons.

skin thickness increasing to 2 mm (0⁵/₆₄ in) at the most stressed locations. The fuselage skin was electrochemically coated and additionally sprayed with clear varnish after assembly; at some locations the fuselage's internal structure was coated with ALG-1 varnish-type yellow primer to prevent corrosion.

Structurally the fuselage consisted of six sections: the flight deck glazing frame (Section F-1), the forward fuselage (Section

F-2), the center fuselage (Section F-3), the center pressure cabin (Section F-4), the rear fuselage (Section F5) and the tail section (Section F-6). The fuselage sections were bolted together by means of flanges.

The *flight deck glazing frame* (fuselage frames 1-2) was a dome-shaped casting made of magnesium alloy, with a radius of 750 mm (2 ft 5³³/₆₄ in). It featured eight glazing panels; the lower

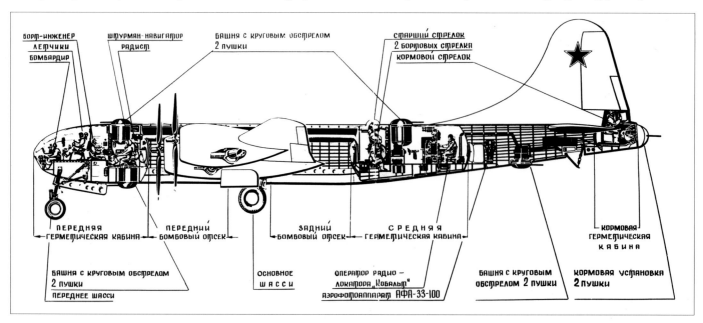

БОРТ-ИНЖЕНЕР ЛЕТЧИКИ БОМБАРДИР

ШТУРМАН-НАВИГАТОР РАДИСТ

БАШНЯ С КРУГОВЫМ ОБСТРЕЛОМ 2 пушки

СТАРШИЙ СТРЕЛОК 2 БОРТОВЫХ СТРЕЛКА КОРМОВОЙ СТРЕЛОК

ПЕРЕДНЯЯ ГЕРМЕТИЧЕСКАЯ КАБИНА

ПЕРЕДНИЙ БОМБОВЫЙ ОТСЕК

ЗАДНИЙ БОМБОВЫЙ ОТСЕК

СРЕДНЯЯ ГЕРМЕТИЧЕСКАЯ КАБИНА

КОРМОВАЯ ГЕРМЕТИЧЕСКАЯ КАБИНА

БАШНЯ С КРУГОВЫМ ОБСТРЕЛОМ 2 пушки ПЕРЕДНЕЕ ШАССИ

ОСНОВНОЕ ШАССИ

ОПЕРАТОР РАДИО-ЛОКАТОРА "КОБАЛЬТ" АЭРОФОТОАППАРАТ АФА-33-100

БАШНЯ С КРУГОВЫМ ОБСТРЕЛОМ 2 пушки

КОРМОВАЯ УСТАНОВКА 2 пушки

Another cutaway drawing showing a Tu-4 built after May 1950 (with ten NR-23 cannons).

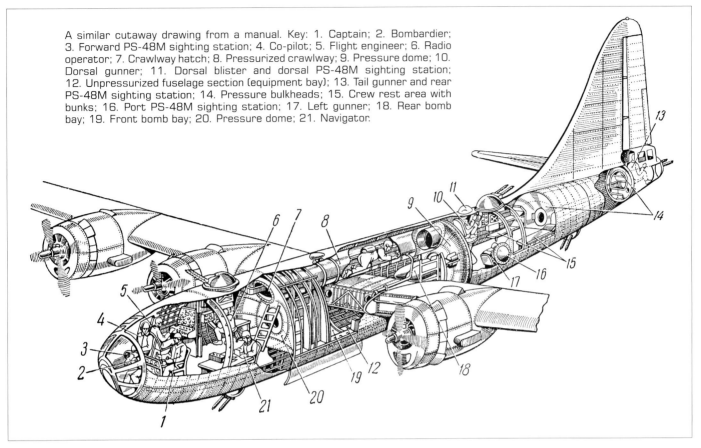

A similar cutaway drawing from a manual. Key: 1. Captain; 2. Bombardier; 3. Forward PS-48M sighting station; 4. Co-pilot; 5. Flight engineer; 6. Radio operator; 7. Crawlway hatch; 8. Pressurized crawlway; 9. Pressure dome; 10. Dorsal gunner; 11. Dorsal blister and dorsal PS-48M sighting station; 12. Unpressurized fuselage section (equipment bay); 13. Tail gunner and rear PS-48M sighting station; 14. Pressure bulkheads; 15. Crew rest area with bunks; 16. Port PS-48M sighting station; 17. Left gunner; 18. Rear bomb bay; 19. Front bomb bay; 20. Pressure dome; 21. Navigator.

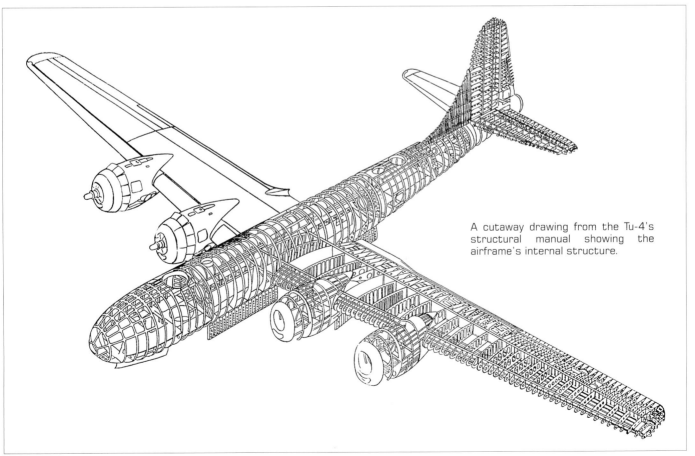

A cutaway drawing from the Tu-4's structural manual showing the airframe's internal structure.

center panel of quasi-oval shape tapering towards the bottom was an optically flat window made of triplex silicate glass for bomb-aiming. The other panels (four sector-shaped ones, two triangular ones and a crescent-shaped panel) were made of Plexiglas.

The *forward fuselage* (frames 2-13) was basically the forward pressure cabin terminating in a rear pressure dome with a pressure door in the center and a crawlway hatch above it. The cabin accommodated six crew members (captain, co-pilot, bomb-aimer, flight engineer, navigator and radio operator); the forward-facing navigator's station was on the port side, with the aft-facing flight engineer's station opposite and the aft-facing radio operator's station behind them. Part of the forward pressure cabin roof was glazed (again with a cast magnesium alloy frame and Plexiglas being used), featuring four rows of lateral and dorsal rectangular or trapezoidal windows (including two upward-sliding direct vision windows for the pilots); there were also two single rectangular windows further aft between frames 10-11 (at the flight engineer's and navigator's stations), the port one incorporating an emergency exit. The nosewheel well was located under the flight deck between frames 3-9, a pressure hatch in the wheel well roof providing access to the cabin. Two circular cutouts for the forward dorsal and ventral cannon barbettes were made on the centerline at the rear.

The *center fuselage* (frames 13-37) was unpressurized, being mated with the wing center section; the upper portion between frames 22-28 was manufactured as a separate subassembly to permit insertion of the wing carry-through box. It incorporated two bomb bays located fore and aft of the wing carry-through box which were closed by clamshell doors actuated by MBL-1 electromechanical drives (*motor bombolyuka* – bomb bay [door drive] motor). A circular cutout for the ventral radome was located between the bomb bays. A pressurized crawlway of 760 mm (2 ft 5⁵⁹⁄₆₄ in) diameter passed along the top of Section F-3 (above the bomb bays and the wing carry-through box), connecting the forward and center pressure cabins.

The *center pressure* cabin (frames 37-46) had front and rear pressure domes with pressure doors, the front one having a cutout for the crawlway above it; the rear pressure door was for crew access. The cabin accommodated the dorsal, port and starboard gunners (working the dorsal and ventral remote-controlled barbettes) and the radar operator whose workstation was at the rear, facing forward. The cabin featured three Plexiglas sighting blisters (two lateral and one dorsal) and a circular cutout for the rear dorsal cannon barbette.

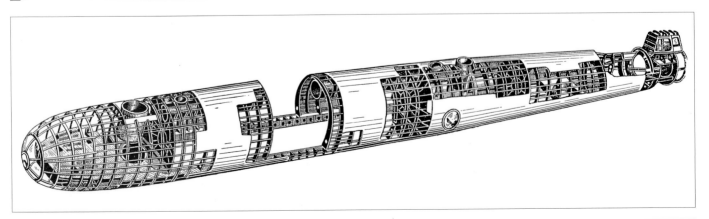

Opposite page, top: The Tu-4's forward fuselage and nose gear unit.

Opposite page, center: An upper view of the Tu-4's flight deck glazing.

Opposite page, bottom: A drawing of the fuselage from the Tu-4's structural manual. Note the cutout between frames 22-28 for inserting the wing carry-through structure.

This page:
Above right: The rear ventral barbette with twin NR-23 cannons. The exhaust port of the M-10 APU is visible ahead of it.

Right: The center pressure cabin section, showing the dorsal and port sighting blisters and the rear dorsal barbette.

The center/rear fuselage; note the rectangular emergency escape hatch high on the port side near the APU. The 'smokestack' ahead of the dorsal blister is apparently an object in the background.

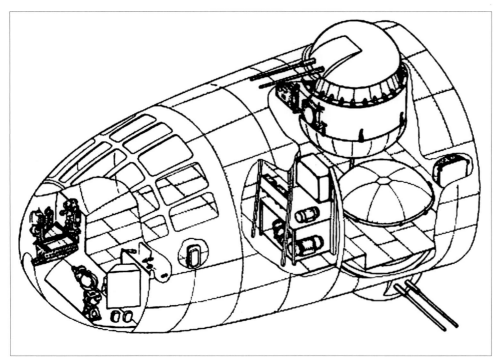

The *rear fuselage* (frames 46-57) was unpressurized, accommodating various equipment, including the APU and the photo cameras for which a centerline camera port was provided. An inward-opening rectangular entry door for the center and rear pressure cabins was located low on the starboard side, with an emergency exit hatch high on the port side across from it. Further aft was a circular cutout for the rear ventral cannon barbette and a cutout for the retractable tail bumper.

The *tail section* (frames 57-61) was the tail gunner's pressurized compartment; the forward pressure dome was at frame 59 and incorporated a door for normal entry/egress. The tail gunner's station glazing (four side windows and one rear window) was Plexiglas on the forward

Above: The forward pressure cabin (sections F-1/F-2). The forward cannon barbettes encroach on flight deck space quite a lot. The forward PS-48M sighting station was operated by the bombardier.

Right: A drawing from the structural manual showing the captain's and navigator's seats and the equipment on the port side of the flight deck (including oxygen bottles).

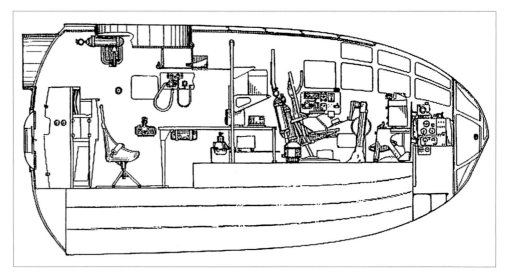

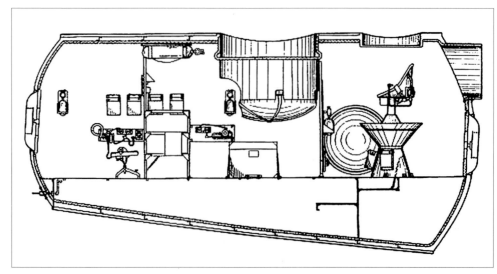

Another drawing from the same manual showing the port side of the center pressure cabin. The dorsal gunner's revolving seat was on a sort of podium.

pair of side windows and silicate glass elsewhere. An escape hatch was located in the port forward side window. The tail section incorporated attachments for the tail turret.

The pilots were protected against bullets and cannon shell fragments by triplex bulletproof glass panels (positioned aft of the Plexiglas glazing), forward steel armour plates, armoured seat backs and hinged rear armour screens. The gunners and radar operator in the center cabin were protected by an armoured door and an armoured box housing the weapons control system processors under the floor, while the tail gunner was protected by a rear steel plate and three triplex bulletproof glass panels. To make the working conditions more comfortable, all three pressure cabins and the crawlway

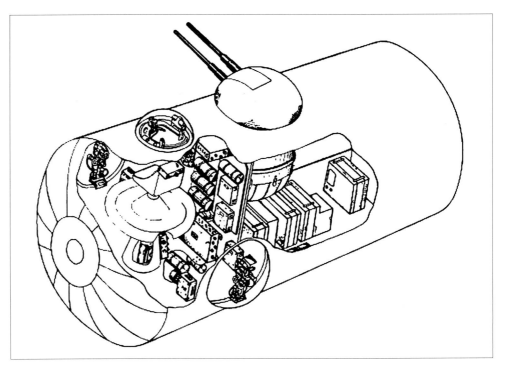

Above: This drawing shows the equipment racks in the center pressure cabin (some of them under the floor), as well as the lateral and dorsal PS-48M sighting stations.

Left: This drawing shows the starboard side of the flight deck, including the co-pilot's and flight engineer's workstations. The other workstations are not shown.

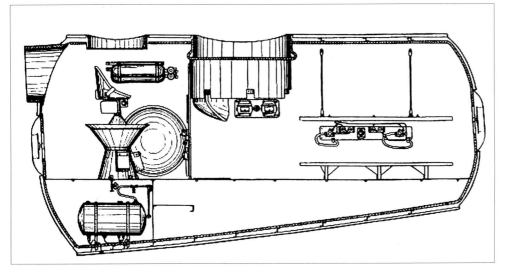

One more drawing from the same manual showing the starboard side of the center pressure cabin with oxygen equipment and the crew rest area. The tank under the floor held alcohol for propeller de-icing.

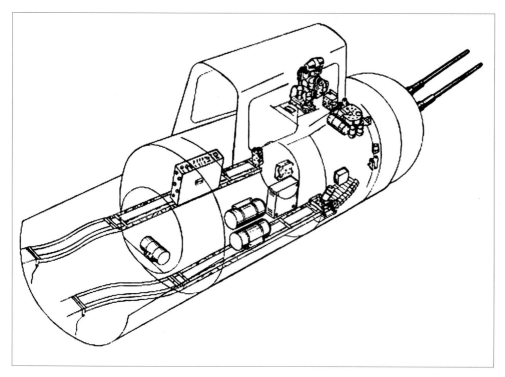

Above: The tail gunner's station, showing the rear PS-48M sighting station and the ammunition belt feed sleeves for the rear NR-23 cannons.

Below: The port wing trailing edge with the flaps retracted.

Bottom: The port wing with the flaps at 45° maximum deflection, showing the recesses for the flap uplocks. Note how the inboard nacelles' tail fairings deflect with the flaps.

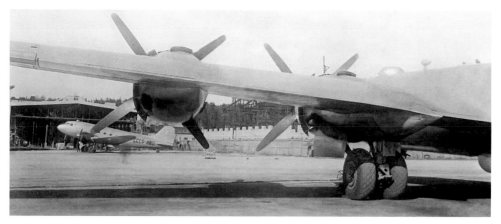

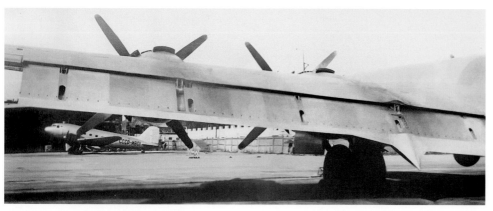

between the front and center cabins were faced with ATIM thermal insulation/soundproofing mats.

Wings: Cantilever mid-wing monoplane; sweepback at quarter-chord 0°, leading-edge sweep 7°1', dihedral 4°29' from the roots, incidence 4°, aspect ratio 11.5, taper 2.36. The wings utilised an RAF-34 airfoil with a thickness/chord ratio of 22% at the root.

The wings were of two-spar stressed-skin construction; the structural materials used were D-16T, D-16AT and D-16ABTN duralumin. The wings were built in three sections – center section/inner wings and detachable outer wings. The *center section* was mated to the center fuselage (Section F-3), the spars being attached to frames 22 and 28. It comprised a center portion (consisting of two halves joined at the fuselage centerline), a detachable leading edge (built in three sections on each side, which were located inboard of, outboard of and between the engine nacelles) and a trailing-edge portion incorporating flaps; it featured 32 ribs and 24 stringers. The skin thickness varied from 0.8 to 2.5 mm ($0^{1}/_{32}$ to $0^{63}/_{64}$ in) on the lower panels and 4-5 mm ($0^{5}/_{32}$ to $0^{13}/_{64}$ in) on the upper panels. The spars, ribs, stringers and heavy-gauge skin formed a tough torsion box carrying the engine nacelles and main landing gear units; it also housed the main fuel tanks.

Each *outer wing* was built in four pieces – the torsion box, a detachable leading edge, a detachable tip fairing and a full-span aileron incorporating a servo tab. The outer wings were joined to the wing center section at ribs 14L/14R and the upper/lower spar caps.

The wings were equipped with one-piece area-increasing Fowler flaps which occupied the entire trailing edge of the wing center section (up to ribs 14L/14R), making up 19% of the wing area. They were of all-metal single-spar construction and incorporated the tail fairings of the inboard engine nacelles. The flaps

Right: The tail unit of a Kazan'-built Tu-4 ('08 Red'), showing the metal strip sealing the manufacturing joint between the fin proper and the dorsal fillet.

Below: Head-on view of the Tu-4's tail surfaces.

Below right: The tail surfaces have rubber de-icer boots on the leading edges.

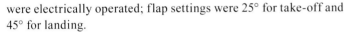

were electrically operated; flap settings were 25° for take-off and 45° for landing.

The aerodynamically balanced ailerons were one-piece single-spar structures with a rear false spar, 32 ribs and fabric skin, except for the leading edge which was skinned with duralumin. The ailerons were carried on six unequally spaced hinges each; aileron travel limits were 7°30' up/8° down. The trailing edge incorporated an all-metal trim tab with a maximum deflection of ±5°.

Tail unit: Cantilever conventional tail surfaces of all-metal stressed-skin construction. The *vertical tail* utilised a symmetrical airfoil section. It comprised a fin, a large detachable fin root fillet and a one-piece rudder. The fin was a two-spar structure with a front false spar, 23 ribs, a forward bulkhead and duralumin skin 0.6 mm (0.023 in) thick. The fin root fillet had 24 frames, a number of stringers, a 'false spar' along the leading edge and duralumin skin. The fin was joined to the fuselage and the stabilisers by bolts.

The aerodynamically balanced and mass-balanced rudder was attached to the fin's rear spar by three hinges; rudder travel limits

were ±18°. It had an all-metal framework with a single spar, 16 ribs, leading-edge and trailing-edge fairings and fabric skin. The root portion incorporated a trim tab attached to a rear false spar, with a maximum deflection of ±5°.

The *horizontal tail* had zero dihedral and featured an asymmetrical inverted airfoil section. The fixed-incidence stabilisers were built as a one-piece riveted structure with two spars, 27 ribs, 22 stringers and duralumin skin. The horizontal tail was attached to fuselage frame 53 by bolts.

The interconnected elevators were symmetrical single-spar riveted structures with 20 ribs and fabric skin. They were carried on two hinges each; elevator travel limits were 25° up/15° down. Each elevator featured a trim tab attached to a rear false spar, with a maximum deflection of ±12°. The elevators were likewise aerodynamically balanced and mass-balanced.

Landing gear: Electrically retractable tricycle type, with twin wheels on each unit. A retractable tail bumper was provided to protect the rear fuselage in the event of overrotation on take-off or a

Top: The port main landing gear unit, seen from the front, showing the dished wheel well doors and the breaker strut.

Above: The port main landing gear unit, seen from the rear, showing the torque link and the small rear door segment attached to the oleo strut.

Top left and above left: The nose landing gear strut was sharply inclined forward when extended. Note the long-stroke oleo and torque link.

Far left: The nose gear strut's lower portion was vertical, being the axle on which the nosewheels turned left and right for steering.

Left: The nose gear steering actuator/shimmy damper was on the starboard side of the strut.

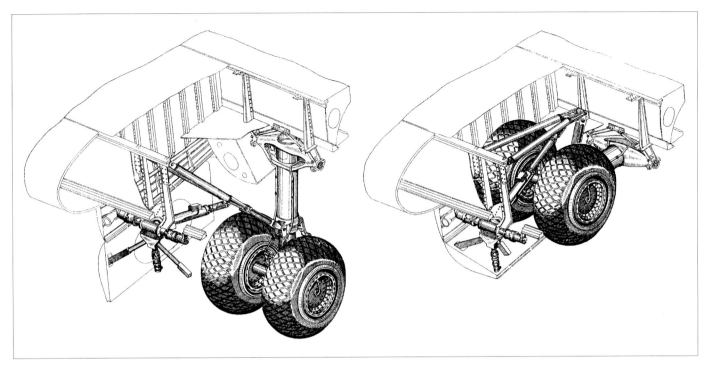

Above and above right: Drawings from the structural manual showing the main gear in extended and retracted position.

Right and below right: Drawings from the same manual showing the nose gear in extended and retracted position.

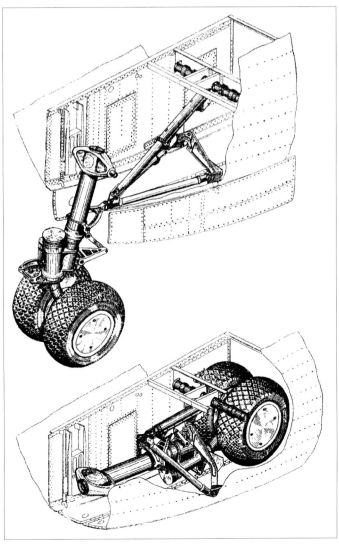

tail-down landing. The nose unit retracted aft, the main units forward into the rear portions of the inboard engine nacelles. All three landing gear struts and the tail bumper were actuated by screwjacks featuring MPSh-1 electric motors (*motor pod"yoma shassee* – landing gear retraction motor) and auto-brake mechanisms, with backup electric actuators for emergency extension/retraction.

The main units had 1,450 x 520 mm (57.08 x 20.47 in) wheels with hydraulic brakes. The steerable nose unit was equipped with 950 x 350 mm (37.40 x 13.77 in) non-braking wheels and a shimmy damper. All units had oleo-pneumatic shock absorbers charged to 7 kg/cm² (100 psi) for the nose unit and 17 kg/cm² (242.8 psi) for the main units. Tyre pressure was 3.6 kg/cm² (51.4 psi) for the nosewheels and 5 kg/cm² (71.4 psi) for the mainwheels.

The nosewheel well was closed by mechanically linked twin lateral doors, the mainwheel wells by twin lateral doors (operated by mechanical linkages) and a small rear door attached to the strut. All doors remained open when the gear was down.

Powerplant: Four Shvetsov ASh-73TK air-cooled 18-cylinder two-row radial engines delivering 2,400 hp at 2,600 rpm for take-off, 2,200 hp at 2,400 rpm at combat rating up to the altitude of 8,000-8,700 m (26,250-28,545 ft) and 2,000 hp at 2,400 rpm in cruise mode (at nominal power). Displacement 58.122 litres (3,546.8 cu in), bore 155.6 mm (6⅛ in), stroke 169.9 mm (6¹¹⁄₁₆ in), compression ratio 6.9:1. The valvetrain was of the pushrod type, with two valves per cylinder and sodium-cooled exhaust valves. The engine had a planetary reduction gearbox with a gearing ratio of 0.375.

The ASh-73TK had a two-stage supercharging system to improve its high-altitude performance. The first stage consisted of two TK-19 exhaust-driven single-stage two-speed superchargers operating in parallel; they were automatically controlled by an RTK-46 electronic governor and could be selected on or off from the flight deck. After passing through an intercooler the air was fed to the second stage, a PTsN single-speed engine-driven centrifugal blower (*privodnoy tsentrobezhnyy nagnetahtel'*). The fuel feed system included VK-4 carburettors with automatic mixture control; the ignition system included dual MD-18 magnetos and twin distributors. The bank of throttles was located at the flight engineer's workstation.

Engine starting was by means of an electrically-actuated inertia starter known as 'unit 263'. If electric power was unavailable, the starters could be spun up and engaged manually.

The ASh-73TK had a length of 2.29 m (7 ft 6³⁄₃₂ in), a diameter of 1.375 m (4 ft 6⅛ in) and a dry weight of 1,339 kg (2,952 lb); some sources state a dry weight of 1,275 kg (2,810 lb). SFC was 350 g/hp·hr (0.77 lb/hp·hr) at take-off power and 315-335 g/hp·hr (0.69-0.74 lb/hp·hr) at maximum continuous power.

Each engine initially drove a V3-A3 four-blade variable-pitch propeller of 5.056 m (16 ft 7 in) diameter turning anti-clockwise when seen from the front. Propeller pitch was automatically adjusted by an R-18B centrifugal governor which was also responsible for propeller feathering; once the propeller had been feathered, it could only be unfeathered manually on the ground. Late-production Tu-4s had V3B-A5 propellers with R-18A governors allowing the propellers to be unfeathered in flight.

The engines were installed in individual nacelles of quasi-elliptical cross-section; the latter were attached to the front spar of the wing center section and to the wing skin by gussets. The nacelles were of stressed-skin semi-monocoque construction. Each engine was mounted on a welded truss-type engine bearer via nine vibration dampers; the bearer was attached to the nacelle structure by six fittings. Each nacelle featured two air intakes; the larger circular intake was for engine cooling, while the crescent-shaped lower intake catered for the superchargers, the intercooler and the oil cooler. Electrically actuated multi-segment outlet flaps were located at the trailing edge of each engine cowling for controlling the cooling air flow.

The powerplant controls were located on both pilots' control consoles and the flight engineer's console. All control inputs were transmitted by means of cables.

An M-10 auxiliary power unit was installed on the port side of the unpressurized aft fuselage to provide emergency power. The M-10 was a 10-hp two-stroke engine driving a GS-5000 generator.

Control system: Conventional mechanical flight control system with full dual controls, the control columns featuring U-shaped control wheels. Control inputs were transmitted to all control surfaces, including trim tabs, by cables and pulleys. The flight controls were manual (that is, unpowered). The system included an AP-5 electric autopilot providing stabilisation in all three channels.

Fuel system: 22 flexible fuel cells (bag-type tanks) with a total capacity of 20,180 litres (4,439.6 Imp gal) were housed in the wing

Opposite page, top: The port inboard engine nacelle of a Tu-4. The cooling air flow control flaps are closed. The outer TK-19 supercharger is visible.

Opposite page, bottom: The port outboard engine nacelle of a Tu-4. Note the cloth cover on the mainwheels to protect them from the elements.

Right: The port engine nacelles of a decommissioned Tu-4, with propellers and access panes missing. Note the partially open air flow control flaps.

center section torsion box. The tanks were divided into four groups, one for each engine. Groups II and III catering for the inboard engines each held 5,310 litres (1,168.2 Imp gal), comprising tanks Nos. 1L-4L and 1R-4R respectively, while groups I and IV catering for the outboard engines each held 4,780 litres (1,051.6 Imp gal), comprising tanks Nos. 5L-11L and 5R-11R respectively. The fuel used was 100 octane aviation gasoline (Avgas).

Fuel was fed to the engines by electric transfer pumps or by gravity. Late-production Tu-4s featured BE-4M electric fuel level meters (from c/ns 2207301, 1840647 and 230120) and SRB-6 fuel flow meters (from c/ns 2207301, 1840647 and 230102).

For long-range missions three auxiliary bag-type tanks, each holding 2,420 litres (532.4 Imp gal), could be installed in the front bomb bay. The port and starboard tanks were installed rigidly and attached by bolts, while the center tank could be jettisoned in case of need. The fuel from these tanks was likewise transferred by electric pumps.

The M-10 APU had its own fuel system with a 15-litre (3.3 Imp gal) bag-type tank.

Oil system: Each engine had its individual oil system with a 320-litre (70.4 Imp gal) bag-type oil tank housed inside the engine nacelle. However, the maximum amount to be filled in each tank was 280 litres (61.6 Imp gal); total usable oil capacity in the main tanks was thus 1,120 litres (246.4 Imp gal). A 420-litre (92.4 Imp gal) bag-type auxiliary oil tank was also provided and the oil from it was transferred to the main tanks in flight as required. Type 729A oil coolers were located in the lower portions of the nacelles, the airflow and hence oil temperature being controlled by a flap at the rear of the oil cooler air duct. Oil temperature was adjusted by an ARTM-46 controller (*avtomaht regooleerovaniya temperatoory mahsla* – automatic oil temperature governor, 1946 model).

The TK-19 superchargers had their own oil system with one rigid oil tank holding 9.5 litres (2.09 Imp gal) for each pair of superchargers.

Electric system: The Tu-4 featured more than 150 electric servo systems and actuators. 28V DC power for these was supplied by six engine-driven 9-kilowatt GS-9000M generators (two on each outer engine and one on each inner engine); these were replaced by GSR-9000 generators from c/ns 225301, 184132 and 230101 onwards. R-20R or R-25A voltage regulators were installed. Backup DC power was provided by a 12A-30 (24 V, 28.5 A·h) silver-zinc battery and a 5-kilowatt GS-5000 generator driven by the M-10 APU. Two ground power receptacles were provided to the right of the nosewheel well.

26 V/400 Hz and 115 V/400 Hz AC power for some equipment items was supplied by two PK-750F converters and two MA-1500K converters (one of the latter served the IFF system and the other powered the radar set).

The electric system used single wiring throughout, with BPVL type conductors or BPVLE type shielded conductors. There were three electric circuits: main DC (powered by the engine-driven generators), emergency DC (powered by the storage battery and the APU) and AC (powered by the PK-750F combined AC converters).

Interior lighting included overhead dome lights, fluorescent lights, PS-45 and PSG-45 directional lamps, an ARUFOSh ultraviolet light system for illuminating the instrument panels, and cabin wander lights. Exterior lighting comprised BANO-45 port and starboard navigation lights (*bortovoy aeronavigatsionnyy ogon'*), retractable LFSV-45 landing lights in the undersurface of both outer wings, VBSOS-45 and KOS-45 identification lights and PSSO-45 formation lights.

Hydraulic system: Two hydraulic systems (main and backup) worked the wheel brakes and the nosewheel steering mechanism. Hydraulic pressure for normal braking was supplied by a hydraulic accumulator charged by an electrically driven pump ('unit 265M'); a hand-driven hydraulic pump was used in an emergency. Hydraulic pressure was 56-70 kg/cm² (800-1,000 psi). The systems used GMTs-2 grade hydraulic fluid.

De-icing system: Rubber de-icer boots on the wing, fin and stabiliser leading edges; compressed air for operating the de-icer boots was supplied by engine-driven vacuum pumps. The system worked automatically after being selected on. Alcohol de-icing on the propeller blades, with a 90-litre (19.8 Imp gal) tank beneath the floor of the center pressure cabin, two filters and two SN-1 pumps.

Fire suppression system: An OSU fire suppression system using carbon dioxide (hence OSU for *ognetooshitel' statsionarnyy ooglekislotnyy* – 'permanently installed CO_2 fire extinguisher') was used to fight engine fires. The system's two bottles charged with CO_2 were installed in the nosewheel well, feeding the gas to distribution manifolds around the engines. Three portable CO_2 fire extinguishers were provided in the pressure cabins for fighting any fires breaking out there.

Air conditioning and pressurisation system: Three ventilation-type pressure cabins; the forward and center cabins were connected by the pressurized crawlway and the center and aft cabins by an air duct. The cabins were pressurized by air bled from the inner engines' superchargers, which passed through air/air heat exchangers buried in the wing roots; a variable pressure differential depending on the flight altitude was maintained by an RDK-47 pressure regulator (*regoolyator davleniya v kabinakh*). Cabin temperature was controlled by RTVK air temperature regulators (*regoolyator temperatoory vozdukha v kabinakh*) which opened or closed the flaps of the heat exchangers as necessary. Additionally, Model 900A electric heaters installed in the forward and center cabins, plus a Model 1010 electric heater in the tail gunner's station, were provided to warm the air and demist the glazing.

Oxygen system: The permanent oxygen system comprised 15 individual oxygen supply stations, each with a KP-16 breathing apparatus, an IK-15 oxygen flow meter, an MK-13 oxygen pressure gauge (*manometr kislorodnyy*), indicators and a connector for charging a portable breathing apparatus. 18 oxygen bottles were connected by a system of piping; the nominal pressure was 30 kg/cm² (428.5 psi) and the oxygen supply was sufficient to keep a 15-man crew alive in totally decompressed cabins at 7,000-8,000 m (22,965-26,250 ft) for 4 hours 12 minutes. Additionally, 14 KP-19 portable breathing apparatus were provided; finally, each crewman's parachute pack included a KP-23 breathing apparatus enabling the crewman to survive a descent from high altitude after bailing out.

Avionics and equipment:

a) navigation equipment: SP-48 ILS comprising the ARK-4 or ARK-5 Amur automatic direction finder (*avtomaticheskiy rahdiokompas* – ADF), RV-2 *Kristall* (Crystal) low-range radio altimeter (0-1,200 m/3,940 ft), RV-10 high-range radio altimeter (100-15,000 m/330-49,210 ft) and MRP-45 marker beacon receiver (*markernyy rahdiopriyomnik*) replaced by the MRP-48 *Dyatel* (Woodpecker) from c/ns 221401, 184402 and 230101 onwards. Tu-4s from c/ns 225501 and 230101 onwards, as well as c/ns 184232 and 184133, featured the SP-50 Materik ILS comprising the SD-1 DME, the KRP-F localiser receiver (*koorsovoy rahdio-*

priyomnik) and the GRP-2 glideslope beacon receiver (*glissahdnyy rahdiopriyomnik*). An NK-46B navigation computer was fitted; an AK-53P astrocompass enabled celestial navigation.

b) communications equipment: 1RSB-70 communications radio for two-way air-to-ground communication with fixed and trailing wire aerials. SCR-274N two-way HF command radio for two-way air-to-air communication, later replaced by the RSB-5 VHF radio with a range of 120 km (74.5 miles). 12RSU-10 *Koora* VHF command radio (named after a Georgian river; later redesignated RSU-5) with blade aerial, replaced by the RSIU-3 radio from c/ns 221901, 184301 and 230101 onwards. An AVRA-45 emergency radio (*avareeynaya rahdiostahntsiya*) was fitted. Early aircraft had an SPU-14 intercom replaced by the SPU-14M on later aircraft.

c) cockpit instrumentation: US-7-OO airspeed indicator (*ookazahtel' skorosti* – ASI), VD-15A altimeter indicators, AGK-47B artificial horizons, UPE-46 electric turn and bank indicator (*ookazahtel' povorota elektricheskiy*), VAR-30-3 vertical speed indicators (*variometr* – VSI), GPK-46 pneumatic directional gyros (*gheeropolukompas*), AB-52 drift sight, KI-11 magnetic compasses (*kompas indooktivnyy*), DIK-46 remote flux-gate compass (*distantsionnyy indooktivnyy kompas*), UVPD-3 cabin altitude/pressure indicator (*ookazahtel' vysoty i perepahda davleniya*), AVR-M and AChKhO chronometers, TNV-45 outside air thermometer (*termometr naroozhnovo vozdukha*) and TV-45 cabin air thermometer, UZS-46 wing flap position indicators, UYuZ-4 cowling flap position indicators, MG manometers, TE-2 engine tachometers, 2MB-2 fuel pressure gauges, 2MM-15 oil pressure gauges, 2MV-18-P vacuum meters, 2TUE-46 multi-purpose electric thermometers, 2TTsT-47 cylinder head thermometers, ME-4 electric oil meter (*maslomer elektricheskiy*) and BE-4 electric fuel meter (*benzinomer elektricheskiy*). The barometric instruments received data from TP-45 or (from c/n 224202 onwards) TP-49 pitot heads.

d) IFF equipment: Magniy IFF interrogator and Bariy transponder (replaced by the Magniy-M and Bariy-M respectively on late production batches).

e) radar equipment: Kobal't or Kobal't-M bomb-aiming/navigation radar with a 360° field of view installed in a hemispherical retractable radome under the center fuselage.

f) photo equipment: For bomb damage assessment the basic bomber version initially featured an AFA-33/100 aerial camera for vertical photography, an AFA-27/T hand-held camera for oblique photography and a KS-50B cine camera. From c/ns 223001 and 184309 onwards the bomber version could be equipped with one of three still cameras: an AFA-33/100, an AFA-33/75 with a shorter focal length for daytime photography or an NAFA-3S/50 for night photography, plus an AKS1-50 cine camera. The cameras were installed in an unpressurized bay in the rear fuselage and controlled by the captain.

g) interior equipment: Apart from crew seats, the Tu-4 featured a crew rest area with bunks on the starboard rear side of the center pressure cabin. The flight deck glazing and other windows were provided with curtains (for blind flying training and for eye protection) and stained glass filters. Thermos flasks and toilet buckets were provided for the crew.

The captain's workstation of the Tu-4, showing the U-shaped control wheel, the bank of throttles on the outboard side (in order to allow the bombardier unrestricted access to his workstation). The latter two features were present on later Tupolev aircraft having a glazed navigator's station (up to and including the Tu-134 twinjet short-haul airliner of 1963). The placard in the lower right-hand corner reads: 'Warning! Remember to replace the fuse in the landing gear retraction circuit before flight' (it was standard operational procedure to remove this fuse when the aircraft was parked, preventing accidental retraction on the ground).

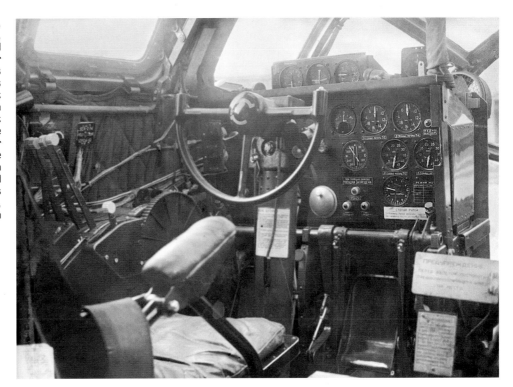

The co-pilot's workstation had a slightly different set of instruments but the same bank of throttles. Note the quilted ATIM heat- and soundproofing blanket on the flight deck wall.

Armament:

a) offensive armament: Free-fall bombs were carried internally in two bays. Depending on the mission the bombs were loaded into the aft bomb bay or both bays (for extra long range missions the forward bomb bay was occupied by auxiliary fuel tanks). Initially a mixed complement of bomb racks was fitted, but from Kazan'-built Tu-4 c/n 225701 onwards a standardised set of bomb racks was introduced; this allowed FAB-250M46 and FAB-500M46 bombs to be suspended on two rows of shackles, doubling the number of 250- and 500-kg (551- and 1,102-lb) bombs carried by the Tu-4. Bomb loading was performed using a BL-47 bomb hoist.

Normal and emergency bomb release could be performed by the bomb-aimer, the captain or the radar operator. The bombs could be dropped singly, in series of one, two or four, or in a salvo by means of the ESBR-45 electric bomb release mechanism (*elektrosbrahsyvatel'*); the bombs' fuses could be set for detonation or non-detonation. A special signal light was provided, allowing the lead aircraft in a formation to give the 'bombs away' signal to the

Top: The captain's side console with the throttles, the elevator trim tab handwheel and the flap control lever.

Above: The aft-facing flight engineer's workstation with banks of engine instruments and other gauges and controls, an extra set of throttles, the propeller pitch controls and the landing gear control lever.

Top: The co-pilot's side console with an intercom switch box aft of the throttles.

Above: Another view of the flight engineer's workstation, with minor differences showing this is an aircraft from a different production batch, or perhaps a different factory. The flight deck emergency exit is just visible on the left.

Above: An electrics control panel in the flight deck.

Above right: The radar operator's workstation in the center pressure cabin, with equipment racks and a small radarscope.

Right and below right: The radome of the Kobal't radar in semi-retracted and deployed positions.

Below: The radio operator's workstation.

Far left: The antenna of the Kobal't radar (seen here in inverted position on a test bench).

Left: The port side blade aerial of the 12RSU-10 VHF radio.

wingmen while maintaining radio silence. The bomb bay doors were operated by the bomb-aimer.

In VMC the bomb-aimer used an OPB-4S optical bombsight installed in the foremost part of the forward pressure cabin. This was superseded by the OPB-5SN model from Kazan'-built Tu-4 c/n 221901 and Kuibyshev-built Tu-4 c/n 184402 onwards, whereas all Moscow-built Tu-4s had the new bombsight. Data from the bombsight were fed into the AP-5 autopilot, enabling the bomb-aimer to take over control of the aircraft during the bombing run. In IMC bombing was done using the display of the Kobal't radar for targeting, the bomb-aimer and the radar operator working together; the radar operator could also perform the entire operation himself if necessary.

b) defensive armament: Early Tu-4s had the PV-20 defensive weapons system comprising eleven 20-mm (.78 calibre) Berezin B-20E cannons in remote-controlled twin-cannon barbettes (two dorsal and two ventral) and a tail barbette with three cannons. The

five barbettes were positioned so as to give 360° coverage, excluding 'blind spots' in the sectors of fire. They were located outside the pressurized areas of the airframe and remote-controlled by an electric servo system using synchros.

Aiming was done by means of five PS-48 sighting stations: forward (in the flight deck), dorsal, port and starboard (at the three sighting blisters in the center pressure cabin) and rear (at the tail gunner's station). All sighting stations except the tail gunner's had two barbette control modes (primary and auxiliary), allowing the gunners to operate other barbettes than those they were assigned to if one of the crew was incapacitated. In primary mode the bomb-aimer operated both dorsal barbettes and the forward ventral barbette; the rear dorsal barbette was operated by the dorsal gunner, the rear ventral barbette by the lateral gunners and the tail barbette by the tail gunner. In auxiliary mode the forward dorsal barbette could be operated from the dorsal sighting station, the tail barbette and the forward ventral barbette from the lateral stations.

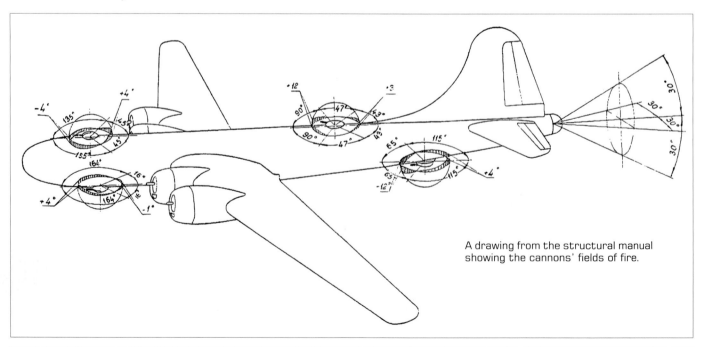

A drawing from the structural manual showing the cannons' fields of fire.

The dorsal and ventral barbettes had a 360° field of fire in azimuth and +88°/–3° in elevation for the dorsal barbettes (and vice versa for the ventral ones); the tail turret covered a 59° cone (that is, ±29°30' in both azimuth and elevation).

The ammunition was stored in ammo boxes located inside the dorsal and ventral barbettes or, in the case of the tail barbette, in the rear fuselage. The total ammunition supply was 4,700 rounds – 400 rounds per gun for each twin-cannon dorsal/ventral barbette and 500 rpg for the three cannons in the tail turret; however, a figure of 4,680 rounds is also quoted.

From c/ns 223201, 184309 and 230102 onwards the Tu-4 featured the PV-23 Zvezda defensive weapons system with ten 23-mm (.90 calibre) Nudelman/Rikhter NR-23 cannons (two in each station – a DK-3 tail barbette was fitted). Cannon aiming was done by means of PS-48M sighting stations.

The total ammunition supply was reduced to 3,150 rounds. The cannon were cocked by an electrically-controlled pneumatic mechanism and then recharged by recoil action, which allowed the heavy-calibre weapons to have a high rate of fire and be relatively lightweight. The NR-23 weighed 39 kg (86 lb) and fired 200-gram (7.06-oz.) projectiles (the complete rounds weighed 340 g/12 oz.); the rate of fire was 800-950 rounds per minute and muzzle velocity was 680 m/sec (2,231 ft/sec). The ammunition belts to the dorsal and tail turrets were fed electrically. The weapons control system included electric barbette actuators and the aforementioned servo system which controlled these actuators by means of servo amplifiers and amplidyne generators. The system featured PVB-23 processors which automatically introduced target lead and compensated for parallax (caused by the distance between the gunner and the barbette) and shell ballistics.

The Tu-4's payload options				
Payload	Rack type	Bomb type (conventional)	Quantity of bombs (forward bay/aft bay)	Warload, kg (lb)
a) mixed set of bomb racks				
Option 1	KD4-546A KD3-246A	FAB-250M43	24 (12/12)	6,000 (13,230)
Option 2	KD4-546A KD3-246A	FAB-250M46	24 (12/12)	6,000 (13,230)
Option 3	KD4-546A	FAB-500M43	12 (6/6)	5,830 (12,850)
Option 4	KD4-546A	FAB-500M44, FAB-500M46	12 (6/6)	6,000 (13,230)
Option 5	KD4-546A	FAB-1000M44*	8 (4/4)	7,120 (15,700)
Option 6	KD4-248	FAB-1000M44*	8 (4/4)	7,120 (15,700)
Option 7	KD4-248	FAB-1500M44	.8 (4/4)	11,840 (26,100)
Option 8	KD4-248	FAB-3000M44	4 (2/2)	11,930 (26,300)
b) standard set of bomb racks †				
Option 1	KD4-547 SD3-248	FAB-50	48 (20/20 + 4/4)	2,000 (4,410)
Option 2	KD3-547 SD3-248	FAB-100	48 (20/20 + 4/4)	5,900 (13,010)
Option 3	KD3-547 SD3-248	FAB-250M46	48 (20/20 + 4/4)	10,600 (23,370)
Option 4	KD4-547 SD3-248	FAB-500M46	28 (12/12 + 4/4)	12,000 (26,455)
Option 5	KD4-547 SD3-248	FAB-250M43 or FAB-250M44	24 (10/10 + 2/2)	6,000 (13,230)
Option 6	KD4-547 SD3-248	FAB-500M43 or FAB-500M44	14 (6/6 + 1/1)	6,800 (14,990)
Option 7	KD4-248	FAB-1000M44	8 (4/4)	7,120 (15,700)
Option 8	KD4-248	FAB-1500M46	8 (4/4)	11,840 (26,100)
Option 9	KD4-248	FAB-3000M46	4 (2/2)	11,930 (26,300)
Option 10	BD5-50	FAB-6000M46	4 (2/2)	11,930 (26,300)
Option 11		FAB-250M44	24 (12/12)	6,000 (13,230)
Option 12		FAB-500M44	14 (7/7)	6,802 (14,995)
Option 13		FAB-1000M44	8 (4/4)	7,120 (15,700)

* Shortened version
† The bombs were suspended in two rows for options 1-9 and in one row for options 10-13

Tu-4 specifications

Length overall (less cannon barrels)	30.177 m (99 ft 0¾ in)
Wing span	43.047 m (141 ft 2⅝ in)
Horizontal tail span	13.108 m (43 ft 0⅛ in)
Height on ground	8.46 m (27 ft 9¼ in)
Landing gear track	8.676 m (28 ft 5 in)
Landing gear wheelbase	10.444 m (34 ft 3 in)
Ground angle	+2°
Maximum rotation angle	9°30'
Root chord	5.18 m (16 ft 11¹⁵⁄₁₆ in)
Tip chord	2.25 m (7 ft 4³⁷⁄₆₄ in)
Mean aerodynamic chord (MAC)	4.001 m (13 ft 1¹³⁄₆₄ in)
Wing area	116.17 m² (1,250.44 sq ft)
Aileron area (total)	12.01 m² (129.27 sq ft)
Flap area (total)	30.79 m² (331.42 sq ft)
Vertical tail area	22.01 m² (236.91 sq ft)
Fin area (including root fillet)	15.94 m² (171.58 sq ft)
Rudder area	6.1 m² (65.66 sq ft)
Horizontal tail area	30.94 m² (333.04 sq ft)
Stabiliser area	20.251 m² (217.75 sq ft)
Elevator area (total)	10.683 m² (114.87 sq ft)
Airframe weight	15,196 kg (33,501 lb)
Powerplant weight	14,270 kg (31,460 lb)
Weapons systems weight:	
offensive (bomb racks, release controls etc.)	407 kg (897 lb)
defensive (cannons, barbettes etc.)	1,890 kg (4,166 lb)
Operating empty weight	35,270 kg (77,760 lb)
Take-off weight:	
normal	47,500-47,600 kg
	(104,720-104,940 lb)
high gross weight configuration	54,500; 60,900; 61,500; 65,000;
	66,000 kg (120,150; 134,260;
	135,580; 143,300; 145,500 lb) *
Landing weight:	
normal	48,000 kg (105,820 lb)
maximum	60,000 kg (132,280 lb)
Payload, normal:	12,330 kg (27,180 lb)
crew (11)	990 kg (2,180 lb)
cannon ordnance	1,060 kg (2,340 lb)
fuel	3,480 kg (7,670 lb)
engine oil	800 kg (1,760 lb)
bombs	6,000 kg (13,230 lb)
Payload, high gross weight configuration:	19,230 kg (42,390 lb)
crew	990 kg (2,180 lb)
cannon ordnance	1,060 kg (2,340 lb)
fuel	8,150 kg (17,970 lb)
engine oil	1,030 kg (2,270 lb)

bombs	8,000 kg (17,640 lb)
	(eight FAB-1000 bombs)
CG range	25.7-30.1% MAC
Maximum speed:	
at sea level	420 km/h (261 mph)
at 10,000 m (32,810 ft)	558-578 km/h (346-359 mph)
Landing speed	172 km/h (106 mph)
Rate of climb: ‡	
at sea level	4.6 m/sec (905 ft/min)
at 1,000 m (3,280 ft)	4.6 m/sec
at 2,000 m (6,560 ft)	4.6 m/sec
at 3,000 m (9,840 ft)	4.6 m/sec
at 4,000 m (13,120 ft)	4.5 m/sec (885 ft/min)
at 5,000 m (16,400 ft)	4.5 m/sec
at 6,000 m (19,685 ft)	4.4 m/sec (866 ft/min)
at 7,000 m (22,965 ft)	4.1 m/sec (807 ft/min)
at 8,000 m (26,250 ft)	3.7 m/sec (728 ft/min)
at 9,000 m (29,530 ft)	3.1 m/sec (610 ft/min)
at 10,000 m (32,810 ft)	2.0 m/sec (393 ft/min)
at 11,000 m (36,090 ft)	0.8 m/sec (157 ft/min)
at 11,200 m (36,750 ft)	0.5 m/sec (98 ft/min)
Time to height: ‡	
to 1,000 m	3.6 minutes
to 2,000 m	7.2 minutes
to 3,000 m	10.9 minutes
to 4,000 m	14.5 minutes
to 5,000 m	18.2 minutes
to 6,000 m	22.0 minutes
to 7,000 m	25.9 minutes
to 8,000 m	30.1 minutes
to 9,000 m	35.1 minutes
to 10,000 m	41.6 minutes
to 11,000 m	53.3 minutes
to 11,200 m	58.0 minutes
Service ceiling	11,200-12,000 m
	(36,745-39,370 ft)
Climb time to 5,000 m (16,400 ft)	18.2 minutes
Maximum range	5,600 km (3,479 miles)
Range with maximum payload	5,000 km (3,105 miles)
Combat radius (Tu-4K missile carrier)	250-300 km (155-186 miles)
Take-off run	960-2,210 m (3,150-7,250 ft) †
Landing run	1070-1750 m (3,510-5,740 ft) †
G limit:	
with a 47,600-kg all-up weight	4.6
with a 55,000-kg (121,250-lb) AUW	4.05
with a 65,000-kg AUW	3.56

Crew rescue equipment: All crew members were provided with PLK-45 parachutes fitting into the seat pans. Two inflatable dinghies were provided to save the crew in the event of ditching; they were stowed in a bay above the wing center section closed by a cover.

* With different bomb loads and fuel quantities

† Depending on weight

‡ At maximum continuous power, AUW 47,850 kg (105,490 lb)

Chapter 6

The Tu-4 in Soviet service

As noted earlier, the advent of nuclear weapons brought about a need to bolster the Soviet Air Force's bomber component and made the Soviet government admit it had made a mistake by 'demoting' the ADD to an Air Army (the 18th VA) during the war. Now this mistake was put right when the DA was (re)established in April 1946. The 'new' Long-Range Aviation consisted of three Air Armies that were initially designated as the 1st, 2nd and 3rd VA.

Actually DA crews had begun preparing for conversion to the Tu-4 as early as 1946 when the first production B-4 was still under construction. Pending delivery of the Tu-4s, several bomber units re-equipped with B-24s and B-25Js; the two types were used for mastering the technique of piloting heavy aircraft with a tricycle undercarriage.

The western military districts were the first to receive the Tu-4, being closest to the potential adversary; this was normal practice in the USSR at that time when introducing new combat materiel. The first Long-Range Aviation regiment to receive the type was the 203rd *Orlovskiy* GvBAP (*Gvardeyskiy bombardirovochnyy aviapolk* – Guards Bomber Regiment) of the 45th *Gomel'skaya* BAD (*bombardirovochnaya aviadiveeziya* – Bomber Division), which was part of the aforementioned 50th VA. The division was headquartered at Balbasovo AB near Orsha and assigned to the Belorussian Military District. Like the aforementioned 890th BAP, the 203rd Regiment was based at Balbasovo AB; the honorary appellation *Orlovskiy* had been conferred on the unit for its part in liberating the Russian city of Oryol during the

The flight line of one of the first Tu-4 regiments, with early-production Kazan'-built Tu-4s. '39 Black' (c/n 220804) has canvas covers on the wings and the horizontal tail.

Left: One more view of the same aircraft. All of the unit's Tu-4s are from the Kazan' production line. Note the posts marking the line not to be crossed when the hardstand was being guarded.

Right: Another section of the same flight line. Note the distinctive angular font of the serials applied to early Kazan'-built Tu-4s. The font of the c/ns changed to a rounded one after Batch 5 as shown opposite.

Great Patriotic War. The 45th Division had received its honorary appellation for its part in liberating the city of Gomel'.

It is fitting to note here that the regiments and divisions re-equipping with the Tu-4 had previously operated smaller aircraft; hence they were renamed by adding the 'Heavy' designator. Thus, for example, the 45th BAD became the 45th TBAD, while the 203rd GvBAP became the 203rd GvTBAP.

Conversion training in the 203rd Regiment commenced in June 1948. The first crew to accomplish the training course comprised

crew captain Guards Lt.-Col. V. V. Ponomarenko, co-pilot Guards Capt. P. A. Voronin, navigator Guards Lt.-Col. K. P. Ikonnikov, engineer in charge of the tests D. I. Kantor, powerplant engineer-in-charge V. F. Moskovskiy, flight engineers Guards Capt. N. S. Grishko and Guards Capt. V. R. Kishchenko, technician Guards Lt. (SG) V. T. Krasnov, radio operator Guards MSgt. S. P. Daniuk and electrics engineer Guards Engineer-Lt.-Col. O. B. Aronov. This experienced crew had previously flown the Pe-8, B-17 and B-24; the Tu-4 in question was the sixth production aircraft ('1 Black', c/n

Left: In the early days the Tu-4s of a given unit were normally parked on the flight line, without any dispersals. Since each regiment had four squadrons (until 1951, when the number of squadrons in a regiment was reduced to three), the row of aircraft stretched as far as the eye could see.

Right: The same Tu-4s '32 Black' and '39 Black' seen from the rear; note the canvas 'bags' enclosing the guns in the tail turrets so that only the muzzles protrude.

Left: Tu-4s '19 Black', '24 Black', '23 Black' etc. seen from the rear. The lower half of the fuselage appears to be coated with matt paint, the colour division line curving upwards ahead of the stabilisers.

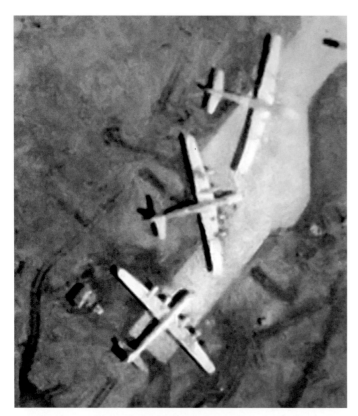

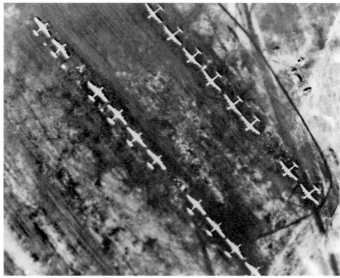

Above: An intelligence photo of a Long-Range Aviation base with a double row of Tu-4s parked at an unpaved airfield. At least three aircraft are out on a mission, judging by the gaps in the lines.

Left: These photos of a dispersal area with individual stands occupied by Tu-4s were taken by a Yak-25R tactical reconnaissance aircraft.

Right and below left: A different airbase occupied by a Tu-4 unit, with the bombers parked on individual stands along a taxiway. These shots were taken by the Yak-25R tactical reconnaissance aircraft prototype during trials, using the oblique camera, and come from the test report. The bombers parked on the grass are presumably unserviceable. Note the Yak-12 liaison aircraft parked next to the bomber in the foreground in the right-hand photo.

Below: Another shot of the same dispersal area taken by the Yak-25R, this time using the vertical camera. Note the varying type of pavement slabs on the taxiway and the hard-stands.

220201). Several more crews were trained later; they took part in the joint testing of the Tu-4. The 203rd GvTBAP completed conversion training at the end of 1948 but retained its B-24s well into 1949 alongside the Tu-4s because there were not enough Tu-4s at first.

Meanwhile, back in the autumn of 1946 the 890th BAP had been excluded from the 45th TBAD and had moved to Kazan', becoming the 890th *Bryanskiy* OUTAP (*otdel'nyy oochebno-tre-nirovochnyy aviapolk* – Independent Training Air Regiment) reporting directly to the DA HQ, and Col. Vladimir V. Abramov

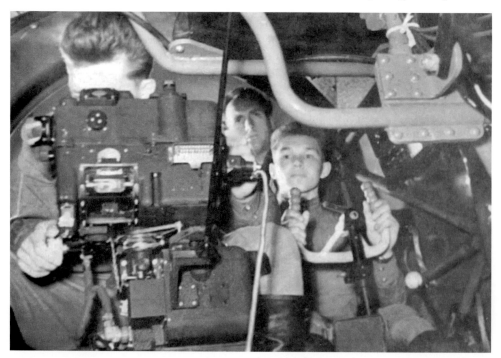

Above: The command staff of a DA unit during the conversion training course to the Tu-4.

Left: The watcher watched: trainees in action in the Tu-4 flight simulator. The 'captain' looks on as the 'bombardier' works with the OPB-4S sight, while the instructor watches them both.

Opposite page: The Tu-4s were updated in the course of their service; here, '25 Blue' (c/n 2805703) has a different (forked) ILS aerial under the nose.

The crew of an 840th TBAP Tu-4 receives a briefing prior to a mission at Sol'tsy AB.

Tu-4 crews receive a briefing during an exercise at Ashuluk AB in the Astrakhan' Region.

was appointed regiment CO. The pilots of this regiment had accumulated much experience with the restored Flying Fortresses and Liberators collected from all over Eastern Europe, which enabled them to be the first regular Air Force pilots to fly the Tu-4 (as distinct from GK NII VVS test pilots). Thus was born an operational conversion unit easing the DA aircrews' transition to the new strategic bomber. The regiment catered for training Tu-4 crews until 1955, when the jet-powered Tu-16 began to enter service.

In addition to their main job of coaching the aircrews of first-line bomber units, the personnel of the 890th OUTAP was involved in the Tu-4's pre-delivery tests and state acceptance trials. Maj. (Retd.) Nikolay V. Chernolikh, a former navigator in the unit, recalled that in March 1948 a crew comprising crew captain Capt. Aleksandr Kh. Romanov (a deputy squadron commander in the 890th OUTAP), co-pilot Lt. (SG) Dmitriy P. Vaoolin, navigator Lt. Nikolay V. Chernolikh and flight engineer Capt. Vasiliy G. Vlasov was tasked with accepting a brand-new Tu-4 at the factory and conducting manufacturer's tests and state acceptance trials. (*Sic* – Chernolikh does not mention any other crewmembers, not even a radio operator. He refers to the aircraft as 'Tu-4 No.15', which must be c/n 220305 and was very probably coded '15 Black'.) The author's style in the following quote has been retained:

'After making a check flight our crew ferried our aircraft to Ramenskoye airfield and got down to routine test work – Chernolikh wrote in his memoirs – *The enthusiastic and creative atmosphere at the airfield inspired us to conduct the test flights with high quality. The ground crews and aircrews were doing their utmost. Andrey Nikolayevich Tupolev would be present at the airfield almost each day.*

On 12th June 1948, in parallel with the test work, the Tu-4s started making formation flights along the parade route, practic-

Left: An 840th TBAP Tu-4 taxies out past three sister ships, including '93 Black'.

Below left: The aircrew members of a Guards heavy bomber unit are instructed prior to a celestial navigation training sortie. The one on the left is Guards Capt. Titarenko.

Below: The flight line of a heavy bomber regiment equipped with early Tu-4s having no undernose ILS aerials.

Right: A Tu-4 taxies at an unpaved airfield. A Polikarpov UT-2 trainer, probably used as a liaison aircraft, is visible on the left.

Below right: A black-overalled member of the ground crew poses in front of a Tu-4 on the military apron of Gor'kiy's Strigino airport.

Bottom right: A Tu-4 retracts its undercarriage after take-off.

ing for the Air Fleet Day flypast. In addition to our Tu-4s, numerous production aircraft that had been already mastered by Air Force pilots – MiG-15s, Yak-15s etc – were making practice flights for the flypast at the same airfield. [...] In recognition of the effort they had made and the flying skill displayed by the participants of *the 25th July 1948 flypast, many of them, including the Tu-4 captains, were awarded state orders. Our crew captain Capt. Romanov received the Red Star Order. The Commander of the Moscow MD Air Force Lt.-Gen. V. I. Stalin issued a formal commendation to all participants of the flypast.*

In August-September our crew did a lot of flying and completed the test program, the hardware giving little cause for complaint. On 13th October our crew was to make the tenth and final long-range flight at 10,000 m [32,810 ft]. The mission was flown with instructor pilot Col. V. V. Ponomarenko along a highly important route from Moscow to the Crimea Peninsula and back. On the outbound leg of the flight the direction finder failed unexpectedly. All attempts of radio operator/chief technician Frenkel' to fix it were fruitless – the AC converter [catering for the DF] had failed and we did not have a spare converter on board. The remainder – that is, the greater part – of the

Top: '41 Black', a Kazan'-built Tu-4 (c/n 220901), at a grass airfield during an exercise.

Another Kazan'-built Tu-4 (c/n 221902). At first the bombers wore very large tactical numbers.

An unusual aspect of a Tu-4; the picture was probably taken through the upper sighting blister of a sister ship.

route above the clouds was flown without the benefit of the DF and the search radar.

Being low on fuel, we did not fly the usual landing circuit but came straight in to land. When the tanks were emptied to check the remaining amount, it turned out that we had only 200 litres [44 Imp gal] of petrol left. Our endurance was 11 hours 25 minutes – an impressive figure by the day's standards.

Since our long-range flight was the finishing touch to the manufacturer's tests and state acceptance trials, the successful completion of the mission was immediately reported to Comrade I. V. Stalin.

On 16th October we made a positioning flight to the airfield where the preparations of the 203rd Orlovskiy Guards Bomber Regiment for the 7th November 1948 flypast over Red Square were going full steam ahead. The very next day we were participating in formation flights along the parade route.

Capt. A[leksey]. D. Perelyot's Tu-4, with Commander of the Moscow MD Air Force Lt.-Gen. V[asiliy]. I. Stalin aboard, led the way during the flypast, with fighter escort. The long line astern formation composed of flights of three Tu-4s was led by 45th Gomel'skaya Air Division CO Maj.-Gen. S. K. Nabokov; our aircraft was the left wingman in his flight. (Sic; according to other sources, Maj.-Gen. Nabokov was CO of the 48th BAD – Auth.)

For the participation in the parade, and for the skills attained in mastering new aviation hardware, all the participants of the flypast, including our crew, received commendations from Minister of Defence Marshal N[ikolay]. A. Bulganin.

For the ensuing 20 days we reported the test results attained with our aircraft (Tu-4 No.15). Then we returned to our airfield, where we immediately joined in the 890th OUTAP's high-priority and important work of training the aircrews of Long-Range Aviation units to fly new hardware – the Tu-4. The nearly two-year manufacturer's tests and state acceptance trials were over. The 203rd GvTBAP started taking delivery of the aircraft and commenced service tests, and the type was progressively fielded by the DA.'

In 1947 another unit of the 45th TBAD – the 52nd GvTBAP – took up residence at Balbasovo AB, moving from Migalovo AB near Kalinin (now Tver'), Russia, and started converting to the B-25J. The conversion was completed by January 1948 and the B-25s were operated quite actively; in August they were augmented by another interim type, the B-24. The actual conversion of the crews to the Tu-4 took place in Kazan' (at plant No.22 and in the 890th OUTAP). The unit received its first Tu-4s in 1949.

When the third unit of the 45th TBAD – the 362nd TBAP, likewise based at Balbasovo AB – had taken its training, in late 1948 it was the turn of the 13th *Dnepropetrovsko-Budapeshtskaya* GvBAD of the 43rd VA to start the transition to the Tu-4. This division, which had earned its titles for liberating the cities of Dnepropetrovsk (the Ukraine) and Budapest in the Great Patriotic War, was headquartered at Poltava (a regional center in the Ukraine), comprising the co-located 185th GvBAP, the 202nd BAP in nearby Mirgorod (Poltava Region) and the 226th BAP (also at Poltava), all equipped with IL-4s. In November 1948 the division's ground personnel was dispatched to the Kazan' aircraft factory to familiarise themselves with the Tu-4. Between January and April 1949 the flight crews took conversion training at the factory and in

Above: The aircrew and ground crew of Kuibyshev-built Tu-4 '05 Blue' poses for a photo in a winter setting.

Below: A crew receives a briefing in front of Moscow-built Tu-4 '35 Red' being refuelled by an articulated fuel bowser.

Top and above: '08 Red', a fairly late-production Kazan'-built Tu-4 (c/n 226002) on a snowbound hard-stand sometime in the 1950s.

Left: Three-quarters rear view of the same aircraft, with signs of operational wear and tear very much in evidence.

Top right: The same aircraft is prepared for a sortie, with the front bomb bay doors open and the boarding ladder in position.

Above right: A Tu-4 undergoes minor repairs at an Air Force aircraft repair workshop in Konotop in the Ukraine. Minor defects are painted up with brushes. Note the cannons in the forward dorsal barbette at extreme upward deflection.

Right: Here, apparently more serious work is done on the same aircraft involving the use of spray guns, as evidenced by the ZiF-55 mobile compressor on the hardstand.

the 890th OUTAP; the 185th GvTBAP and 226th TBAP led the way, followed by the 202nd TBAP in April 1949. Again, the B-25J was used as a transitional trainer. The navigators and bomb-aimers took conversion training courses *in situ* under the guidance of tutors and instructors from the training center; flight training took place at the training center in Ivanovo, central Russia.

The first six Tu-4s destined for the 185th GvTBAP arrived at Poltava in July 1949; by mid-month the crews assigned to the bombers had started flying them. In early September 1949 the crews captained by the 185th GvTBAP's command staff – the regiment CO and the squadron commanders – started practicing flights along designated routes involving weapons training; soon the other crews followed suit. The 184th TBAP in Priluki, Chernigov Region, also switched to the Tu-4 in due course. All in all, 72 aircraft captains and 67 complete crews in the 13th GvTBAD received their Tu-4 type ratings; the average flight time in the course of conversion training was 14 hours. However, the mastery of Tu-4 combat techniques in the division continued all the way to 1952.

One more bomber unit stationed in Belorussia converted to the Tu-4 in 1949. This was the 330th BAP based at Bobruisk (Mogilyov Region), which had previously flown Yer-2s; this was probably also a unit of the 45th TBAD.

As one might imagine, the Tu-4 was plagued by teething troubles for the first few years, and reliability was rather poor. For instance, in August and September 1949 the DA crews flew a total of 5,306 sorties in the Tu-4, logging a total of 3,139 hours. During these two months 895 manufacturing defects and hardware failures were detected, including 94 for the airframe, 58 for the engines, 233 for the armament and 510 for the mission equipment. As a result, the mean time between failures (MTBF) was no more than 3.5 hours.

Cold-weather operations turned up a whole spate of defects. In late February 1950 Soviet Air Force C-in-C Pavel F. Zhigarev sent a long list of defects occurring in the Tu-4 during the winter of 1949-50 to Vice-Minister of Aircraft Industry Pyotr V. Dement'yev. The latter forwarded the list to Andrey N. Tupolev, requesting that remedial action be taken. The list contained 28 items:

'Aircraft: (that is, airframe and aircraft systems – Auth.)

1. Type R-5612-0/2 hydraulic line hoses burst due to insufficient strength and the rubber's poor resistance to frost;

2. Leaks develop in all rubber seals of the hydraulic equipment, the wheel well door actuating rams and bomb bay door actuating rams due to the rubber's poor resistance to frost;

3. Air and hydraulic fluid seep from the landing gear oleos due to the rubber seals' poor resistance to frost;

4. The current in the flap actuation increases sharply (up to 300-320 amperes) during flap extension due to the lubricant on the flap screw jacks congealing;

5. Brake shoes stick to the brake drums during operations from snow-covered airfields;

6. The reinforced springs in the wheel brake mechanisms due to the brake chambers losing elasticity;

7. The perimeter seals of the pressure hatches become stiff due to the rubber's poor resistance to frost;

8. The glazing of the pressure cabins ices up both on the ground and in flight;

An early Kazan'-built Tu-4 with the unusual serial '1000 Red' (sequentially it should have been '10 Black').

9. The engine of the M-10 [auxiliary power] unit runs cold in flight, making it impossible to start before landing;

Propulsion group:

10. The cylinder head temperature drops to the minimum permissible value (140°C [284°F]) in level flight and during descent [even] with the cowling flaps fully closed;

11. Engine oil pressure aft of the forward oil pump falls in flight and during ground runs due to the oil freezing in the pipe from the front casing to the oil pressure distributor;

12. Oil cooler tubes burst in flight and during taxying due to the oil congealing because of improper functioning of the ARTM automatic oil temperature regulator;

13. Reinforced rubber hoses in the pipelines from the oil pump to the centrifugal filter and from the filter to the oil cooler burst due to the oil congealing;

14. The engine fire extinguishers are put out of action by ice forming on the [CO_2] distribution manifold and blocking the orifices in it;

15. The instruction in the manual concerning oil dilution does not provide for operation of the aircraft at ambient temperatures below −30°C [−22°F];

16. The design of the warm engine covers is inadequate, causing large heat losses when the engines are warmed up pre-flight and causing the engines to cool quickly after being ground-run, especially in high winds;

Armament:

17. The electric motors of the compressors catering for the recharging of the cannons in the tail barbettes burn out because the compressor piston freezes in the cylinder due to moisture entering the cylinder with outside air via the filter;

Tu-4 '1000 Red' taxies on the power of the outer engines alone.

18. The Type RVA, RPS and RS relays in the cannon barbettes' remote control boxes fail due to hoar frost forming on the contacts;

19. The lubricant in the range selector mechanism of the sighting stations freezes, with the result that the grip of the sighting station's mount turns when the range selector handwheel is spun;

20. The ballistic computer's electric motors burn out because the lubricant in the computer's mechanisms freezes, overloading the motor;

21. The contacts of the amplifiers in the synchro system ice up, with no electric contact as a result;

22. The movable safety bar on the KD4-546F bomb cassette freezes and sticks to the base, which means it cannot be moved from 'safety on' to 'safety off' position;

23. The UN-1 electric motor actuating the bomb bay door locks freezes up;

Two more views of the same aircraft. Note the battery cart near the nose gear and the Li-2T and IL-12D transports in the background.

Left: Lt.-Gen. Vasiliy I. Stalin in the captain's seat of a Tu-4 – without even a flying helmet.

Below: Another shot of Vasiliy Stalin flying a Tu-4 – this time in proper flying attire.

Bottom: Vasiliy Stalin in command of one of the Tushino flypasts.

Special equipment:

24. The MPSh-3 landing gear actuators fail because water seeps through the cover of the helical gear and freezes, and because the limit switches ice up;

25. The RL-14A minimum relays fail all over the place because the contacts ice up and the command relays fail when the condensation forming on them during cooling of the engine and oil tank freezes;

26. The RL-14A relays become maladjusted due to weakening of the springs;

27. The vinyl tubes and wraps enclosing the wiring bundles and individual wires become stiff and brittle;

28. The switches and buttons on the control panels fail because their movable contacts ice up.'

Here we have to go back in time a bit. As a counter-intelligence ploy, pursuant to the Chief of General Staff's directive No.Org/1/120030 dated 10th January 1949 the 1st, 2nd and 3rd Air Armies were renumbered as the 50th, 43rd and 65th VA respectively, effective from 15th February that year. Another eight years later the 65th VA was reorganised to become the 5th VA headquartered in Blagoveshchensk in the Far East.

The Soviet government and military top brass attached considerable importance to fielding the Tu-4 in large numbers as quickly as possible – mainly due to the type's intended role as a nuclear bomb delivery vehicle and so measures were taken to intensify crew training for the Tu-4. Hence the Officers' Aircrew College of the DA at Dyagilevo AB near Ryazan' in central Russia joined in the training process; this establishment later became the well-known 43rd Combat Training & Aircrew Conversion Center.

Another unit involved in the training of Tu-4 crews was the 34th UAP (*oochebnyy aviapolk* – Training Air Regiment) based at Lebedin AB in the Soomy Region of north-eastern Ukraine, which was part of the Nezhin Flying School (this base is no longer in existence). The unit received its first *Bull* in the autumn of 1949, having previously operated the IL-4 and B-25. Anatoliy Kravchenko, who served in the 34th UAP as a ground crew member in the rank of Sergeant at the time, recalled that the arrival of the first Tu-4 was a grand occasion; the unit's entire personnel assembled on the flight line to greet the aircraft, with a brass band for added effect. '*The giant aircraft created a strong impression. When the aircraft taxied in, everyone came bustling out onto the flight line to watch. Even though we knew that it had been copied from the American B-29 down to the last rivet, everyone was aware that this bomber laid the foundations for the development of completely new combat aircraft in Soviet aviation.*

This marked the beginning of a new era in the aircraft maintenance system as well. Quite apart from the fact that the aircrew now included a flight engineer and a flight technician, the ground crew was significantly expanded. (The Tu-4 had a ground crew of ten – *Auth.*) *In addition to the crew chief, each aircraft was serviced by a port wing technician and a starboard wing technician, each with several mechanics under his command. Each squadron had maintenance teams specialising in radio equipment, armament, and electrics and special equipment (ESE); the ESE team comprised electrics, instrument and photo equipment technicians.*

The regiment quickly built up its fleet and started flying almost immediately. Typical missions were the same ones which the crews

Top: Guards Maj. Tsygankov (left), the maintenance chief of Sqn 1 of a Tu-4 unit, inspects the No.1 engine's left-hand TK-19 supercharger before a sortie.

Above: Several Tu-4s at an unpaved tactical airfield during an exercise.

Top right: A very early Tu-4 (c/n 220201) in service with a first-line unit

Above right: A pair of Tu-4s in cruise flight.

Right: A Tu-4 shows off its high wing aspect ratio during a low-level pass.

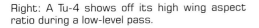

About 20 early-production Tu-4s on the flight line of a Long-Range Aviation airbase.

were to perform in first-line units. These included take-offs and landings, flights around the airfield circuit and bombing practice at nearby target ranges. The bombing practice involved so-called "concrete bombs" – dummies filled with cement powder.' Anatoliy Kravchenko was then a 22-year-old conscript in the Soviet Armed Forces, but the regiment's Chief of Staff Lt.-Col. Kachanov sug-

gested that he should consider a career in the Air Force when his conscription term was over. Kravchenko accepted and, after taking appropriate training, became a photo equipment mechanic in the 34th UAP's Sqn 4, with a job of servicing the Tu-4's BDA cameras. In 1952 he was promoted to Lieutenant Junior Grade and became a photo equipment technician in the same squadron. Kravchenko

Another view of the same flight line.

recounted that a camera technician's job offered just about the least career opportunities in the DA. Up to 1953 the ground staff of DA units had included the post of photo equipment engineer in the rank of Major, but then it was abolished, and being a senior photo equipment technician in the rank of Lieutenant Senior Grade was as far as you could go. This was blatant injustice towards many officers who, after 20-25 years of sterling service, were forced to retire from active duty in the rank of Lt. (SG).

One of the top-priority tasks facing the DA was the enhancement of poor-weather flying capability in daytime and at night. The importance of this aspect was so high that on 15th January 1950 the Council of Ministers issued a special directive calling for the outfitting of airbases with instrument landing systems. A system of aircrew qualification ratings was introduced (Pilot 3rd Class, Pilot 2nd Class, Pilot 1st Class, Merited Military Pilot; the same system applied to navigators). In accordance with this system, airmen with higher qualification ratings received various awards and bonuses (including financial ones) and were eligible for faster promotion for mastering the technique of flying in IMC. The document introducing the rating system also introduced qualification badges to be worn on the airmen's uniforms. The incentive worked: by the end of 1950 no less than 74% of the DA's pilots had received qualification ratings. (Later, however, the top command decided this was too much of a good thing; in 1955 the number of IMC flight hours needed to be eligible for an award was doubled, and in 1959 the awards for IMC flying were cancelled altogether.)

Deployment of the Tu-4 necessitated equipping the airbases with inner and outer marker beacons, localisers and glideslope beacons, relay stations for long-range control, light beacons and well-equipped air traffic control towers. The 52nd GvTBAP crews were the first to master the OSP-48 ILS in late 1949; the system permitted landings with the cloudbase at 200 m (660 ft) and 2,000 m (6,560 ft) horizontal visibility. As early as 1950, however, the regiments began to receive Tu-4s equipped with the more advanced SP-50 Materik ILS which made it possible to reduce the weather minima to 100 x 1,000 m (330 x 3,280 ft).

The DA command gave much attention to the aerial gunnery service which was responsible for training the bomber crews in anti-fighter defence techniques. The chiefs of the bomber divisions' and bomber corps' aerial gunnery service included both battle-hardened veterans of the past war – Vladimir A. Vyazovskiy, Aleksandr F. Petrov, Mikhail T. Ryabov, Sergey V. Shcherbakov and Grigoriy I. Nesmashnyy (all HSUs) – and newly trained officers: V. N. Arsen'yev, G. M. Bytsenko, N. I. Glookhoyedov, V. L. Kolchin, G. A. Kirsanov, N. D. Morrison, V. M. Savin, I. Ye. Cherkashin and P. S. Shelest.

This was also a time when bomber crews intensively practiced low-level and ultra-low-level flying as a means of penetrating enemy air defences. This was a highly complex job, and the absence of ground proximity warning systems enhancing flight safety (the Tu-4's radio altimeters were rather inaccurate at first) did not make it any easier. Here a major contribution to the training effort was made by war veteran pilots and navigators – Maj.-Gen. A. V. Koshkin, Maj.-Gen. V. I. Chernyshov, Maj.-Gen. V. A. Bykhal, Maj.-Gen. V. T. Taranov, Maj.-Gen. F. S. Yalovoy, Col. V. S. Vakhnov, Col. I. A. Shchadnykh, Col. V. A. Kolodnikov,

Col. V. V. Golubkov, Col. L. M. Karpov, Col. A. V. Cherkasov, Col. A. F. Kravchenko, Col. B. S. Korbut, Col. V. D. Vanchoorin and Col. K. F. Batishchev.

Meanwhile, more and more units were transitioning to the Tu-4. In 1950 it was the turn of the 200th *Brestskiy* GvTBAP, the 210th GvTBAP and the 111th TBAP, all based at Bobruisk. The tradition of showing heavy bombers at festive events also continued. In 1950 it was the turn of the 185th GvTBAP and 226th TBAP to take part in the flypasts on Aviation Day and during the May Day and 7th November parades. The Tu-4s of the 52nd GvTBAP participated in such parades on three occasions. This show of force was meant not so much for the home public but rather for the foreign observers. Behold the power of the Soviet war machine! Hear ye! Fear ye!

Re-equipment of bomber units with the Tu-4 continued apace in 1951-52. In particular, in early 1951 the 57th TBAD was redeployed to Belorussia; this division comprised the 170th Smolenskiy GvTBAP at Bykhov (Mogilyov Region), the 171st GvTBAP at Baranovichi (Brest Region) and the 240th GvTBAP at Bykhov, all equipped with Tu-4s. Also in 1951, the 132nd *Berlinskiy* BAP of the 116th BAD converted from the Tu-2 to the Tu-4 and was transferred to the 326th *Ternopol'skaya* BAD, moving to Tartu, Estonia, where the division was headquartered.

The 840th BAP of the 326th BAD (reporting to the 3rd VA from 1946) started converting to the type from the Tu-2 in May 1951. In November the regiment moved to its new home – Sol'tsy AB near the town of the same name in the Novgorod Region, north-western Russia; by December the unit had received its first ten Tu-4s, becoming the 840th TBAP. Sol'tsy AB also hosted another 326th TBAD unit equipped with the Tu-4 – the 345th *Berlinskiy* TBAP; each regiment ultimately had 30 aircraft in three squadrons.

Here it is worth again quoting the memoirs of Anatoliy Kravchenko, who ended up in the 345th TBAP/2nd AE after the DA regiments shifted from four-squadron to three-squadron organisation in 1953 and Sqn 4 of the 34th UAP, where he had served, was disbanded. '*There was, of course, a major difference in operations support between a training regiment and a first-line regiment. Firstly, bombing practice was done using real high-explosive bombs. Secondly, there was night photography, of which we had had no notion back at Lebedin. For night photography FotAB flash bombs were loaded into the bomb bays. The squadron had about five [NAFA-75] night cameras and about as many [AFA-33] daylight cameras for the entire complement of aircraft. They were used mainly for bomb damage assessment. The bombers proper were identical but we would install daylight or night cameras in special camera ports, depending on the mission. Changing the cameras was no easy job because the daylight camera weighed 78 kg [172 lb] and the night camera weighed 45 kg [99 lb]. Together with the [photo equipment] mechanic we had to move them from aircraft to aircraft all the time. The Tu-4s flew night sorties frequently, at least two or three days a week, alternating them with daytime sorties. Also, as distinct from the training regiment, each crew was able to fly any time, day or night. Hence the responsibility associated with the job in the 345th TBAP was much greater. For example, during exercises the crews used additional cameras recording the radar imagery; they were linked to the display of the Kobal't*

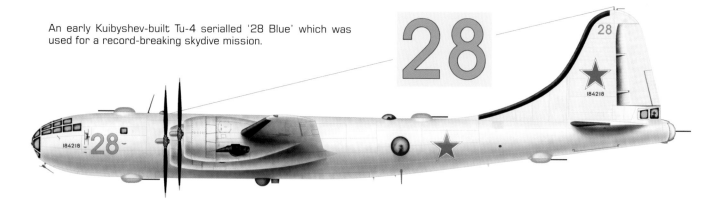

An early Kuibyshev-built Tu-4 serialled '28 Blue' which was used for a record-breaking skydive mission.

radar, and we worried the most about the pictures provided by these cameras. It was all too easy to have disciplinary action taken against you at the post-exercise debriefing if the photo lab turned out poor-quality photos. Our job was to prepare and maintain the cameras, while developing and interpreting the photos was the responsibility of a separate aerial photography service.

Interestingly, during exercises our regiment occasionally attacked maritime targets in addition to the usual ground targets. The cameras provided some excellent pictures of the radar imagery in these cases. The cameras used 28-cm [approx. 11^{1}/$_{32}$ in] film from which paper prints were made later. In the case of ground targets the photos showed contours of the targets which were analysed by a specially trained decoding officer, and a detailed caption would be made. Of course, the radar operators were also able

to interpret these pictures. One of them, A. Ye. Filatchev, was a first-rate person who became my lifelong friend. In addition to radar imagery, the decoders interpreted regular photos taken by the daylight cameras, gluing them to produce photo maps, which likewise featured comments to all objects depicted on them. As a rule, the photos were taken from high altitude; the quality was generally good, though there were cases when BDA was not performed due to thick overcast.'

Another unit which transitioned to the Tu-4 at this point was the 251st TBAP commanded by Col. Pavel A. Taran (Twice HSU), which was part of the 15th TBAD and was based at Belaya Tserkov' ('White Church') in the Kiev Region of the Ukraine. This division, like the aforementioned 13th GvTBAD, came under the control of the 2nd OTBAK (*otdel'nyy tyazholyy bombardirovochnyy avia-*

Tu-4 '28 Blue' during preparations for a record-breaking skydive mission between 10th and 12th September 1952.

korpus – Independent Heavy Bomber Corps) when the latter was established in 1960. One more unit known to operate the Tu-4 was the 229th *Roslavl'skiy* TBAP based at Konotop in the Soomy Region of north-eastern Ukraine.

In 1952-55 the 13th GvTBAD continued improving the crews' flying skills. Tu-4 crews flew daytime and night maximum-range sorties in IMC during operations in detachment, squadron and regiment strength. The crews became proficient in the bombardment of targets obscured by clouds or darkness, as well as of moving maritime targets, using the Kobal't radar and the OPB-4S bomb sight. Additionally, the crews were trained to make evasive manoeuvres when the bomber was caught by air defence searchlights. In 1950-51 each Tu-4 crew logged an average 80-100 flight hours per annum; in 1952-54 this figure rose to 120-139 hours.

All in all, during the first post-war decade the *Bull* saw service with 20 heavy bomber regiments. Some of them were reshuffled in the course of reorganisations; thus, in January 1951 the 52nd GvTBAP – previously a 9th BAK unit – came under the control of the 79th GvTBAK and moved from Balbasovo to Machoolishchi AB immediately south of Minsk.

Wartime bomber pilots, many of whom became commanders of DA regiments and divisions, were the first to master the operational use of the Tu-4. Well-known airmen who had earned the HSU title in combat – Aleksandr I. Molodchiy, Vasiliy V. Reshetnikov, A. V. Ivanov, Serafim K. Biryukov, Vladimir M. Bezbokov and Pavel A. Taran – spearheaded the aircrew conversion effort.

Apart from regular training sorties, the Tu-4s performed various special missions. Thus, 52nd GvTBAP crews obtained radar images of several strategic objectives in the USSR and participated in the trials of the SP-50 ILS, filming automatic landing approaches using this system. The unit's Tu-4s also acted as practice targets when the guidance radars of new AA artillery systems were being tested. In February-October 1954 appropriately equipped Tu-4s were used for air sampling to determine radiation levels in the wake of nuclear tests. In March-June 1954 a detachment of 226th TBAP Tu-4s commanded by Lt.-Col. Kirsanov was temporarily deployed to China on a special assignment. Missions flown by Tu-4s included reconnaissance flights of more than 12 hours' duration along the northern borders of Turkey, Iran and Afghanistan in order to provoke the air defence radars located there into 'painting' the aircraft and revealing themselves. Many airmen of the 13th GvTBAD received combat awards for successfully mastering the new hardware and for fulfilling various special assignments.

Mastering the Tu-4 was no simple matter for the aircrews. The *Bull* was larger and far more complex than the Pe-8, Yer-2 and IL-4 that it came to supersede (they were ultimately phased out in 1945, 1946 and 1949 respectively). The Tu-4 was the first Soviet aircraft to have IFR capability. The plethora of complex avionics and equipment entailed considerable difficulties. It was necessary to radically improve the training level of the personnel, especially the radio operators. Simply learning the operational procedures was not enough; practice and a steady hand were required. For example, adjustment of the Kobal't radar's receiver was effected by tuning to the maximum current in the *Kristall* (Crystal) detector; if the operator turned the vernier just a little too briskly he missed the optimum value and the reception quality suffered. To comment on

Tu-4s '21 Black' (c/n 220501), '66 Black' and '12 Black' await the next mission at an unpaved tactical airfield during an exercise.

this situation, a rhyming phrase was coined, loosely translating as 'Kristall current is too weak – operator's fault I seek'.

Control of the defensive armament was also organised along new lines. Now it was possible to aim and fire any of the turrets making use of any one of the PS-48 sighting stations. On the other hand, adjustment and ranging of the system was a complicated and labour-intensive matter.

The IFF system was a can of worms in its own right, albeit this was not so much the system's fault but rather a result of the operational procedures. Since the IFF equipment was classified (for obvious reasons), additional security measures were enforced in the first-line units operating the Tu-4; as often as not, these measures boiled down to additional warnings and bans which were hammered into the heads of the personnel. To prevent the IFF codes from falling into enemy hands if the aircraft was shot down or captured, the IFF transponder and interrogator featured detonators which were

A still from an early Soviet colour film showing a Tu-4 making a low pass.

Seen over the heads of the troopers, aircrews are briefed before a training sortie, with a Tu-4D transport in the background; the flight deck and the engines are under wraps.

The crew of '14 Red', a Kuibyshev-built Batch 46 Tu-4D operated by the 566th VTAP, are instructed before a sortie from Seshcha AB.

Kuibyshev-built Tu-4 with the post-1955 tactical code '15 Red' (note the absence of star insignia on the fuselage) at Kiev-Borispol' in the 1950s – possibly a Tu-4D. Note that only the tail turret has been retained, the dorsal and ventral barbettes being deleted.

triggered either by a button on the captain's instrument panel or by an inertia or crash switch. The detonator circuit was to be armed when flying over enemy territory by inserting a jumper plug into a special socket. Well, the instructions of the security officers were such a pain in the butt that occasionally the pilots messed up the fairly straightforward procedure, inadvertently blowing up the transponders; the detonator's charge was small and the explosion could not be heard above the roar of the engines. Occasionally the inertia switch would be triggered by a hard landing, with the same result. Furthermore, an indicator flag was provided to show when the detonator was armed, but it was located inconveniently and, what's more, often broke off due to vibrations. Removing the transponder for inspection/maintenance was a complex and dangerous procedure because of its explosive contents. The IFF system was a constant source of problems, and sometimes the crew simply switched it off – even though this was strictly prohibited.

A 566th VTAP Tu-4D coded '18 Red' taxies out at Seshcha, with another following close behind.

Above: The flight line of a military airlift unit equipped with Tu-4Ds in the 1950s; the aircraft are fitted with P-90 pods.

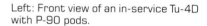

Left: Front view of an in-service Tu-4D with P-90 pods.

Left: The ground crew wheels a P-90 pod loaded with a ZiS-2 anti-tank gun into position under the port wing of a Tu-4D. The pods wore the same tactical code as the aircraft they were carried by.

Below left: The starboard P-90 pod is lifted into position under the wing of Tu-4D '16 Red', using BL-47 bomb hoists. Once the pod had been secured, the ground handling wheels were detached.

Below: This view shows the canvas flaps enclosing the parachute system in the rear end of the P-90.

Right: This Tu-4D coded '18' is in 28-seat troopship configuration. The paratroopers are standing alongside, awaiting the order to board.

Below right and bottom right: Three Tu-4Ds in Vee formation stage an airborne assault.

The ground personnel also had to shoulder more than a fair share of hard work when servicing the aircraft, its engines and systems, getting to grips with totally new hardware within very tight time limits. For example, many of the bolts designed to sustain high stresses were screwed in under substantial pressure and consequently could be removed only by drilling them out, which was no easy thing because they were made of extremely hard chrome-molybdenum steel. When the skin panels of the wing center section were removed, it was necessary to prop up the outer wings with trestles – otherwise the wings could collapse under their own weight. For the first time the technicians had to deal with remote control systems and electronic equipment the likes of which had not been seen in the Soviet Union. If the bomber sat parked for an extended time it had to be placed under wraps, and putting the heavy canvas covers in place, using tall and flimsy stepladders, was a back-breaking job. Many of the DA's generals and officers, including I. V. Markov, V. G. Balashov, N. D. Grebennikov, Yu. G. Mamsurov, I. K. Sklyarenko, P. I. Vostrikov, G. T. Fomin, N. U. Timoshok, N. Ye. Yefimov, A. V. Omel'chenko and V. M. Semyonov, did a lot to facilitate and expedite the routine maintenance of the Tu-4.

Yet, not only difficulties were in store for those who mastered the Tu-4s. It was far more capable than any wartime Soviet bomber had been, having a greater warload and longer range. The potent defensive armament had three times the weight of fire as compared to the same number of heavy machine-guns.

The Kobal't radar radically altered the tactics of bomber aviation, making it possible to deliver bomb strikes in adverse weather and at night from altitudes between 3,000 m (9,840 ft) and the aircraft's service ceiling. The radar detected big industrial premises at a distance of 100 km (62 miles) and determined the target's co-ordinates with an accuracy margin of ±2° with regard to direction and ±100 m (330 ft) with regard to distance; its maximum detection range was 400 km (248 miles). The radar could also be used for navigation by comparing the image on the radarscope to a map.

Most of the instruments and equipment items were conveniently placed and easily accessed. The pressurized cabins afforded a degree of comfort that had been unheard-of previously; they were even provided with Thermos bottles for warm food and with sleeping bunks for off-duty crew. Considering that missions could last ten hours or more, due care was taken even of the crew's need to relieve themselves – there were several vessels provided

with tight covers to prevent the smell from escaping. On the whole, the aircraft immediately commanded genuine respect.

The Tu-4 was intended for delivering massive bomb strikes with conventional bombs against key military and industrial targets located deep in the enemy's rear area and directly against enemy troops, both singly and in groups, day and night, in any weather. The use of bases in East Germany and in the countries of Eastern Europe was envisaged for this purpose; in the course of military exercises the Tu-4s actually redeployed to forward operating locations in Eastern Europe and were thus ready to make bombing raids against NATO troops stationed in Europe.

As was the case with the B-29s, the Tu-4s would scramble on alert and fly towards the borders of the Eastern Bloc. Single machines or groups of Tu-4s were sent to patrol the borders of the nations that were presumed to be the Soviet Union's adversaries in the future Third World War. The weapons options carried by the Tu-4s included 1,500-kg (3,310-lb) FAB-1500 and 3,000-kg (6,610-lb) FAB-3000 high-explosive bombs. Weapons of mass destruction carried by the Tu-4 included 500-kg (1,102-lb) KhAB-500-280S M-46 and 250-kg (1,102-lb) KhAB-250-150S M-46 chemical warfare bombs (*khimicheskaya aviabomba*) filled with a jellified mixture of mustard gas and lewisite having a per-

A Kuibyshev-built Tu-4D with the post-1955 tactical code '26 Red'.

sistence of at least 72 hours. However, the main weapon was to be nuclear bombs of various yield, which were under development at the time. The 'nuclear' page in the Tu-4's biography is described separately in Chapter 7.

As mentioned earlier, the Tu-4 introduced IFR capability in the Long-Range Aviation, albeit only a small proportion of the bomber fleet had refuelling receptacles and Tu-4 tankers were scarce. It is known that the 251st TBAP at Belaya Tserkov' had a few IFR-capable Tu-4s and a few tankers (the latter were operated by Sqn 2). The crews started practicing IFR in 1953, and Maj. (Retd.) Viktor S. Soozikov, who was then a tanker crew member in the 251st TBAP, flew a total of 178 refuelling sorties in the Tu-4 – first as refuelling system operator (RSO) and then as flight engineer. The receiver aircraft led the way, deploying a steel cable tipped with a drogue-stabilised locking weight from the starboard wing-tip. The tanker assumed echelon starboard formation, placed its port wing over the cable and moved to the right so that a ring clip attached to the hose engaged the cable; the weight at the end locked into the clip and extracted the hose. Then the receiver aircraft rewound the cable until the hose locked into the refuelling receptacle, forming a loop; when the RSO reported that contact had been established, the tanker's flight engineer switched on the fuel transfer pumps. The fuel transfer rate was 2,000 litres (440 Imp gal) per minute and, as several tons of petrol had to be transferred, the operation lasted several minutes. When the transfer was completed the RSO scavenged the hose with compressed nitrogen and rewound it in reverse order.

'Utmost precision in piloting the aircraft was required, and the crews' manoeuvres were calculated literally down to the centimetre and to the second – Soozikov reminisced – Otherwise, if the tanker banked to port the cable would be severed by the No.1 propeller; if it banked to starboard, the ring could not engage the cable and contact was impossible. Throughout the contact procedure, the fuel transfer and until the hose had been rewound, the tanker's RSO communicated with the captain and the flight engineer over the intercom, and with the receiver aircraft's RSO over the radio.

In May 1955 the flight technicians were assigned the RSO's functions. Until the aircraft reached the rendezvous area they sat at their usual position between the pilots; then, immediately before the contact they would crawl from the first cabin to the second cabin via the crawlway, assuming the port beam gunner's position

on the tanker and the starboard beam gunner's position on the receiver aircraft. When refuelling was completed, they returned to their regular position.

I have to repeat that this IFR system was very demanding as regards piloting skill, and quite often the cable or even the hose would be broken. There were cases of spontaneous disengagement; then the 30-m [98-ft] hose would start flailing erratically, and there was a real danger that it might strike the elevator and stabiliser. In the crew captained by Lt.-Col. V. Ikonnikov there was a case when the RSO failed to scavenge the hose properly after the fuel transfer, and when the hose was rewound the [residual] petrol spilled into the bomb bays (the Tu-4 had two of these). An explosion occurred, blowing off the doors of both bomb bays; the fuselage skin came loose half way up, and only Ikonnikov's skill and presence of mind averted a fatal crash. In another incident which occurred in the 229th Regiment, our brothers-in-arms, the hose

Opposite page, top: Tu-4D '26 Red' (c/n 1840347) in post-1955 markings, with what appears to be PDMM-47 paradroppable bags in the foreground.

Opposite page, above: An uncoded Tu-4D (c/n 223204) in pre-1955 markings in a winter setting.

This page, top and above: The same aircraft comes in to land at a tactical airfield in the High North.

Right: A Tu-4D creates a local snowstorm as it runs its engines at a northern airfield.

A fine air-to-air shot of a Tu-4 on an overwater sortie.

could not be rewound after the contact and it was decided to cut the cable (the one actuating the hose – Auth.). As a result, the hose started flailing wildly. In order to get rid of it, the RSO would have had to walk all the way through the No.2 bomb bay, running the risk of treading on the bomb bay doors, cut the 6-mm [0¹⁵⁄₆₄ in] cable with a pair of nippers, then return to his station and only then jettison the hose. Not the most pleasant procedure one might think of, to say nothing of the fact that it is not easy to bite through a 6-mm

Wheels caught in mid-retraction, a Tu-4 departs on a routine mission.

steel cable with nothing but nippers. The long and short of it is that the crew made a landing with the hose deployed at Mozdok, which was the alternate airfield.' (The city of Mozdok is situated in what was then the Chechen-Ingush Autonomous SSR; now it is in the Russian Federation's constituent Republic of Ingushetia – *Auth.*)

Soozikov mentions a rather embarrassing episode in the summer of 1955, when a pair of Tu-4s flew at 2,000 m (6,560 ft) over a military town near Borispol' township just as the local officers were having marksmanship practice with their handguns at the shooting range. Being accustomed to low-flying aircraft, nobody paid much attention until an object started falling earthwards with an unmistakable high-pitched screech. Everybody ran for cover, believing the aircraft had lost a bomb. However, on close inspection the object, which hit the ground with a thump, turned out to be a strange-looking pear-shaped nickel-plated thingamajig with a small parachute marked 'Please return to military unit 22634'. This was the field post office code of the 251st TBAP, and the object was, of course, the locking weight from an IFR-capable Tu-4 bomber whose cable had been severed by the tanker.

'In February 1955 the Long-Range Aviation command took the decision to conduct test flights of Tu-4s with a duration of more than 20 hours involving IFR – Soozikov continues – We were prepared for these flights with all due care. A week or so before the mission, the aircrews began receiving enhanced rations at the mess; their flight suits were exchanged for brand-new ones, regardless of how long they had worn the previous ones. In addition to the enhanced rations and the on-board emergency rations, each of us was issued a kilogram [2.2 lb] of chocolate. Special covers were made for the on-board Thermos flasks to keep the contents warm longer; all the crew members were issued pistols, army knives and compasses. These details give some idea of the importance attached to these flights.

On 7th February 1955 our tanker crew made a positioning flight to Tolmachovo airfield in Novosibirsk (now the city's principal airport – Auth.) as part of the support group commanded by Twice HSU Guards Col. Pavel Taran. Another crew redeployed to Belaya airbase near Irkutsk. We had a spell of severe frost at the time, so each four or five hours we had to remove the warm engine covers, start up and warm up the engines, then shut down and replace the covers. This job was sheer hell, especially considering the size of the bomber and the weight of the icy covers...

At length, on 11th February 1955 the aircraft of Capt. Musatov set out on a mission from Borispol' airfield (now Kiev's main international airport – Auth.). Our aircraft captained by Maj. Vasenin rendezvous'd with it over Barabinsk. The contact was performed without any problems and we transferred 10,000 litres [2,200 Imp gal] of petrol. At the next designated point Musatov's aircraft was met by another tanker, which also transferred 10,000 litres of petrol. Finally, Capt. Serov's aircraft took off from Belaya AB and transferred another 2,500 litres [550 Imp gal]; this last top-up was not really necessary and was performed for the sake of practice only. Capt. Musatov's aircraft landed at Vozdvizhenka AB (in the Primor'ye Territory of the Soviet Far East – Auth.) after a non-stop flight that had lasted more than 24 hours.

On 11th April 1955 our group comprising the same crews deployed [again] to Tolmachovo. On 15th and 23rd April we flew in support of two more ultra-long-range flights (non-stop from Borispol' to Krasnoyarsk and back, with two fuel top-ups en route), each of which lasted about 24 hours. Of course, we flew along other routes as well...'

Not bombers alone

Somehow the missile carrier version of the *Bull* has come to be associated primarily with the Naval Aviation. Yet, in 1954-55 the DA established a special division equipped with Tu-4Ks. This division, which was commanded by Maj.-Gen. Vladimir P. Dragomiretskiy, was designated 116th TAD (*tyazholaya aviadiveeziya* – Heavy Air Division) because, technically, it was not a bomber division but a missile strike division.

The 116th TAD comprised two missile carrier regiments – the 12th TAP (*tyazholyy aviapolk* – Heavy Air Regiment) and the 685th TAP – and an independent air squadron equipped with MiG-15SDK missile emulator aircraft. The latter were used for practice purposes (training missile launch operators) in order to avoid wasting the costly KS-1 missiles. All three units were based at Ostrov AB in the Pskov Region of north-western Russia; the name of the base means 'island' but, ironically, the base was nowhere near an island of any kind. A nuclear weapons storage facility was built in the woods near Lake Gorokhovoye ('Pea

Above: Naval Aviation bomber crews in winter flight suits pose with Tu-4 '09 Red' in the late 1950s. Note the unusual alphanumeric c/n (2266609B-07).

Below: Soviet officers in winter uniforms pose with a late-production Kuibyshev-built Tu-4.

The aircrew and ground crew of a Tu-4 captained by F. R. Zabelin before a sortie from Shagol AB near Chelyabinsk in 1958. The men don't seem to mind at all that a Tu-4 is engine running right behind their backs.

Lake') a short way from the base. The entire area was a maximum security zone – it was strictly guarded on a 24/7 basis and no one could enter or leave without special authorisation. In compensation for the inconveniences caused by the heightened security, the personnel received higher salaries, promotions and other perks.

The Tu-4R PHOTINT/ELINT and ECM version was another important variant of the *Bull* on the Long-Range Aviation inventory. As per 1st January 1955 the DA had five specialised units equipped with Tu-4Rs, each having 18 aircraft – a little short of the specified complement of 22. These included the 199th *Brestskiy* GvODRAP (*Gvardeyskiy dahl'niy razvedyvatel'nyy aviapolk* – Guards Independent Long-Range Reconnaissance Air Regiment, equivalent to the USAF's Strategic Reconnaissance Wing) based at Nezhin (Chernigov Region) in central Ukraine (Kiev MD), the 290th GvODRAP at Zyabrovka AB (Gomel' Region) in Belorussia and the 22nd TBAD's 121st *Sevastopol'skiy* GvODRAP at Machoolishchi AB near Minsk. Both of the Belorussian DD units received the type in 1953. The 199th GvODRAP came under the control of the 2nd OTBAK in 1960. Additionally, several bomber regiments (including the 202nd TBAP in Mirgorod) re-equipped with Tu-4Rs but retained their 'bomber' designations. Also, many bomber units had a few Tu-4Rs for ECM support; for example, the 345th TBAP had two such aircraft.

According to some sources, in late 1951/early 1952 a squadron of Tu-4Rs operated above the Sea of Japan and the East China Sea,

monitoring the deployment of United Nations reinforcements to Korea during the Korean War. In a more peaceful vein, in the early 1950s the Tu-4Rs were used to make a complete photo map of the Soviet Union; this operation was controlled personally by Lavrentiy P. Beria.

The Tu-4R had a few operational problems unique to it. The camera lenses iced up at high altitudes; the photo film turned brittle and lost sensitivity because of the cold. Some of the problems were tackled while others remained unsolved.

On 26th April 1958 a pair of 121st GvODRAP Tu-4Rs participated in a unique operation whose aim was to check the possibility of operating jet bombers from ice fields in the Arctic Ocean as a way of offsetting the threat posed by US bombers flying near the northern borders of the USSR. The bombers in question were Tu-16As from the 52nd GvTBAP, which had converted to the *Badger-A* in the meantime. According to the plan, the Tu-16As coming in from a forward operating location at Tiksi were to land on the ice floe hosting the Soviet drifting Arctic research station SP-6 (*Severnyy polyus* – North Pole) and take off again immediately in a reciprocal direction, one after another; next, the Tu-4Rs would land and offload supplies for the SP-6. However, the best-laid plans go astray – the first of three Tu-16As participating in the experiment, '04 Red' (captained by 52nd GvTBAP CO Guards Col. Anton A. Alekhnovich) hit a pothole during the take-off run and veered off the laboriously prepared ice runway, colliding with a parked civilian (Polar Aviation) Il'yushin IL-14T *Crate* transport that had arrived earlier on a support flight. Both aircraft were damaged, and the other two *Badgers* aborted the landing and returned to Tiksi. The Tu-4Rs (captained by Guards Maj. Alfyorov and Guards Maj. Aleksey Akoolov) did land and take off safely afterwards, taking the crew of the wrecked bomber with them. The SP-6 was then at 81°15' N and 147°42' E.

Tu-4D transports/troopships were placed on the strength of the Air Force's airlift regiments; for example, the 339th VTAP (*voyenno-trahnsportnyy aviapolk* – Military Airlift Regiment) based near Vitebsk, Belorussia, re-equipped with the type on 6th December 1956. Another airlift unit that switched to the type was the 566th *Solnechnogorskiy* ATP (*aviatrahnsportnyy polk*, later 566th VTAP), which was then based at Vypolzovo in the Kaliningrad Region and equipped with Il'yushin IL-12D *Coach* assault transports and Tsybin Ts-25 assault gliders. The conversion training took place between 1st March and 15th April 1956 in Kakhovka in the Kherson Region of the Ukraine; the unit commenced flying the Tu-4D in May and received its complement of *Bulls* in June. By the end of the year 31 crews had taken conversion training. After that, the 566th ATP returned to Vypolzovo but again deployed to Kakhovka in the summer of 1957. On 26th August 1958 the 566th VTAP moved to a new base – Seshcha AB in the Bryansk Region of central Russia; however, in 1959 it started converting to the then-latest Antonov An-12 four-turbo-prop transport. Until the early 1960s the Tu-4Ds were actively used for carrying troops and combat materiel.

This was the only version of the *Bull* actually used in combat by the VVS, namely during the Hungarian anti-Communist uprising which started in late October 1956. An air bridge was set up to bring in Warsaw Pact troops and crush the rebellion. Operating alongside other Soviet Air Force transport aircraft, Tu-4Ds airlifted VDV units to Budapest-Férihegy airport and to nearby Hungarian Air Force bases; on 4th November these troops were the first to storm the rebellious Budapest on their ASU-57 self-propelled guns and then set about restoring 'revolutionary order'. During the airlift the Tu-4Ds operated mainly from the town of Chop in the Zakarpat'ye Region of the Ukraine near the Soviet-Hungarian border. There were cases when they were fired upon by the rebels on final approach; many aircraft were damaged by small arms fire but none were lost. On the return trip the Tu-4Ds carried the bodies of Soviet servicemen killed in action. One can only imagine the psychological impact the sight of coffins being offloaded at Chop had on paratroopers waiting their turn to board the aircraft. 'See Budapest and die.'

In fact, the troopships were not the only *Bulls* involved in the crushing of the rebellion. On the night of 4th November several standard Tu-4s, each toting 3,000 kg (6,610 lb) of bombs – two FAB-500s and eight FAB-250s – took off from Borispol' AB near Kiev with a mission to bomb Budapest where the rebels had their headquarters. In order to mislead NATO intelligence assets tracking their flight, the bombers took a circuitous route to the target instead of heading straight to Budapest. It was just as well that they did, because fortunately wiser counsels prevailed and the mission was called off at the last moment when the bombers were passing the Romanian city of Moreni.

In naval service

Tu-4 bombers were operated not only by the Air Force but by the Naval Aviation (AVMF – *Aviahtsiya Voyenno-morskovo flota*) as well. The AVMF units stationed on the Baltic Sea were the first to convert to the new bomber. In February 1955 the 57th TBAD comprising two regiments of Tu-4s (the 170th GvTBAP and the 240th GvTBAP) was transferred to the Red Banner Baltic Fleet air arm. It was renamed the 57th MTAD (*minno-torpednaya aviadiveeziya* – Minelayer and Torpedo-Bomber Division), the regiments changing their designations accordingly to 170th GvMTAP and 240th GvMTAP (*Gvardeyskiy minno-torpednyy aviapolk* – Guards Minelayer and Torpedo-Bomber Regiment). The flying personnel were well trained, but the Naval Aviation was not particularly in need of the Tu-4 – the aircraft had already became obsolete. Therefore it was decided that the division should become the first to convert to new aircraft. Some of the Tu-4s were transferred to the AVMF's 33rd Combat Training & Aircrew Conversion Center at Kool'bakino AB in Nikolayev, the Ukraine, some were written off, and the command staff were dispatched to production plants in Kazan', Kuibyshev and Voronezh to study the Tu-16.

(As already mentioned, DA units equipped with the Tu-4 were also used for bombing maritime targets and setting naval mines. The 52nd GvTBAP trained for this purpose in 1951.)

In June 1953 the Independent Instructional & Training Unit No.27 tasked with evaluating the Kometa weapons system was set up at Gvardeiskoye AB near Kacha on the Crimea Peninsula. As early as December 1953 the unit made the first practice launch of a KS-1 missile, the decommissioned freighter M/V *Kursk* serving as the target. Pursuant to Soviet Navy General HQ directive No.52380 dated 30th August 1955, the Independent Instructional & Training Unit No.27 was transformed into the 124th TBAP of the Black Sea Fleet's 88th MTAD. Still based at Gvardeiskoye AB, the 124th TBAP operated 12 Tu-4Ks, two MiG-15SDK missile emulator aircraft, a UTI-MiG-15 *Midget* trainer, a Li-2S staff transport and two Po-2 liaison aircraft. Afterwards the MiG-15SDKs were replaced by similarly converted MiG-17SDKs. The use of missile simulator aircraft allowed the Tu-4K crews to prepare for real-life launches of KS-1 missiles quicker, making it possible to identify and rectify crew errors and avoid wasting live missiles. On reflection, the latter point was apparently of no major concern, judging by the fact that 82 live KS-1s were launched in 1955-56.

In 1955 the 124th TBAP spent three months on active duty away from base (at a North Fleet airbase), making a total of 40 missile launches. Next, in late 1956 a North Fleet unit likewise began its conversion to the Tu-4K, by which time enough of these aircraft had been delivered to the North Fleet to equip a detachment.

Practice launches of KS-1 missiles showed that the guidance system was still immature and vulnerable to ECM. Launching two missiles at a time was a problem because of the difficulty of having them both follow the target illumination radar's beam; however, once the missile had come close enough, its seeker head could achieve a lock-on against virtually any target. Initially all KS-1 launches were performed at 60-65 km (37-40 miles) range by single aircraft flying at 3,000-4,000 m (9,840-13,120 ft) and 340-360 km/h (211-223 mph); in the absence of ECM, 'kill' probability averaged 80%.

Above: A Tu-4A with the forward bomb bay open.

Below: Technicians use a special ladder for maintenance of a Tu-4 coded '22 Blue' (c/n 2805002), with sister ship '21 Blue' visible beyond. The red objects in between are fire extinguishers.

In the course of Tu-4K operations someone suggested a joint strike tactic: first, the Tu-4Ks would launch their missiles, keeping the enemy air defences busy, whereupon IL-28N bombers would deliver nuclear bomb strikes at minimum intervals. However, considering the huge difference in the bombers' performance (the IL-28 was much faster but had much shorter range), organising such joint operations would be a problem and the idea was deemed impracticable. As early as 1957 the 124th TBAP – by then redesignated the 124th MTAP DD (*minno-torpednyy aviapolk dahl'nevo deystviya* – Long-Range Minelayer and Torpedo-Bomber Regiment) – started converting to the Tu-16KS *Badger-B* having the same Kometa weapons system.

The Tu-4K never fired in anger. It deserves mention that there is an oft-cited myth: at a session of the Politbureau during the Korean War period some 'hawk' allegedly suggested using two regiments of Tu-4Ks armed with the first 50 production KS-1 missiles against US Navy and Royal Navy aircraft carriers involved in the war. The idea was allegedly rejected for fear of the local conflict in Korea escalating into the Third World War. In reality, considering the timeline, the Kometa weapons system simply entered service too late to take part in the Korean War.

Bulls, thrills and spills

Several interesting episodes from the service career of the Tu-4 can be related here.

As recounted earlier, on 3rd August 1947 the first three Tu-4s (still designated B-4 at the time) made the type's public debut at the Aviation Day flypast in Tushino. A curious episode occurred involving these aircraft. Air Chief Marshal Aleksandr Ye. Golovanov, who had been Commander of the Long-Range Aviation during the war, was the co-pilot of the lead bomber captained by Nikolay S. Rybko (aircraft R-01). Golovanov, a brilliant pilot possessing a perfect command of instrument flying and navigational calculations, was 'in command of the parade' aboard the aircraft. Nevertheless, a mistake in the calculations amounting to just one minute compelled the three bombers to literally dive under a formation of fighters flying ahead of them in order to avoid a collision; thus, to quote a subsequent wisecrack from the airfield personnel, for the first time an air display featured a flypast in 'sandwich formation'. Nevertheless, all the top government officials agreed that 'the overall impression was favourable' and the incident entailed no reprisals.

After the war, documentary films were shot during the Tushino air displays; these films were later shown nationwide for propaganda purposes. The *Ves'nik Vozdooshnovo Flota* (Air Fleet Herald) magazine (later closed down but now reborn – *Auth.*) published a detailed account of the parade and a detailed review of the film. According to some memoirs, the first such documentary was in black and white, and Iosif V. Stalin ordered the film shooting to be repeated in colour. So the flypast had to be 'repeated' over Kubinka AB, Monino AB and Tushino. In order to make the film 'ring true', the film sequences had to include 'enthusiastic crowds of spectators'. These were lured to Tushino, using a sale of some short-supply goods as a pretext.

Organisation of air displays and parades in the late 1940s was invariably associated with the name of Lt.-Gen. Vasiliy I. Stalin,

Commander of the Air Force component of the Moscow Military District. Promoted to the rank of General at the age of 25, the Soviet leader's younger son was not easy to deal with, being endowed with a bad temper, and had a drinking problem.

At the end of 1946 Vasiliy I. Stalin – then in the rank of Major-General – was appointed deputy of Lt.-Gen. N. A. Sbytov, the then Commander of the Moscow MD Air Force. Shortly thereafter Sbytov was forced into retirement and Stalin Jr. became Commander. His first 'undertaking' designed to enhance the combat efficiency of the Air Force was the construction of a new headquarters building for the Moscow MD Air Force at the Central airfield (Moscow-Khodynka). A constant stream of trains from Germany brought construction materials and costly embellishments (marble, sculptures, carpets etc.). Built from the ground up within six months, the HQ was rather more reminiscent of a palace.

When inspecting air units and formations of the district in Kalinin, Kubinka, Torzhok, Tyoplyy Stan and other garrisons, Vasiliy Stalin flew there in a Tu-4 converted to a command post of the Moscow MD Air Force. As related by the son of his aide-de-camp Viktor Polyanskiy, the fuselage housed a lounge, a study and a bedroom, all lavishly trimmed. Outwardly the aircraft was in no way different from production Tu-4s; it flew as a flagship at air displays in the late 1940s and early 1950s. Press correspondents often photographed Vasiliy I. Stalin sitting in the captain's seat of this aircraft. In his capacity of Commander of the Moscow MD aviation, he conducted 14 parades over the Red Square; usually, nearly 200 aircraft took part in these parades – several fighter regiments, two tactical bomber regiments and one DA regiment. The flypast was usually concluded by the Tu-4 bombers, but during the last 'parade' years they immediately followed the flagship.

After having flown over Moscow all the aircraft returned to their bases, and the flagship landed at the Central Airfield; from there V. Stalin was whisked in a limousine to the Kremlin where he mounted to the right wing of Lenin's Mausoleum in Red Square and reported to the Air Force Commander-in-Chief about the completion of the flypast. On the evening of that day the Government gave a reception in the Kremlin in the honour of the parade's participants (mainly, the parade's commanders).

As related by Polyanskiy, in his later years Stalin Jr. almost stopped flying (and he was, in the opinion of many, quite a skilled pilot), his alcoholism going from bad to worse, but sometimes he attempted to fly his Tu-4 in an inebriated condition – though, in fairness, not during preparations for official flypasts. He gave full throttle to the engines and made the bomber climb sharply or enter a dive, forgetting that it was not a fighter he was piloting. If reasoning did not work, the other crew members used brute force to remove him from the pilot's seat, wisely considering that it was better to suffer a punishment from 'the boss' than to be killed in a crash.

Many people have a vivid recollection of the story told by test pilot Ivan Ye. Fyodorov (HSU) about a 'demonstration flight' over the Kremlin that he had performed together with Vasiliy Stalin. One fine day Stalin arrived in his aircraft at the LII airfield and, goaded by the desire to show off his flying skill to his father, suggested to Fyodorov that he should fly over the Kremlin at very low altitude. Being well aware of the possible consequences, Fyodorov

asked Vasiliy to write a flight document stating the details of the forthcoming flight – the route, the altitude etc. Vasiliy Stalin did so, and they set out for the flight. Barely half an hour after their landing at Ramenskoye, secret service agents turned up at the airfield; they quickly found out who the pilot was, put the arm on Fyodorov and set about extracting from him the 'evidence' as to who had 'masterminded an attempt against the Soviet leader's life'. The 'flight document' written by Stalin Jr. failed to convince them and was promptly torn to pieces; indeed, it added more zeal to their interrogation. Fyodorov was saved only because Vasiliy Stalin was still at the airfield. When the pilot's friends gave him a hint about what was going on, he entered the room where the interrogation was taking place, asked Fyodorov: *'What are you doing here?'*, half-embraced him by the shoulders and led him away. The MGB officers dared not intervene.

Unlike the majority of his subordinates, Vasiliy Stalin failed to earn the coveted Gold Star medal that went with the Hero of the Soviet Union title. As for his simple human 'guiding star', it eclipsed during the 1952 May Day parade in Moscow in which a large formation of combat aircraft (Tu-4s, IL-28s, MiG-15s) was to participate. The weather report that day was absolutely prohibitive – dark clouds were hanging low over the city, it was drizzling. The weather was no better at the airfields where the Air Force units taking part in the parade were based. The Air Force C-in-C cancelled the flying display, but Vasiliy Stalin gave him a phone call from his command post, requesting permission to undertake the flypast, should the weather take a turn to the better. The C-in-C washed his hands of the affair, permitting him to act at his own discretion. As a result, without waiting for the weather to improve, Stalin took a risk and ordered all the units to take to the air and head for Moscow in instrument flying mode. That was a decision unprecedented in peacetime.

The result was deplorable. Some units missed Red Square altogether, others flew across it instead of lengthwise; still others received a message from the flagship aircraft ordering them to abort the mission and return to base. The spectators in Red Square did not see the aircraft at all – they could only hear the roar of jet engines above the clouds. This is how the story of this 'parade' was related by chief of the Main Command Post Col. B. A. Morozov and GK NII VVS test pilot Lt.-Gen. Stepan A. Mikoyan and retold by S. Gribanov. *'Vasiliy Stalin himself was the first to take off; he made a pass exactly over Red Square – Morozov recalls – He was followed by a division of Tu-4s. But the bombers' formation stretched out during the approach pattern: the regiment bringing up the rear was some 30 seconds late and it was difficult to catch up with those that were ahead. They were flying very low, just some 600 m [1,970 ft]. At that moment [...] a group of Il'yushins* (IL-28 tactical bombers – *Auth.*) *led by [Lt.-]Gen. [Sergey F.] Dolgooshin (the CO of a bomber division based at Migalovo AB, Kalinin – Auth.) put in an appearance. What do we do now? If I let them fly on, collision is imminent – the Il'yushins are flying at greater speed. I took the decision to stop Dolgooshin's group. I ordered, "Turn to port!" and directed the group towards Monino...'* As a result, the formation broke up; one of the *Beagles* clipped the top of a tall pine tree at low altitude and brought home some twigs, another IL-28 broke its pitot tube. Worst of all, two IL-28s collided

Below left: Another view of the same aircraft in the 1970s, with one of the museum's hangars in the background.

Opposite page: The same aircraft at a different location in the open-air display of the Monino museum in the 2000s. The c/n has been removed on the orders of some security-mad official.

near Migalovo AB and crashed, killing their crews (according to other sources, only one of them crashed, the other bomber sustaining damage but landing safely).

The 'debriefing' conducted in the study of Minister of Defence Nikolay A. Bulganin ended in a slight fright for both Vasiliy Stalin (and all the others involved) – he earned no more than a reprimand. However, at the reception held after the parade, in everybody's presence Iosif V. Stalin called Vasiliy 'a fool, a snotty kid' and the like. in very vivid expressions of which the Soviet leader was past master. Then he ordered Vasiliy to leave the study and instructed the Air Force C-in-C to dismiss him from the post of Commander of the Moscow MD Air Force.

A little-known page in the Tu-4's biography is its operation from *ad hoc* ice airstrips. These had to be used because of the scarcity of regular airfields in the High North. Tu-4 flights along the Soviet Union's northern coastline for the purpose of pinpointing suitable locations for forward airstrips began in 1950; they were commanded by Gen. Serebrennikov and the DA's Chief Navigator Viktor M. Lavskiy. The crews had to survey the Arctic Ocean and the coastline, determining the possibility of using the Kobal't radar for this. The original intention was to equip the Tu-4 with skis for operation from ice airstrips, but this was never done. A civilian Polar Aviation Li-2T transport was captained by the famous Polar Aviation pilot Ivan I. Cherevichnyy. His task was to find icefields suitable for landing strips and mark the latter. Once this had been done, the Tu-4s captained by Vagapov and Simonov would come in to land. On one occasion the Tu-4s paid a visit to the Arctic research station SP-2.

Unfortunately, like any major combat aircraft type, the Tu-4 had its share of accidents and incidents in Soviet Air Force service. The first known case was on 25h May 1949 when a 890th OUTAP aircraft (c/n 220704, probably serialled '34 Black') had its rudder lock up suddenly while flown manually (with the autopilot disengaged). Not knowing the cause of the problem, the crew resourcefully used differential power for directional control, managing a

safe landing; it turned out that one of the autopilot servos had jammed.

The first three operational losses occurred in the summer of 1951 in quick succession. The first aircraft captained by Lt. (SG) Kazantsev crashed on 26th July near the town of Dzerzhinsk in the Gor'kiy Region (now Nizhniy Novgorod Region) after departing Staryy Bykhov AB on a weather reconnaissance sortie. After entering heavy cumulus clouds at 8,000 m (26,250 ft) the aircraft encountered turbulence so severe that it was tossed about with bank angles reaching 80°. The Tu-4 stood up to this abuse, but apparently the pilots had lost attitude awareness in the clouds (the artificial horizon must have toppled) and hence unwittingly put the aircraft into a power dive. The bomber disintegrated after exceeding the dynamic pressure limit, and four of the 12 crewmembers, who were thrown clear of the aircraft, lived to tell the tale.

The second aircraft, which was lost on 8th August near Pavlovskiy Posad in the north-east of the Moscow Region, crashed in similar circumstances. The bomber, which had departed from Bobruisk in Belorussia, ran into severe turbulence and lost part of the horizontal tail. Breaking out of the clouds, the crew attempted an emergency landing, but when the landing gear was extended the resulting pitch-down force was too much, causing loss of control – the damaged elevators could not provide enough control authority.

The third Tu-4 crashed on 18th August near Mironovka village in the Piterskiy District of the Saratov Region, southern Russia, after departing from Kuibyshev-Bezymyanka (the factory airfield the factory airfield of plant No.18) on a pre-delivery test flight. The tail unit broke off in turbulence and the bomber dived into the ground, killing all 15 occupants – captain Vasiliy T. Booryonkov, co-pilot V. F. Lyubchenko, navigator V. A. Ivanov, radio operator M. M. Volkov, flight engineer M. S. Shesternev, flight mechanic I. S. Kiselyov, radar operator V. I. Poleyev, electrics engineer Ye. I. Ternikov, senior gunner P. V. Nikitin, gunners A. P. Ivanov, P. N. Lopatin and V. A. Dmitriyev, engineer in charge of the tests

V. P. Politov, his assistant N. V. Kozhevnikov and factory representative Yu. A. Golovanov.

Two days later, on 20th August, the Politbureau convened to discuss these crashes and determine what remedial action was necessary. A month later, on 17th September, Vice-Chairman of the Council of Ministers Nikolay A. Bulganin, War Minister Marshal Aleksandr M. Vasilevskiy and Air Force Deputy C-in-C Lt.-Gen. Filipp A. Agal'tsov reported thus at a session of the Politbureau:

'Although there have been no fatal accidents [with the Tu-4] in the Air Force and MAP for similar reasons between 1948, when the Tu-4 was inducted, and July 1951 (there have been two fatal accidents for other reasons – one was caused by pilot error during a night landing and the other by engine failure), pilots with the greatest experience [with the type] (Korovitsin, Mar'yan, Balenko) have reported at a briefing that they have had cases when, after encountering turbulence at altitudes of 8,000-9,000 m [26,250-29,530 ft] and the [indicated] airspeed of 275 km/h [170.8 mph] prescribed for this altitude by the flight manual, the Tu-4 would stall, dropping a wing sharply. In so doing the indicated airspeed would increase to 570 km/h [354 mph]. Recovery to level flight in these cases had been very difficult, involving each pilot's individual technique because the manual contained no recommendations for such a situation.

Research conducted by TsAGI has shown that at indicated airspeeds of 260-270 km/h [161-167 mph] and altitudes of 8,000-10,000 m [26,250-32,810 ft], in vertical gusts of 10-12 m/sec [20-24 kts] the Tu-4 may stall and drop a wing, the speed increasing past the maximum design speed. The same research has shown that the stall can be averted by increasing the level flight speed.'

Accordingly, on 15th September 1951 the Commander of the DA issued an order increasing the Tu-4's indicated airspeed at 8,000-10,000 m to 300-310 km/h (186-192 mph) in any weather. This resolved the problem, albeit at the expense of a 3-6% reduction in range and a 500-800 m (1,640-2,620 ft) reduction in service ceiling. At the same time, MAP was tasked with reinforcing the Tu-4's

horizontal tail (the modification was to be made fleetwide by July 1952) and installing an AGK-47B back-up artificial horizon.

There was one more fatal crash in 1951 when a 203rd TBAP Tu-4 and its crew of 12 were lost during a practice flight from Balbasovo AB (the exact date and flight details are unknown). Other incidents in 1951 were less dramatic. For example, 1 hour 15 minutes into a practice bombing sortie Tu-4 c/n 184306 suffered a failure of the No.4 engine's crankshaft at 4,500 m (14,760 ft); the mission was aborted and the bombed-up aircraft force-landed at an alternate airfield. Another Kuibyshev-built *Bull* (c/n 184326) suffered a failure of the No.1 engine while cruising at 8,000 m. Shutting down the engine, the crew decided to carry on with the mission, reaching the target range and dropping the bombs as planned, but again a diversion to an alternate airfield had to be made.

Three Tu-4s were lost in 1954. The first of these, a 184th GvTBAP bomber captained by Capt. Dolgov, crashed on 16th March soon after take-off from Priluki AB, coming down 40 km (24.85 miles) from the base; the entire crew of 12 perished. On 21st August a Tu-4R from the 199th ODRAP captained by the regiment CO Col. Anatoliy P. Roobtsov (HSU) suffered an engine fire and broke up in mid-air; seven of the 13 crewmembers died, including Roobtsov. It was the same story with the third aircraft, a 251st TBAP bomber captained by Capt. N. Novikov (the exact date of the crash is unknown). When the in-flight fire could not be extinguished, Novikov ordered his crew to bail out – only he and an inspector pilot from the 15th TBAD HQ remained aboard, attempting an emergency landing. Unfortunately the stricken bomber exploded on final approach, killing the brave pilots.

A further *Bull* captained by Capt. Veryagin was lost with all hands on 14th February 1955. Disaster struck again on 23rd April 1956 when Tu-4 c/n 2302604 captained by Capt. G. I. Shabanov was coming in to land at Seshcha AB near Bryansk, central Russia. Due to a defective attachment nut that had gone AWOL, the

propeller of the No.2 engine had worked loose and eventually broke away when the bomber had descended to 250 m (820 ft). Only two of the 12 crewmen survived the crash. On 1st November 1956, a few days before the abortive raid on Budapest, a Tu-4 was lost when the pilot of a MiG-15 fighter making a practice attack misjudged his speed and collided with the bomber, all 13 airmen in the two aircraft losing their lives.

Several fatal accidents were caused by the tail gunners whose responsibilities included operating the M-10 APU. To start up the 'putt-putt' the gunner had to get up and leave his station, which was no easy task in the cramped conditions. In so doing he would often accidentally grab hold of the trim tab control cables, causing uncommanded elevator deflection. When this came to light the cables were enclosed by protective covers.

There were also ground incidents involving the type. On 2nd November 1956 an armourer hooking up an FAB-500 bomb to a Tu-4 forgot to engage the safety mechanism of the BL-47 bomb hoist and the 500-kg (1,102-lb) bomb fell on top of him, crushing the luckless man to death.

Eclipse

The type's service career turned out to be rather brief because the development of turbine-powered bombers in the Soviet Union rendered the piston-engined Tu-4 obsolete. Korean War experience was undoubtedly a contributing factor; in Korea Soviet fighter pilots flying the MiG-15*bis* made mincemeat of the Tu-4's American progenitor, but it was obvious that in an encounter with western jet fighters the Tu-4 would not stand much chance either. In 1954 the 52nd GvTBAP started converting to the Tu-16 jet bomber; the units of the 13th GvTBAD followed suit in January 1955 in keeping with the Minister of Defence's order No.00230-54. From 1956 onwards the Long-Range Aviation units began to receive the intercontinental Tu-95 as a replacement for their Tu-4s.

As per 1st January 1955 the DA had a total of 30 heavy bomber regiments distributed between 12 divisions. Of these, ten divisions were fully equipped with Tu-4s, having between 63 and 94 aircraft each; the remaining two were equipped with Tu-16s – or rather underequipped, having no more than 54 aircraft each instead of the required 130. Of these, one Tu-4 division with 63 aircraft and one Tu-16 division with 43 aircraft were so-called special divisions armed with nuclear weapons. At this time, DA crews averaged 125 flight hours and 30 bomb drops per annum; thus, they were much better off than the tactical bomber crews who had only 55 flight hours and 11 bomb drops per annum.

By 1st January 1958 the DA's bomber fleet had changed qualitatively, numbering 1,120 Tu-16s and 778 Tu-4s. Because the new Soviet leader Nikita S. Khrushchov had a predilection towards missile systems (especially intercontinental ballistic missiles) to the detriment of manned combat aircraft, many bomber regiments (including the 251st TBA) were transformed into ICBM units; others, such as the 229th TBAP, became military airlift units. Outlining the Long-Range Aviation's development prospects, Soviet Air Force C-in-C Air Chief Marshal Konstantin A. Vershinin and DA Commander Air Marshal Vladimir A. Soodets suggested that by 1st January 1961 the DA's bomber fleet should have the following composition:

'- new aircraft types – 405 aircraft (15.5%);
- obsolete Tu-16s – 1,381 aircraft (52.7%);
- Tu-4 piston-engined bombers – 255 aircraft (9.2%);
- shortage of aircraft [versus required full strength] – 578 aircraft (22%).'

By the early 1960s the *Bull* remained in service only with the military airlift units of the Air Force, in training establishments and as flying testbeds in the system of the Air Force and MAP; the Tu-4USh navigator trainers were the last to go. Finally, in 1964 the Tu-4 was officially withdrawn from the Soviet Air Force inventory.

Some of the Tu-4s were simply broken up for scrap; others were converted into Tu-4M target drones and shot down. A few ended up as ground targets at test and practice ranges; among other things, decommissioned Tu-4s were used for testing the effectiveness of nuclear weapons. One Tu-4 airframe became part of a ground test rig used for testing the effectiveness of aircraft cannons against heavy bombers; it was chosen for this purpose because of its American origins, on the assumption that other American heavy bombers would show similar results. The tests showed that a single hit of a 37-mm (1.45 calibre) round fired by a Nudelman N-37 could cause enough damage to the wing center section to disable the bomber.

Unfortunately only a single example of the Tu-4 survived in the Soviet Union (and post-Soviet Russia); on the other hand, it is the only completely original Tu-4 in existence today (the ones preserved in China are not – see Chapter 8). The aircraft in question, a Kuibyshev-built example manufactured in 1952 (c/n 2805103), was one of the bombers that took part in the famous cancelled raid against Budapest in 1956. In 1958 the aircraft, by then wearing the tactical code '01 Red', was struck off charge, making its last flight to Monino airfield on 7th October with pilot Altukhov in the captain's seat; it had logged 1,540 hours 5 minutes total time and 2,004 cycles since new. The aircraft became an instructional airframe at the local maintenance workshop; when the latter was closed down, the Tu-4 was donated to the Soviet Air Force Museum (now the Central Russian Air Force Museum) in Monino south of Moscow, becoming its first exhibit.

Tu-4 colours

Unlike the B-29, which initially wore an olive drab camouflage scheme with pale grey undersides (though this was extremely short-lived), the Tu-4s in Soviet service invariably sported a natural metal finish. A red, blue or black serial (or, after 1955, tactical code) which had the size of about one-third of the fuselage diameter was painted on the fuselage behind the forward cannon barbettes. The tactical codes in a given air formation ran in sequence, the colour of the numbers could be different for different regiments. The c/n was applied in smaller black numerals on both sides of the forward fuselage and on the fin. Propeller blades were black with yellow tips. Until 1955 the Soviet star insignia were applied to the rear fuselage, the underside of the wings and the fin. After 1955 the insignia were removed from the fuselage but added to the wing upper surfaces. Sometimes the stars were edged in white between the star itself and the red outline (usually this gap was not filled with colour). In the 1960s the tactical code was repeated in small digits at the top of the fin for quick identification on the flight line.

Chapter 7

With nukes on board

The development and testing of nuclear weapons in the USA, which culminated in the atomic bombings of Hiroshima and Nagasaki on 6th and 9th August 1945 respectively, compelled the Soviet government to take measures aimed at achieving nuclear parity with the USA and discouraging a possible American nuclear attack. Ten days after the Nagasaki bombing, Iosif V. Stalin ordered the establishment of a special committee under the State Defence Committee. This committee, which was headed by People's Commissar of the Interior Lavrentiy P. Beria, co-ordinated the new-born Soviet nuclear program.

On 9th April 1946 the Council of Ministers issued a directive ordering the establishment of the KB-11 design bureau of the Soviet Academy of Sciences' Laboratory No.2. KB-11 was headed by Pavel M. Zernov, with Academician Yuliy B. Khariton as Chief Designer. It was based in the town of Sarov (Gor'kiy Region), which has been off limits to foreigners ever since and which was known by various cover names over the years, including Arzamas-75 and Arzamas-16. (The real town of Arzamas is located further north in the same region, between Gor'kiy – now renamed back to Nizhniy Novgorod – and Sarov.)

On 21st August 1946 the Council of Ministers issued directive No.1286-525ss which, among other things, required KB-11 to develop a nuclear charge. The latter was cryptically referred to in the directive as *reaktivnyy dvigatel' spetsiahl'nyy* ('special reaction engine' or 'special jet engine'), or RDS for short, for security reasons. In later years the RDS acronym became widely known, generating alternative explanations, such as *reaktivnyy dvigatel' Stalina* ('Stalin's jet engine'), *Rosseeya delayet sama* ('Russia does it (read: makes the atomic bomb) on her own') etc.

Development of the first Soviet nuke – the RDS-1 implosion-type charge (alias *izdeliye* 501 or 'nuclear device 1-200') having a yield of 22 kilotons – began in 1947. Even with the benefit of intelligence on the US nuclear weapons program (the Manhattan Project) furnished by Soviet secret agents, this was a highly complex task that required a lot of fundamental research and took time; the first two RDS-1s were manufactured in 1949. The first RDS-1 was exploded on 29th August 1949 at the proving ground built for this purpose 170 km (105 miles) west of Semipalatinsk, Kazakhstan, and known as the War Ministry's Practice Range No.2 (UP-2 – *oochebnyy poligon*). This first test was a ground test; yet, development of an air-droppable version of the RDS-1 had been going on

since 1948. Predictably, the Tu-4 was chosen as the delivery vehicle, since it had the longest range of all Soviet bombers; hence the size of the future bomb was chosen with the dimensions of the *Bull*'s bomb bay in mind. The weapon was 3.7 m (12 ft 1⁴³⁄₆₄ in) long, with a diameter of 1.5 m (4 ft 11³⁄₆₄ in), and weighed about 4,600 kg (10,140 lb).

Ballistic tests of the first Soviet nuclear bomb began in the first half of 1948. Initially dummy bombs were dropped by a Tu-4 at a test range near the town of Noginsk (in the north-east of the Moscow Region) which was run by the 4th Directorate of GK NII VVS; the bomber was flown by LII test pilots Aleksey P. Yakimov and Stepan F. Mashkovskiy. These test drops, and other research conducted by KB-11 jointly with TsAGI, showed that the bomb was not stable enough as it travelled on its ballistic trajectory. The manufacturer had to optimise the design by streamlining the bomb casing, shifting the weapon's CG and altering the inertia forces.

Further ballistic tests were then performed at the Air Force's 71st Test Range at Bagerovo in the eastern part of the Crimea Peninsula, 14 km (8.7 miles) west of the city of Kerch. This outfit had been established pursuant to a Council of Ministers directive issued on 21st August 1947 (exactly a year after the one that kicked off the Soviet nuclear program!) for the purpose of conducting aerial nuclear tests and testing the delivery vehicles of nuclear munitions – that is, bombers because no ICBMs existed yet. It was to be commissioned in the summer of 1949, when the first Soviet nuclear test was to take place; this was a major effort and included complete reconstruction of the wartime auxiliary airfield at Bagerovo, which was in a state of total disrepair.

Actually the designation '71st Test Range' – *sem'desyat pervyy poligon* – was something of a misnomer. In reality this was a test and support establishment which, in addition to an instrumented test range with a target for the inert bomb drops, included three Air Force units. The latter were the 35th BAP, sometimes referred to as the 35th OSBAP (*otdel'nyy smeshannyy bombardirovochnyy aviapolk* – Independent Composite Bomber Regiment) equipped with Tu-4s (and later with jet bombers and fighter-bombers), the 513th IAP equipped with Lavochkin La-9 *Fang* piston-engined fighters and, later, MiG-17 jet fighters, and the 647th SAPSO (*smeshannyy aviapolk spetsiahl'novo obespecheniya* – Composite Special Support Air Regiment). This latter regiment operated a mixed bag of aircraft (which changed over

An RDS-3 atomic bomb on a purpose-built ground handling dolly is towed to the flight line under the watchful eye of a security officer.

Another view of the same bomb and dolly behind a GAZ-67B jeep, with a waiting Tu-4A visible in the background.

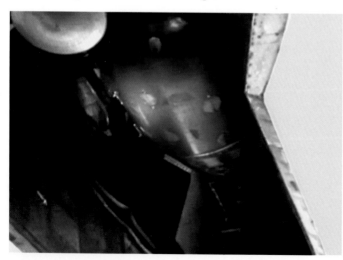

The RDS-3 in the bomb bay of a Tu-4A.

the years) – Po-2 utility biplanes, Yakovlev Yak-12 *Creek* utility monoplanes, Li-2, Il'yushin IL-14, An-8 and An-12 transports, An-24 *Coke* airliners, Yak-25RRV *Mandrake* and IL-28R *Beagle* reconnaissance aircraft, Mil' Mi-6 *Hook* heavy-lift helicopters. It was tasked with air sampling in the wake of nuclear tests, aerial photography, transport and liaison duties. Only instrumented inert bombs were dropped at Bagerovo; nobody in their right mind would test a live nuclear bomb in the Crimea which was, to use a common Soviet cliché, *the all-Union resort!*

Tests of inert RDS-1 bombs began in the spring of 1948; the first examples were equipped with tracers to permit trajectory measurements by means of cine camera footage, while later examples had telemetry sensors. The chief purpose at this stage was to find the optimum location for the barometric sensor which was to detonate the bomb at the preset altitude. No fewer than 30 test drops from a Tu-4 were required before KB-11 had achieved the required accuracy and made the trajectory measurements necessary for developing the *izdeliye* 501's automatic detonation control system. Unlike the first bombs, which were truly dummies, some of the later inert bombs were equipped with data recorders, which had to be salvaged and deciphered; this meant the remains of the bomb had to be excavated – manually, as the bomb buried itself 6-7 m (19-23 ft) into the soft ground on impact.

Next, the inert bombs were outfitted for testing the electric detonators which, in the live version, were to ignite the conventional explosive charge compressing the fissionable material to trigger a chain reaction. The bomb's multiple detonators were fitted with smoke cartridges, each of a different colour, to check if they fired in the correct order. On one occasion, however, the detonators fired when the aircraft was still on the ground. Aborting the take-off, the pilots opened the bomb bay and the crew evacuated hastily as multi-coloured smoke billowed from the aircraft. The base's crash rescue service promptly arrived on the scene, preventing a catastrophic fire, but after this incident the smoke cartridges were discarded in favour of a data link system. The ballistic tests of the RDS-1 bomb and the tests of its automatic systems were duly completed in the spring of 1949.

Tests of the Tu-4A nuclear-capable bomber and its specialised equipment were duly completed in 1951. On 17th May 1951 the Air Force Commander-in-Chief signed an order appointing a State commission for holding ground and flight tests of two *Bulls* which had been converted by OKB-156 to Tu-4A standard. The commission was chaired by Maj.-Gen. Gheorgiy O. Komarov (HSU), the first commander of the 71st Test Range; it included representatives from the Tupolev OKB and KB-11.

Meanwhile, KB-11 was not sitting idle. The RDS-2 second-generation implosion-type charge was developed in 1949; like the RDS-1, it had a plutonium core but the yield was much greater – 38 kt (some sources quote a figure of 42 kt). The bomb was dimensionally similar to the predecessor and weighed about 3,100 kg (6,830 lb). As early as 10th June 1948 the Council of Ministers had issued a directive requiring OKB-11 to develop the RDS-3, RDS-4 and RDS-5 atomic bombs and the RDS-6 hydrogen bomb. The RDS-3 differed from the RDS-2 in the type of fissionable material – the core consisted of 25% plutonium-239 and 75% uranium-235 in order to cut costs (plutonium was expensive and scarce). This design was treated sceptically at first because uranium has a higher critical mass than plutonium, and the opponents believed the chain reaction might be incomplete or not begin at all. However, the sceptics were proved wrong.

The question of how to test the bombs arose; should it be a ground test or was the time ripe for an airborne test? A group of scientists headed by Yuliy B. Khariton demanded a ground test, arguing this would allow the physics of the nuclear explosion to be studied in greater detail; the leaders of the Soviet nuclear program, notably Academician Igor' V. Kurchatov (head of the Nuclear Physics Institute), insisted on an airborne test. Eventually a compromise was reached: the RDS-2 would be tested in ground mode and the RDS-3 would be air-dropped a few days later.

Yet, the test of the RDS-2 on 24th September 1951 did involve the Tu-4A. Actually two 35th BAP bombers were involved (along with other 71st Test Range aircraft), redeploying to Zhana-Semey airfield in the East Kazakhstan Region – the nearest one to the Semipalatinsk proving ground. The larger aircraft were flown there, while smaller aircraft and ground support equipment were delivered by rail – the place lacked even the most basic equipment.

The primary Tu-4A captained by Lt.-Col. Konstantin I. Oorzhuntsev (HSU) was scheduled to take off at 0711 hrs local time; yet the take-off was postponed three times because of poor weather. When the bomber finally took off at 1011 hrs and headed for the target, the weather at the proving ground turned for the worse again. Unable to see the target, Oorzhuntsev made repeated passes over the area and, after running low on fuel, aborted the mission and returned to Zhana-Semey. Hours later, when the weather finally improved, it was the turn of the back-up crew headed by Capt. Konstantin I. Oosachov to try their luck; the back-up aircraft took off at 1416 hrs and reached the target of the first try. At T minus 90 seconds the Tu-4 passed at 10,000 m (32,810 ft) directly over the tower with the bomb at the proving ground (at exactly the same spot where the RDS-1 had been detonated two years earlier), transmitting a radio signal to the command center that triggered the detonator's time delay mechanism. In this way a bomb drop was simulated, confirming the possibility of using the Tu-4 as a nuclear bomb delivery vehicle.

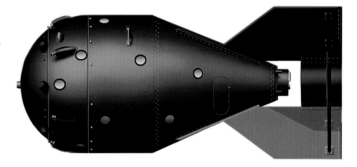

Top right: An artist's rendition of the RDS-3 42-kiloton nuclear bomb. The shape is not unlike that of the Fat Man atomic bomb dropped on Nagasaki. Note the multi-pin connector for a data transfer cable at the rear end of the body.

Above right: The crew of Tu-4A '207 Black', a 35th BAP aircraft, receive the final briefing before a nuclear test.

Right: '207 Black' is seen here a few moments before becoming airborne with a nuclear bomb on board.

Above: A Tu-4A seen from a chase plane en route to the target.

Escorted by an IL-28R recce aircraft, a Tu-4A is about to drop the bomb over the Totskoye practice range on 14th September 1954.

Following Andrey N. Tupolev's recommendations, the crew disengaged the autopilot, depressurized the cabins to equalise the pressure and donned oxygen masks after passing the tower; otherwise, the blast wave could overpressurise the fuselage, causing an explosive decompression, or cause the crew to lose consciousness. The blast wave caught up with the bomber when the latter was 24.2 km (15.04 miles) beyond the 'target'; the pressure of the wave turned out to be three times less than expected, and the Tu-4A suffered no ill effects from it.

The State commission responsible for the tests gave the go-ahead to drop the RDS-3 nuclear bomb but ruled that a series of

A Tu-4A cruises towards the test range.

calculations associated with the blast of a more powerful bomb should be made in the interests of the Tupolev OKB and KB-11. The commission did not focus on the carrier aircraft's flight safety when subjected to the factors of a nuclear explosion; this aspect would have to be dealt with separately.

The first live drop

Live tests of the RDS-3 (*izdeliye* 501-M) were scheduled for 18th October 1951. The bomb was codenamed 'Maria', starting a tradition of assigning Russian women's names to Soviet nuclear bombs. A highly experienced crew was entrusted with performing the first-ever live drop. It comprised crew captain Lt.-Col. Konstantin I. Oorzhuntsev, co-pilot Lt. (SG) Ivan M. Koshkarov, navigator Capt. Vladimir S. Suvorov, bomb-aimer Capt. Boris D. Davydov, radar operator Lt. (SG) Nikolay D. Kiryushkin, radio operator Lt. (JG) Vladimir V. Yakovlev, flight engineer Maj. Vasiliy N. Trofimov, gunner Pvt. Arkadiy F. Yevgodashin, flight technician Lt. Arkadiy F. Kuznetsov and test engineer for the nuclear bomb Lt. Al'vian N. Stebel'kov.

For added reliability, a second Tu-4 was to fly as wingman. The crew of this aircraft comprised crew captain Capt. Konstantin I. Oosachov, co-pilot Lt. (SG) Vasiliy I. Kooreyev, navigator Capt. Aleksey A. Pastoonin, bomb-aimer Lt. (SG) Gheorgiy A. Sablin, radar operator Lt. (SG) Nikita I. Svechnikov, radio operator SSgt Vladimir B. Zolotaryov, flight engineer Lt. Pyotr P. Cherepanov, gunner SSgt Nikolay D. Borzdov, flight technician Lt. Filaret I. Zolotookhin and test engineer Lt. Leonid A. Blagov.

For the live drop of the RDS-3 bomb the 35th BAP Tu-4As and various support aircraft redeployed to Zhana-Semey again. Extremely stringent security measures were taken during the final preparations. The armed, checked and double-checked RDS-3 bomb was placed on a ground handling dolly, carefully draped in tarpaulins to disguise its outline and slowly towed to the hardstand where the Tu-4A was waiting. The hardstand itself was surrounded by a tall fence to prevent unauthorised personnel from watching and had a separate checkpoint at the only entrance which was the taxiway. The hardstand featured a special trench lined with concrete; after the RDS-3 had been uncovered, the dolly was wheeled into this trench and the refuelled aircraft was pushed back into position over it. This was necessary because the Tu-4's ground clearance was insufficient for the dolly with the globular bomb to pass. By then the bomb bay doors had been opened and fitted with the usual BL-47 hand-cranked bomb hoists. To further enhance security the bomb bay area was curtained off by a canvas tent before the RDS-3 was slowly hoisted into the bomb bay. Next, an electrical connector was locked into place on the bomb's rear end, and the approved time delay and critical barometric altitude settings were downloaded to the detonation mechanism, using the aircraft's control panel.

After this, the crew began the pre-flight check. The captain and the navigator inspected the bomb bay, making sure that the bomb was properly secured and all electrical connectors were in place. Together with a representative of KB-11 they activated the bomb's electric locks, checking that the appropriate signal lights were on; the keys were then handed over to the captain in case the aircraft had to make an emergency landing at an alternate airfield.

The captain of Tu-4A '207 Black' in his seat.

An RDS-3 bomb seconds after being released by the aircraft.

After checking the bomb control panel readings to see that they matched the entries in the log, the bomb bay doors were closed and sealed by a MGB officer. Finally the crew captain and the navigator signed their acceptance of the aircraft in the log book.

With the entire crew lined up in front of the bomber, Oorzhuntsev reported his readiness to Maj.-Gen. Komarov and the KB-11 representative. Receiving the go-ahead, the crew climbed in and the aircraft taxied out for take-off. At 0700 hrs Moscow time on 18th October 1951 the Tu-4A lifted off the runway; the first-ever take-off with a live nuclear bomb in the Soviet Union went without a hitch.

Minutes later the back-up Tu-4 captained by Oosachov took off, carrying a dummy FAB-1500 HE bomb. In the event the nuke-armed aircraft's targeting systems failed the pair would change formation; the back-up aircraft would become the leader, proceeding to the target range as planned while transmitting a series of tone-modulated radio signals to the Tu-4A until the designated

bomb release time. The signals, and the moment when the back-up aircraft released the dummy bomb, would tell the crew of the Tu-4A when it was time to drop the nuke.

This time the target was situated 2.5 km (1.55 miles) from the previous location because the radioactive contamination at the latter after the recent test was too strong. This would have prevented post-test examination without unduly putting the lives of the personnel at risk.

The flight was controlled from the Central Command Post (CCP) located 25-30 km (15.5-18.6 miles) from the center of the proving ground where all the people in charge of the tests were assembled. These were Igor' V. Kurchatov (representing the Nuclear Physics Institute), Yuliy B. Khariton, Yakov B. Zel'dovich, Boris L. Vannikov, Pavel M. Zernov (all representing KB-11), Marshal Mitrofan I. Nedelin, Maj.-Gen. Viktor A. Boliatko and, representing the Air Force, Maj.-Gen. Gheorgiy O. Komarov and Engineer-Maj. Serafim M. Kulikov. The CCP was a

The first few seconds after the nuclear explosion: a huge ball of fire rises above the spot (note the so-called Wilson cloud above it)...

...and the tell-tale mushroom cloud forms.

The control box from a Tu-4A, used for setting the nuclear bomb's detonation parameters before the drop.

special structure hardened against the nuclear blast and surrounded by a barbed wire perimeter with a checkpoint. It had been built according to a project developed by the 71st Test Range and featured a special display allowing the command staff to watch the progress of the flight and the bomb drop.

The Tu-4s climbed to 10,000 m (32,810 ft) and proceeded to the target in strict accordance with the schedule; their progress was monitored on a map display at the CCP. En route to the target the bombers were escorted by constantly shifting pairs of La-9 fighters from the 513th IAP; their pilots had orders to shoot the Tu-4 down if it strayed from the designated course. All the while the CCP maintained HF and VHF radio communication with the bombers, using special code tables; for good measure the most important commands were repeated by the proving ground's support service.

The weather was favourable, all systems functioned perfectly, and right on schedule the Tu-4A crew was authorised to make a practice pass over the target while transmitting special radio signals so that the final adjustments could be made to the test range instrumentation. Finally it was time for the live bombing run. The range command post reported 'all systems are go' and the permission to drop the bomb was given. The bomber began sending special signals over the HF and VHF channels: T minus 60 seconds... T minus 15 seconds... The last signal came at 0952:08 hrs when radar operator Kiryushkin pushed the bomb release button. On the display at the CCP, lights came on, one by one, showing the bomb's trajectory as the automatic control system was energised, the

multi-stage safety system was disengaged, the detonator was armed and the barometric altitude sensors activated. This was the first evidence that the bomb was functioning normally. The next and final evidence was earth-shattering; the ground shook and a tremendous boom came from outside, confirming that the Soviet Union's first mid-air nuclear test had been successfully performed.

The 42-kiloton RDS-3 detonated at 380 m (1,250 ft) above ground level (AGL). A huge hemispherical cloud of condensation – a so-called Wilson cloud – formed instantly, spreading out in all directions and obscuring the fireball of the explosion completely for a few seconds until it was dissipated by the heat. Nothing of the kind had occurred previously – during the two preceding tests the Wilson cloud had been ring-shaped, forming at approximately 2,500 m (8,200 ft). Post-test atmospheric and soil sampling (both at the site and along the path of the cloud as it was blown by the wind) showed that radiation levels were 109 times (!) lower than in the case of a ground explosion.

The Tu-4A landed safely at its home base. In his post-flight report Konstantin I. Oorzhuntsev described how the effects of the explosion were felt inside the aircraft. He also reported that no problems were experienced with flying the aircraft manually (as recommended by TsAGI and Andrey N. Tupolev) when the blast wave hit. No equipment failures occurred as a result of the explosion. We'll let bomb-aimer Boris D. Davydov, a participant in the mission, tell the story:

'The weather that day was good enough for me to see the target in timely fashion, take aim and drop the bomb accurately. All systems, including the transmitters sending radio signals which activated the ground equipment, functioned without a hitch. When the bomb was gone and the [bomb bay] doors closed the crew prepared for the flash and the blast wave: we disengaged the autopilot, drew the protective curtains and donned dark goggles, depressurized the cabins and put on our oxygen masks. We used a stopwatch to check the anticipated moment of detonation.

The first thing we knew was a tremendous flash. Then the first and quite powerful blast wave caught up with the aircraft, followed by a weaker second wave and a still weaker third wave. The flight instruments went crazy, the needles spinning round and around. Dust filled the cabin, even though the aircraft had been vacuumed clean before the flight. We watched as the cloud of smoke and debris grew; it quickly mounted to our own flight level, billowing out into a mushroom. There was every thinkable colour and shade to that cloud. I am lost for words to describe what I felt after I had dropped the bomb; I saw the whole world – everything I could see – through different eyes. I guess that was because I had my mind focused on this important mission – a mission I could not fail – for many days before the drop, and it just shut out everything else.

After landing we taxied to our special hardstand and climbed out, still wearing our parachutes and oxygen masks – we were still on pure oxygen from the bottles that came with the parachutes. The ground crew checked us and the aircraft for radioactive contamination; a washing-down station had been set up at the hardstand, and after showering and changing into fresh clothes we were taken by car to the headquarters to file our reports.'

The State commission declared that the Tu-4A carrier aircraft equipped with a bomb bay heating system, modified bombing equipment and other associated mission equipment permitted safe and reliable carriage of the RDS-3 nuclear bomb and accurate delivery of same. By decree of the Presidium of the USSR Supreme Soviet dated 8th December 1951 Lt.-Col. Konstantin I. Oorzhuntsev was awarded the Order of Lenin; Capt. Boris D. Davydov (promoted to Major after the test), Capt. Konstantin I. Oosachov, Lt. Al'vian N. Stebel'kov and Lt. Leonid A. Blagov were awarded the Order of the Red Banner of Combat. All the other crew members of both aircraft involved in the first test, as well as the test engineers and ground crews, also received government awards.

The successful first airdrop test prompted the decision to add nuclear weapons to the Air Force inventory and launch series production of the Tu-4A and the RDS-3 bomb.

In his memoirs, Leonid L. Kerber, a leading equipment specialist in the Tupolev OKB, recalls the atmosphere in which the nuclear tests involving the Tu-4A proceeded. *'Our OKB's representatives at the proving ground were Aleksandr A. Arkhangel'skiy and Aleksandr V. Nadashkevich. Communication with them was maintained via an HF telephone line (such lines were reserved for high-priority government communications – Auth.). On the planned day of the test drop I was summoned to the phone. The call was **from out there** (the proving ground – Auth.). Arkhangel'skiy was on the phone, and I could hear agitation in his voice: there was trouble with the aircraft's electric system; a fuse had blown and the test drop had had to be postponed. Lavrentiy P. Beria, who had arrived at the proving ground, was furious. "Tell us as soon as possible what the problem may be and how to fix it", Arkhangel'skiy told me.*

No sooner had I entered my study than I got a phone call from Nikolay I. Bazenkov, the Deputy General Designer. "Come to my office, and quick!" – he told me.

I ran there. There were three men in Bazenkov's office; they were in civvies, but I know an MGB operative when I see one, and these were clearly from that outfit. (Having been convicted on false charges, like many others, Kerber understandably had hard feelings towards the MGB – Auth.) Taking a quick look into a folder lying on the table in front of him, one of them asked me: "Why has a 50-ampere fuse blown on the Tu-4 aircraft?" In an extremely calm tone I told him: "There are several hundred Tu-4 aircraft around, and in several versions; they have thousands of fuses. Tell me the aircraft's construction number and where the aircraft is, then we will find its electric circuitry layout and try to give you an answer." Without changing his facial expression, the bugger monotonously repeated his question, word for word. Now, I really needed to return to my study and have a look at the layout in order to get to the bottom of the problem, so I turned to Bazenkov and put on a show of defiance. "If you are keeping me in the dark, I see no point in continuing this conversation. I cannot answer the question until I know the aircraft's construction number." With that, I headed for the door. One of the three agents silently got up and followed me. I entered the elevator, and so did the agent. We entered my study, where the layout was already on the table and my electrics specialists had gathered. While we were poring over the layout and discussing the problem, the dumbass agent stood by the door. His job was to make sure that I didn't escape.

Having pinpointed the cause of the problem, we descended [to the floor where Bazenkov's office was] and as we passed by the HF communications cubicle the operator hailed me: "There's again a call for you from the proving ground!" This time it was Engineer-Lt.-Col. S. M. Kulikov, a representative of [Lt.-]Gen. [Mikhail S.] Finogenov who was in charge of the test. Serafim Mikhaïlovich told me: "We've found the problem! The fuse in the bomb bay heating system circuit had been installed consecutively with a 20-ampere fuse, not directly on the main power supply cable as required. It was the 20-amp fuse that blew when we turned on the heating. We'll fix it ourselves if you agree – just give us an appropriate HF cable message endorsed by yourself and Bazenkov."

Two hours later we got another call – everything was working OK. During these hours I tried to get rid of the dumbass agent in order to call home and warn them that there could be a delay, just to save them the worry. Only when the message from the proving ground came in that everything had been checked and OK and the experiment had been scheduled for the following day did the dumbass leave me alone. When our "guests" left, Nikolay Il'yich and I recalled the recent years when we were working in detention within these very walls, and we had a very clear idea that we were walking the edge of an abyss.'

Service entry and more tests

In 1951 the Tu-4A achieved initial operational capability; 18 such aircraft were delivered to the Air Force that year. In keeping with Council of Ministers directive No.3200-1513 issued on 29th August 1951 the War Ministry set about forming a unit equipped with Tu-4As and armed with nuclear bombs at Balbasovo AB. Known initially by the cover name 'No.8 Training Unit' and assigned the field post office number 78724, it was commanded by Col. Vasiliy A. Tryokhin from the 52nd GvTBAP. This unit ousted the resident 45th TBAD from Balbasovo – as early as 7th November 1951 the division HQ and the three constituent regiments (52nd GvTBAP, 203rd GvTBAP and 362nd TBAP) urgently moved to other bases. Adding offence to injury, the 45th TBAD's best crews were hand-picked for inclusion into this so-called 'Atomic Air Group' which had usurped the base.

Next, construction units assigned to the MGB arrived at Balbasovo AB and set to work. The first thing they did was to put up a triple barbed wire fence around the base perimeter and erect watchtowers. Then they proceeded to construct nuclear bomb storage bunkers and special hardstands featuring the aforementioned trenches over which the bombers would be parked to facilitate loading of the big bombs. To prevent leaks of sensitive information, strict security measures were introduced in the garrison and enforced by MGB officers (special permission from the unit's command staff was needed to leave the garrison); this applied to all aspects of the unit's activities. To offset the inconveniences, the personnel had their salaries increased by 50%.

Once the reconstruction of the base was complete, Unit 78724 started taking delivery of Tu-4A bombers. The 'Atomic Air Group' started operations with a mere seven aircraft and crews, but gradually the complement was increased to 22 aircraft. In addition to real nukes, the unit received IAB-3000 and IAB-500 practice bombs (*imitatsionnaya aviabomba* – 'simulation bomb'); these were sim-

Trainees at a special Air Force training course are shown a nuclear bomb and given an explanation of its peculiarities.

ilar in size to HE bombs of the same calibre but were filled with a liquid flammable mixture to replicate reasonably accurately the flash of a nuclear explosion and the tell-tale mushroom cloud. Special training stands were set up and manuals for the use of nuclear bombs were handed down. Instructions were developed for the flight and ground crews, rigidly prescribing the entire operational procedure from the moment the order to deliver a nuclear strike came in to the moment when the bomb release button was pushed.

In 1952 a production bomber and an RDS-3 from one of the first production batches passed check-up tests. In 1953 the Ministry of Medium Machinery (MSM – *Ministerstvo srednevo mashinostroyeniya*) – the agency responsible for the Soviet nuclear program in all aspects, both military and peaceful – launched full-scale production of nuclear bombs, allowing deliveries to the DA to begin. (The nukes manufactured previously – two in 1949, nine in 1950, 25 in 1951 and 40 in 1952) were considered pre-production examples. The RDS-1 was manufactured in five copies only; according to other sources, the first 19 bombs had been built as RDS-1s but then converted to RDS-2 and RDS-3 specification. The RDS-2 was built in quantity but not formally included into the inventory.)

In mid-1953 four Tu-4As were again deployed to Zhana-Semey along with several other aircraft of the 71st Test Range to undertake live nuclear tests at the Semipalatinsk proving ground. In September 1953 they took part in the tests of the new RDS-5 implosion-type tactical nuclear bomb developed by KB-11; this was an experimental derivative of the production RDS-4 'Tat'yana' (8U69 or *izdeliye* 244N), from which it differed in having a composite plutonium/uranium core. As had been the case with the very first test, a pair of Tu-4As ('live' and backup) was involved, with Vasiliy Ya. Kutyrchev and Fyodor P. Golovashko as crew captains; the drops were performed at 9,000 m (29,530 ft). Three tests of different versions of the RDS-5 were held on 3rd September, 8th September and 10th September. On the first occasion the bomb exploded at 255 m (836 ft) AGL with a yield of 5.8 kt; in the other two cases the RDS-5 detonated at 220 m (722 ft) AGL, the yield being 1.6 kt and 4.9 kt respectively.

In addition to the actual drops, two such aircraft fitted with dosimeters were used to measure radiation levels near the mushroom clouds and at long distances from the hypocenter – right up to the Sino-Soviet border. On these missions the crews wore isolating gas masks to protect their lungs; tragically, on one occasion a defective gas mask cost one of the crewmen his life.

The year of 1953 also saw the continuation of flight tests for the purpose of evaluating the safety of take-offs and landings of the Tu-4A with the RDS-3 bomb. After all, a situation when the nuclear strike was aborted for whatever reasons and the aircraft would have to land with the bomb on board could not be ruled out.

Considering that the USA was harbouring plans of a nuclear attack on the Soviet Union (Operation *Dropshot*), in 1953 the Soviet government took the decision to establish several nuclear-capable bomber units within the Long-Range Aviation. In early 1954, pursuant to a directive issued by the DA Commander, a group of 78 persons was hand-picked from the aircrews and ground personnel of the 43rd VA. All of them were highly competent specialists with an excellent record of military discipline, unblemished moral character and no incriminating episodes in their biographies. With a few exceptions, all of them were Great Patriotic War veterans, and all had taken conversion training for the Tu-4. The group was dispatched to Balbasovo AB where Col. Nikolay I. Parygin was to set up the first regiment of nuclear bomb carriers – the 402nd TBAP. The personnel and aircraft of the resident 'Atomic Air Group' (Unit 78724), together with the newcomers, formed the core of this regiment. The task facing them was of truly national importance – for the first time in Soviet Air Force history it was necessary to prepare an air unit for nuclear warfare.

One more noteworthy event took place in 1953. A specially modified Tu-4 captained by S. V. Seryogin was used for testing a 'dirty bomb' – a radiological weapon. Two ballistic missile warheads filled with liquid radioactive materials were carried on underwing pylons.

In September 1954 the newly formed 402nd TBAP was commissioned, with Lt.-Col. V. N. Shevchenko as Chief of Staff and Lt. Col. Ye. N. Bashkatov as Deputy CO (Maintenance). Shortly afterwards, a second operational Tu-4A regiment – the 291st TBAP commanded by Col. N. M. Kalinin – was formed at Balbasovo AB; both units became part of the 160th TBAD, which was again commanded by Col. Vasiliy A. Tryokhin. This bomber division enjoyed a special status, reporting directly to the Air Force Deputy C-in-C, Lt.-Gen. Nikolay I. Sazhin.

Initially the mission preparation time was extremely long – it was nearly 24 hours before the aircraft could become airborne. This may have been due partly to the security procedures which hampered the work at times; in the 160th TBAD garrison alone there were nearly 50 MGB officers. Additionally, the first teams assembling and checking the atomic bombs were not Air Force personnel – they consisted largely of civilian MSM employees.

On the other hand, the crews' training levels were always high. Many of the 160th TBAD's aircraft captains (V. F. Martynenko, Konstantin K. Lyasnikov, Vasiliy Ya. Kutyrchev and others) later tested nuclear bombs at the 71st Test Range in Bagerovo.

In April-July 1954 seven Tu-4As captained by Zharkov, Korshunov, Kisilyov, Lootsik, Polyanin, Stroonov and Kholodov were again detailed for air sampling along the Soviet Union's western border and the Sino-Soviet border, measuring radiation levels in the wake of nuclear tests at Semipalatinsk. For this mission the aircraft captains were awarded the Order of the Red Banner of Combat.

Nuclear exercise

On 14th September 1954 the Soviet Union held the first all-arms exercise involving actual use of nuclear weapons for the purpose of training the troops to operate in a nuclear environment – Exercise *Snezhok* (Snowball). It should be noted that Exercise *Snezhok* was not an unprecedented event; military exercises involving the use of live nuclear weapons were held both before and after it. Specifically, the USA had held Operation *Buster-Jangle* in 1951 (a series of seven tests ranging from 1.2 to 31 kt at the Nevada Proving Ground between 22nd October and 29th November) and Operation *Upshot-Knothole* in 1953 (eleven tests ranging from 0.2 to 61 kt at the same location between 17th February and 4th June). Later, France held the 5-kt test *Gerboise Rouge* (Red Jerboa) on 27th December 1960 and the 1-kt test *Gerboise Verte* (Green Jerboa) on 25th April 1961 in the Algerian Sahara desert.

Exercise *Snezhok* merits a few more details. It took place at the artillery range in Totskoye in the Orenburg Region (in the south of the Urals Mountains). The location was chosen because the terrain there resembled the typical terrain of Western Europe – according to Soviet military theoreticians, the most likely place for the Third World War to begin. The purposes of the exercise were: to test the effects of a nuclear blast on a fortified line of defence, as well as on military materiel and on animals; to verify a possible scenario of an offensive by own forces in direct contact with the adversary without withdrawing them from the line of contact prior to the nuclear strike; and to teach the personnel (enlisted men as well as commanders) how to conduct offensive and defensive operations in a nuclear scenario.

According to the plan of the exercise, the target replicated exactly the defensive positions of a US Army battalion; hence

numerous fortifications were built. The nuclear blast would be followed by an attack by strike aircraft and finally an assault by a mechanised infantry regiment. The exercise involved a total of 45,000 servicemen, 600 tanks and SP guns, 500 artillery pieces, 600 armoured personnel carriers, 320 assorted aircraft and 6,000 automobiles.

The exercise was commanded by Marshal Gheorgiy K. Zhookov, the Commander of the Soviet Army's Ground Forces; however, he was present only at the 'dress rehearsal' (without nukes), not at the actual exercise. The entire Soviet top military establishment (the commanders of all armed services, Military Districts and Air Defence Districts), as well as the invited ministers of defence from 'friendly nations', would be watching from a command post located 2 km (1.24 miles) from the 'forward line of own troops'.

The original intention was that the brand-new Tu-16A would drop the bomb. However, the *Badger* showed disappointing results

The captain of a Tu-4A during a practice sortie.

A pair of Tu-4s en route to the target during a nuclear test mission.

during practice runs: dummy bombs missed the target by as much as 700 m (3,000 ft), creating a considerable danger that the invited guests could be caught by the blast. Hence the proven Tu-4A offering greater accuracy was chosen. The 71st Test Range assigned to this task an air group which included two 35th BAP Tu-4As flown by top-notch crews captained by Capt. Konstantin K. Lyasnikov and Lt.-Col. Vasiliy Ya. Kutyrchev. (The latter's last name is occasionally quoted as Kutyrichev; also, he has been reported in some sources as a 226th TBAP pilot.) In addition to excellent flying skills, these crews had accumulated much experience of flying with dummy nukes during the tests at Bagerovo, and the crew captained by Kutyrchev had participated in two live nuclear tests at Semipalatinsk in 1953. Each Tu-4A had three navigators – V. V. Babets, L. V. Kokorin and V. G. Batov in Kutyrchev's crew, A. N. Kirilenko, K. K. Revin and R. I. Shergin in Lyasnikov's crew. The group further included two IL-28Rs that were to photograph the drop, six MiG-17s providing cover, a Li-2 and a Yak-12. It was commanded by the 71st Test Range's new CO, Maj.-Gen. Viktor A. Chernorez.

On 3rd August, well in advance of the exercise, the group redeployed from Bagerovo AB to Vladimirovka AB, a GK NII VVS facility in Akhtoobinsk (Astrakhan' Region, southern Russia), and started practicing; this base located near the estuary of the Volga River was some 900 km (559 miles) from the target. Starting on 5th August, systematic weather observations were made to determine the prevalent winds in the area; this was no idle interest – on the day of the drop the wind had to be in the south, south-east or south-west for safety reasons. Secure communications channels were organised. Once the ground preparations had been completed, practice flights began on 12th August, involving at least ten drops of inert bombs at the range. Since the MiG-17 lacked the range to escort the Tu-4A non-stop from Vladimirovka to Totskoye, four of the fighters were redeployed to a forward base at Sorochinsk (Orenburg Region); these would take over when the first pair ran low on fuel.

This time it was literally 'X marks the spot': a square measuring 150 x 150 m (490 x 490 ft) was outlined in white on the ground, with a white cross in the middle as the target for the bomb-aimer. Three radio beacons were set up along the bomber's route. Additionally, due to the lack of landmarks along the route several pyrotechnic pickets were set up, which were to ignite orange-coloured smoke cartridges at the right moment to assist the bomber crew. Finally, two angle reflectors were installed 72 km (44.74 miles) from the target as radar landmarks.

Meanwhile, a huge amount of fortifications were built at the range – 383 km (238 miles) of trenches, 394 pillboxes, 140 light shelters etc. All these measures were an absolute necessity, as the participating troops would be just 5 km (3.1 miles) from the hypocenter of the explosion.

Concurrently with the preparation of the bombers, two RDS-3 (some sources say RDS-2) nuclear bombs were prepared. A few days before the exercise, Academician Igor' V. Kurchatov arrived at Vladimirovka AB. He checked the readiness of the bombs and the Tu-4As, as well as the readiness of the crews. Later the pilots recalled warmly the meeting with him, his good humour and his parting words with good wishes for a successful mission. On the

day of the exercise, 14th September, each of the two aircraft had one nuke on board and each crew was ready to perform the task. By the scheduled time of departure both Tu-4As were in maximum readiness with engines running, their crews awaiting the order to take off. It was the crew captained by Lt.-Col. Kutyrchev that received the order; the back-up aircraft remained on the ground.

At 0600 hrs Moscow time it was announced that the bomber was on its way. Meanwhile, the weather started changing for the worse – a stiff wind rose and the sky clouded up. Ten minutes before T-time an atomic alert was sounded; the personnel on the ground took cover, and the crews of armoured vehicles battened down the hatches. The Tu-4A crew performed their mission perfectly: the bomb was released by bombardier Leonid V. Kokorin at 8,000 m (26,250 ft) – the minimum altitude at which the bomber could escape the blast wave – at 0934 hrs. 48 seconds later the bomb exploded at the stipulated height of 350 m (1,150 ft) AGL. This was followed by two simulation nukes for sheer effect.

The nuclear attack was followed five minutes later by a 25-minute artillery attack on the Blue Force positions, then an attack by tactical bombers and MiG-15 jet fighters from the 165th IAP, some of them flying right through the pillar of the mushroom cloud. Scouts sent into the area immediately after this reached the hypocenter 40 minutes after the blast; they established that the radiation level there one hour after the explosion was 50 R/h, decreasing to 25 R/h at 300 m (990 ft) from the hypocenter, 0.5 R/h at 500 m (1,640 ft) and 0.1 R/h at 850 m (2,790 ft). Finally, at 1010 hrs a mechanised infantry assault began. According to some sources, the Red Force personnel advancing through the contaminated zone was issued full nuclear, biological and chemical (NBC) protection suits; other sources say they had only the usual gas masks. (In Soviet/Russian exercises, Red Force is the 'good guys' and Blue Force is the 'bad guys'; in NATO exercises it is vice versa.)

When reviewing and summing up the results of the exercise, Marshal Zhookov gave a high appraisal to the work of the crew of the Tu-4A. As a reward for the successful fulfilment of the mission, Lt.-Col. Vasiliy Ya. Kutyrchev was promoted to the rank of Colonel and awarded the Order of Lenin by a decree of the Presidium of the USSR Supreme Soviet. The other crew members also received State awards.

After the exercise, the 71st Test Range carried on with its work. On 23rd October 1954 an upgraded RDS-3I bomb featuring an external neutron initialisation system was dropped at the Semipalatinsk proving ground by a Tu-4A. The bomb detonated at 410 m (1,345 ft) AGL, with a yield of 62 kt. The weather was cloudy that day, and the low cloudbase made it impossible to observe some phases of the mushroom cloud's development.

One more version of the RDS-5 featuring an external neutron initialisation system was tested on 30th October 1954. This time the bomb detonated at just 55 m (180 ft) AGL, with a yield of 10 kt.

In October 1954 two 226th TBAP Tu-4As captained by L. P. Lootsik and M. L. Bondarenko again undertook air sampling after a nuclear test at Semipalatinsk, flying right through the pillar of the mushroom cloud. The same aircraft also measured radiation levels after the cloud dissipated and followed it as it was borne on the wind all the way to China.

Chapter 8

Chairman Mao's *Bulls*: The Tu-4 in China

The Tu-4 was not only the Soviet Union's first true strategic bomber but also the first Soviet strategic bomber to be exported. The sole export customer was China, which was the Soviet Union's principal ally in the early stages of the Cold War. When the last Chinese Civil War ended in a victory of the Communists led by Mao Tse-tung in 1949 and the People's Republic of China (PRC) was proclaimed, the opposing Kuomintang Nationalists led by Chiang Kai-shek left the mainland and settled on Taiwan, declaring the Republic of China (RoC). Since then, Beijing and Taipei have been in a perpetual state of armed confrontation which periodically escalated into open warfare, the US-backed Nationalists making forays against mainland China while the latter sought to capture Taiwan. The Soviet Union immediately extended military aid to the PRC, helping to build up and train the People's Liberation Army Air Force (PLAAF, or *Chung-kuo Shen Min Taie-Fang-Tsun Pu-tai*; now rendered as *Zhòngguó Rénmín Jiěfangjùn Kòngjùn*) which was created on 11th November 1949.

The situation in the region was taking an alarming turn. The Korean War, in which China was implicitly involved, was on and the conflict with Taiwan showed no prospect of ending. In 1953 a total of 25 Tu-4 nuclear-capable long-range bombers were delivered to the PLAAF free of charge (Stalin and Mao had reached an agreement in principle, providing for a transfer of nuclear weapons technology to China). Some of the bombers were built by plant No.22 (the only Kazan'-built Chinese Tu-4 identified so far is a mid-production aircraft from Batch 50 and thus was obviously transferred from Soviet Air Force stocks, as were many other early PLAAF aircraft); others came from the final Kuibyshev-built batches and were possibly delivered new.

The bombers served with the 4th Independent Regiment based at Shijiazhuang, southwest of Beijing in Hebei Province, and wore four-digit-serials commencing 4. The Chinese crews were trained on site by a team of Soviet instructors commanded by Col. Aleksandr A. Balenko (HSU), a distinguished bomber pilot who had commanded the DA's 22nd GvBAP during the Great Patriotic War. The Chinese Minister of Defence and 'founding father' of the People's Liberation Army, Marshal Chu Teh, kept a tab on the training program. On arrival the Soviet instructors were given a formal reception and banquet at Chairman Mao's residence in Beijing. Apparently the Nationalist intelligence service was earning its keep, because at the reception the officers were shown – in a

'speaking-of-which' manner – a Taiwanese newspaper containing an item about their impending visit. The item featured Balenko's portrait and put a price on his head – 'wanted dead or alive'.

The first hurdle to be overcome was, of course, the language barrier; next, strict discipline among the trainees had to be enforced. What they lacked in special education, the Chinese made up for in steadfastness. The Chinese command also used peculiar incentives to make the trainees learn harder. If the bombs missed their target, the crew would stand accountable and would be required to give a credible answer as to why engine life, fuel and ordnance had been expended unproductively. If they failed to give a satisfactory answer, the entire crew would go to the brig for two hours to think about it.

One of the few breaches of discipline by the Chinese aircrews was that the airmen would consume their in-flight ration – a big bar of chocolate – immediately after the bomber had climbed to the assigned altitude en route to the target. Hence the Chinese command issued an order allowing the crews to eat their chocolate only after the bombs had been delivered – and only if the control

Andrey N. Tupolev shakes hands with Chairman Mao Tse-tung during a Soviet delegation's trip to China in 1959.

Above: One of the 25 Tu-4s transferred to the PLAAF starts its engines at Shijiazhuang. The c/n appears to be 2806207, which would make this aircraft '4003 Red'. Note the open escape hatch serving as an air intake for the APU.

Below: A poor-quality but interesting shot of a Chinese Tu-4 delivering a bomb strike.

tower and the command post at the target range confirmed that the target had been hit. Earn your candy, boys… just like in prep school, for goodness sake!

Conversion of Chinese crews was still in progress when the Tu-4s started their combat operations, delivering bomb strikes against the breakaway island. Interestingly, entries in the flight documents read *'training flight with live bomb release over the target range – the island of Taiwan'*. Initially the crews were captained by Soviet pilots on such assignments; Col. Aleksandr A. Balenko was no exception, flying four sorties against Taiwan. The Soviet instructor staff returned home in 1954.

The Chinese duly rewarded the Soviet instructors for their contribution – for example, Balenko was given a personal car, and upon return to the Soviet Union he was awarded the Order of the Red Banner of Combat, ostensibly 'for prolonged meritorious service'. He retired in 1961 in the rank of Major-General.

The PLAAF intended to use the Tu-4s as nuclear weapon carriers, but the matter took a long turn. In 1957 the Soviet Union granted China a licence to build the Tu-16 twin-turbojet bomber, and it seemed that the days of the Tu-4 were numbered. However, shortly thereupon Sino-Soviet relations went sour due to ideological differences. Added to that, Mao's 'Great Leap Forward' policy of accelerated industrial development (placing the emphasis on quantity to the detriment of quality) and the notorious 'Cultural Revolution' had a profound negative effect on the mastering of new technology in China, including the aircraft industry. As a result, the production and service entry of the *Badger*'s Chinese clone – the Xian H-6 – was delayed, and the *Bull's* operational career in PLAAF service proved to be unexpectedly long.

Even though the Chinese *Bulls* never did carry nukes in service, in May 1965 a PLAAF Tu-4 had the distinction of dropping the first indigenously developed Chinese atomic bomb – thus repeating a page in the Tu-4's Soviet Air Force career 14 years earlier. The weapon had a yield of 40 kilotons.

The *Bulls* were used not only as bombers but also for long-range maritime reconnaissance and transport duties – in similar manner to the Soviet Air Force's Tu-4R and Tu-4D respectively.

Top: Tu-4 '4134 Red' (c/n 225008) re-engined with WJ-6A turboprops sits outside the entrance to the PLAAF Museum at Datangshan. A while later it would be further converted as a drone launcher.

Above: '4134 Red' prepares to fly a mission with WZ-5 (Chang Hong-1) drones suspended under the outer wings.

Right: The same aircraft on display at the PLAAF Museum in company with Shenyang J-5 (licence-built MiG-17F) fighters. Note the white anti-flash colouring of the undersurfaces.

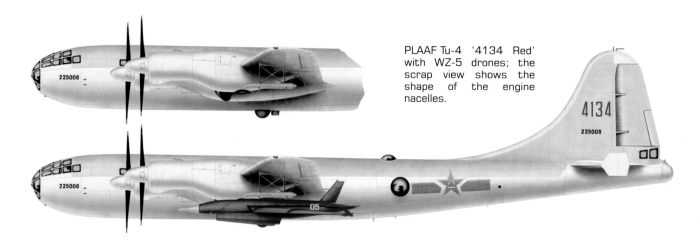

PLAAF Tu-4 '4134 Red' with WZ-5 drones; the scrap view shows the shape of the engine nacelles.

The Chinese set about updating the bomber: they installed the SRO-2 *Khrom-Nikel'* (Chromium-Nickel; NATO codename *Odd Rods*) IFF system, the *Sirena-2* (Siren) radar homing and warning system (RHAWS) and other pieces of equipment copied from Soviet models. However, there were also several locally developed variants of the Tu-4 which mostly were unique to China, having no Soviet equivalents. These are described below.

Tu-4 heavy interceptor version

In Chapter 4 a description has been given of the G-310 heavy interceptor based on the Tu-4. However, that was a development aircraft, whereas in the early 1960s the Chinese actually used the Tu-4 as a heavy interceptor – albeit a less sophisticated one.

In 1951 the US Central Intelligence Agency (CIA) had started performing clandestine reconnaissance/psychological warfare sorties over mainland China from Taiwan, conducting signals intelligence (SIGINT) and dropping anti-Communist propaganda leaflets. These were flown by piston-engined aircraft – initially B-17s and later Lockheed P2V Neptune maritime patrol aircraft – equipped with SIGINT systems and known as 'Black Bats'. While a few B-17s were intercepted and shot down by Chinese fighters during such missions, the P2V proved impossible to intercept, all Chinese radar pickets and alerts from fishing vessels notwithstanding. By carefully choosing the point where they crossed the coastline (known in slang as 'coast-in') the spyplane's crew was able to slip through the belt of AA guns along the coast, and eavesdropping on the Chinese GCI system warned them when fighters were scrambled.

When the PLAAF's then-latest MiG-15*bis* *Fagot-B* and MiG-17F *Fresco-C* jet fighters failed to intercept the intruders, the Chinese tried using three specially modified Tu-2S tactical bombers, but these had no better luck. The reason was that neither type was equipped with an airborne intercept radar. Hence in early 1960 one of the PLAAF commanders suggested adapting the Tu-4

Opposite page: Two more views of Tu-4 '4134 Red' at Datangshan with a pair of WZ-5s under the wings. Note the small hexagonal endplate fins on the stabilisers. The aircraft looks very weathered and some damage to the rudder is evident, but the PLAAF 'stars and bars' insignia have been recently refreshed.

Top right: Close-up of the unique bottle-shaped engine nacelles of the Chinese Tu-4s converted to 'Turbo *Bulls*'.

Above right: Close-up of the port WZ-5 and its lattice-like pylon. The drone on the other side is coded '06 White'.

Right: This photo of '4134 Red' taken at an earlier date shows the dished fairing supplanting the DK-3 tail turret and the slightly kinked trailing edge of the endplate fins.

for this role. The Tu-4 was comparable in speed to the P2V, which could do 555 km/h (345 mph), and had an endurance of more than six hours allowing it to pursue the intruder for a long time. PLAAF C-in-C Liu Ya-lou agreed.

The conversion involved relocating the Kobal't-M radar to a fixed dorsal position aft of the flight deck (the radome supplanted the dorsal forward cannon barbette, reducing the cannon armament to eight NR-23s). The gunners' stations were equipped with infra-red sights having a detection range of up to 3.2 km (2 miles) allowing the target to be acquired visually at night. The bomb bay was converted to a command post with a radar screen and places for an airborne intercept officer, two navigators and two chart plotters. The first bomber thus modified underwent tests at Wukong AB, Shaanxi Province, in the course of a month; the results were encouraging, and three more Tu-4s were converted in this fashion.

The Tu-4 'interceptor' was first used operationally as early as the night of 1st March 1960. A P2V 'Black Bat' overflew Jiangsu, Henan and Anhui provinces, recording no fewer than 216 enemy radar signals and distributing nearly 1,000 kg (more than 2,000 lb) of leaflets along the route. Of course, numerous MiG-17s were scrambled to intercept (unsuccessfully); so was a Tu-4 which followed the intruder across Henan and Anhui provinces but never got within attack range.

Tu-4 turboprop conversions
When the bombers' ASh-73TK engines ran out of service life and the stock of spare engine was depleted, the PLAAF decided to give the remaining *Bulls* a new lease of life by re-engining them. Thus, in the mid-1970s several Tu-4s were fitted with 4,250-ehp Zhuzhou Engine Factory WJ-6 turboprops (*wojiang* – 'turboprop engine,

Opposite page, above: A rare shot of the KJ-1 AEW&C testbed in a test flight. Note the black patches on the flaps aft of the engine jetpipes making the exhaust stains less conspicuous.

Below and below left: The same aircraft on display in the PLAAF Museum at Datangshan. Note the extra ventral radomes.

Top left: The tail of the KJ-1, with a ventral fin augmenting the endplate fins.

Top: The KJ-1's rotodome and its support structure.

Above and above left: Two more views of the KJ-1 at the museum.

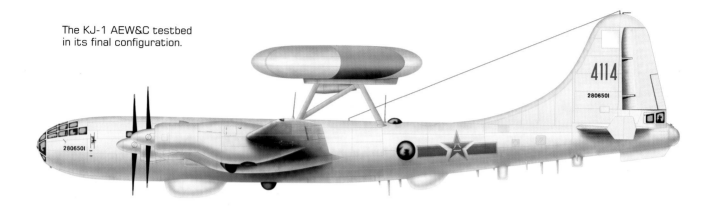

The KJ-1 AEW&C testbed in its final configuration.

Type 6') driving J17-G3 four-bladed reversible-pitch propellers (*jiang* – propeller) of 4.5 m (14 ft 9^{11}/$_{64}$ in) diameter turning clockwise when seen from the front. The WJ-6 and the J17-G3 were locally-made copies of the Ivchenko AI-20K turboprop and the Stoopino Machinery Design Bureau (SKBM) AV-68I propeller; these were fitted to the Antonov An-12BP transport, a derivative of which was built in China under licence as the Shaanxi Y8.

The new engines were markedly smaller in cross-section but much longer than the original ASh-73TK radials, being 3.096 m (10 ft 1^{57}/$_{64}$ in) long, 0.842 m (2 ft 9^{9}/$_{64}$ in) wide and 1.18 m (3 ft 10^{29}/$_{64}$ in) high. This meant that new engine bearers had to be made and fairings had to be fabricated to make the transition from the existing nacelles to the cowlings, resulting in weird-looking nacelles with a pronounced bottle shape. The jetpipes were located on the outer faces of the nacelles beneath the wing leading edge. Another problem was that the AI-20K had almost twice the power of the ASh-73TK, giving stronger yaw in the event of a single-engine failure. To compensate for this, small pentagonal endplate fins were added to the tips of the horizontal tail for added directional stability.

The Tu-4 soldiered on with the PLAAF far longer than with the Soviet Air Force; as many as 15 'Turbo *Bulls*' were reportedly still on strength in 1991! At least two such bombers were further converted into special mission aircraft described below.

Tu-4 – Chinese drone launcher version

At least one of the re-engined Tu-4s, a Kazan'-built example serialled '4134 Red' (c/n 225008), was further converted into a drone launcher aircraft which was broadly equivalent to the Tu-4NM described in Chapter 4, except for the powerplant type. Two racks were similarly fitted under the outer wings; however, they were simple racks attached via truss-type structures, not ejector racks, and the type of drone was different as well. True, the Chinese did receive the La-17 from the Soviet Union and eventually copied it as the CK-1, but this was strictly ground-launched. In contrast, the Chinese Tu-4 drone launcher carried Chang Hong-1 (Long Rainbow-1) reconnaissance drones, aka WZ-5 (*wuren zhenchaji* – unmanned aerial vehicle). The Chang Hong-1 was a reverse-engineered version of the Ryan BQM-34 Firebee, a number of which had fallen into Chinese hands after being downed during reconnaissance missions over China in the late 1960s. Thus, the aircraft could perhaps be described as the Chinese counterpart of the Lockheed DC-130 Hercules, which also carried the Firebee.

Tu-4 '4134 Red' is now preserved at the PLAAF Museum at Datangshan north of Beijing (also referred to as Xiaotangshan).

KJ-1 AEW&C testbed

One more 'Turbo *Bull*', this time a Kuibyshev-built Tu-4 serialled '4114 Red' (c/n 2806501), was converted into an aerodynamics testbed for a prospective airborne early warning and control (AEW&C) system known as the Kongjing-1 ('Sky Cloud') or KJ-1 for short. A conventional rotodome with two dielectric segments was mounted above the wing center section on two N-struts with bracing wires in between; however, no radar was ever installed in it. A while later, a fairly large dielectric canoe fairing was added aft of the nosewheel well, with a smaller canoe fairing, two thimble fairings and several blade aerials aft of the wings. A small ventral fin was added to improve directional stability, augmenting the existing endplate fins.

Upon completion of the test program the AWACS testbed was likewise relegated to the PLAAF Museum. Interestingly, it has been reported that neither of the two *Bulls* exhibited at Datangshan had any non-standard equipment when they were first noted there in 1990 and that they were active aircraft involved in research and development programs when not on display!

Known Tu-4s delivered to China		
C/n	Serial	Notes
225008	4134 Red	Kazan'-built. Re-engined with WJ-6 turboprops; converted into drone launcher aircraft. Preserved PLAAF Museum, Datangshan
2806007	4074 Red	Kuibyshev-built
2806008	4005 Red	
2806010	4104 Red	
2806207	4003 Red	
2806208	?	
2806210	?	
2806301	4001 Red	
2806501	4114 Red	Re-engined with WJ-6 turboprops; converted into KJ-1 AWACS testbed. Preserved PLAAF Museum
2806508	4124 Red	

Chapter 9

Transport derivatives

Tu-70 airliner ('aircraft 70', Tu-12)

In early 1946 the Tupolev OKB began development of a four-engined airliner which received the in-house designation 'aircraft 70'. The B-4 (Tu-4) bomber was chosen as the starting point. The idea of creating an airliner derivative of a combat aircraft then in first-line service seemed extremely appealing in the 1940s and 1950s, especially in the Soviet Union where operating economics did not enjoy the highest priority at the time. (Later the Tupolev OKB pursued this approach with the Tu-104 *Camel* twin-turbojet medium-haul airliner and the Tu-114 *Rossiya* (Russia; NATO codename *Cleat*) four-turboprop long-haul airliner which were spinoffs of the Tu-16 and the Tu-95 respectively.)

Experience gained with the 'donor' bomber in the process of its development, testing and operational use made it possible to design and put into service an airliner based on the said bomber with minimum technical risk and a very high degree of operational

safety. Structural commonality was another important factor, meaning that no drastic changes of manufacturing technology and tooling at the OKB's experimental plant and eventually at the production factory would be required; this, in turn, reduced production costs and helped keep unit costs down. Finally, skilled flight crews which had trained in earnest on the bomber serving as the progenitor would be available, which again enhanced flight safety.

Construction of a full-size mock-up started in February 1946. As already mentioned, on 26th February 1946 the Council of People's Commissars issued directive No.472-191ss, followed by MAP order No.159ss dated 27th March 1946, which required the OKB to complete a set of manufacturing documents and transfer these to the Kazan' aircraft factory. The MAP order included the following item:

'[Chief Designer and Director of plant No.156 A. N. Tupolev shall:]

An air-to-air of the Tu-12. Note the varying type of windows; the paired rectangular windows are for the de luxe compartments and the cafe area, the large circular ones are for the main cabin, and the small ones are for the galley (amidships) and the toilet facilities.

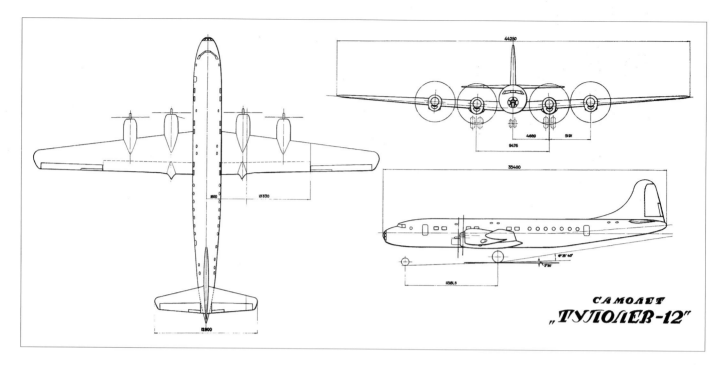

Самолет „ТУПОЛЕВ-12"

3. *Design a passenger version of the B-4 four-engined aircraft having the following performance:*

- *maximum speed at sea level, 470 km/h [292 mph];*
- *maximum speed at 10,500 m [34,450 ft], 560 km/h [348 mph];*
- *normal all-up weight, 47,000 kg [103,620 lb];*
- *maximum all-up weight, 56,000 kg [123,460 lb];*
- *range at maximum AUW, 5,000 km [3,105 miles];*
- *service ceiling, 10,900 m [35,760 ft];*
- *seating capacity, 47 passengers.*

A single prototype shall be built and submitted for [state acceptance] trials on 1st May 1947.'

The aircraft was officially designated Tu-12 the first aircraft thus designated. A decision on whether to proceed with series production would depend on the test results.

To speed up prototype construction it was decided to use, inasmuch as possible, the components of B-29-15-BW '365 Black' which had been dismantled in order to study the design; usable parts salvaged from the wreckage of B-29A-1-BN 42-93829 also came in handy. By early 1946 the Tupolev OKB's experimental plant (MMZ No.156) and the OKB's Kazan' branch had completed the task of reassembling the components of '365 Black'. B-29 parts used for building the Tu-12 included the outer wings, engine nacelles, flaps, main gear units and tail surfaces. The wing center section was new, since the Tu-12 was a low-wing aircraft as opposed to the mid-wing B-29 and had 0.81 m (2 ft 7⁵⁷⁄₆₄ in) greater span. The wings were built in seven pieces instead of three – additional production breaks were introduced at the fuselage sides and just inboard of the outer engine nacelles.

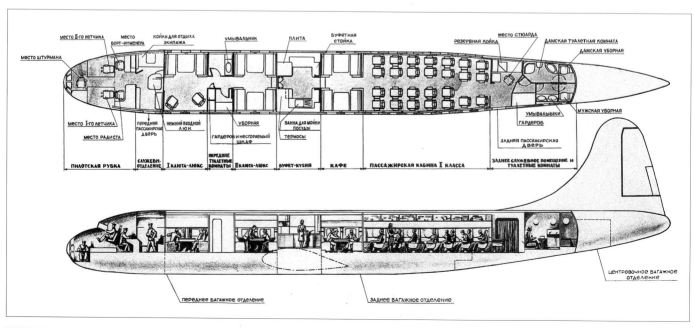

Left: A three-view drawing of the Tu-12 from the ADP documents.

Right: The sole Tu-12 nearing completion at the OKB's prototype construction facility, MMZ No. 156, in Moscow.

Below right: Another aspect of the Tu-12 in the assembly shop of MMZ No. 156, almost hidden in a maze of work platforms.

Below left: The interior layout of the Tu-12's 40-seat version from the ADP documents. Left to right: flight deck with a five-man crew (navigator in the extreme nose, captain and aft-facing radio operator to report, first officer and aft-facing flight engineer to starboard); front vestibule with entry door to port, crew rest area to starboard and ventral entry hatch; first de luxe compartment; forward toilet (port) and washroom (starboard); second de luxe compartment; galley (with thermos jug and kitchen sink to port and stove and buffet stand to starboard); cafe compartment; 1st class cabin; rear vestibule with entry door to port, flight attendant's seat and reserve bunk to starboard; wardrobe and men's room to port, ladies' room to starboard. The side view shows the two underfloor main baggage compartments and the auxiliary baggage compartment (filled as required for CG reasons) aft of the rear pressure dome.

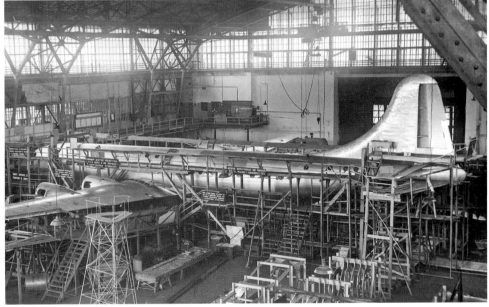

Below: The Tu-12 at the premises of MMZ No. 156 after roll-out.

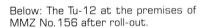

The Tu-12 (with appropriate nose titles) is rolled out at Moscow-Khodynka, where it was reassembled prior to the maiden flight. A Yer-2 bomber, an IL-4 bomber and a pair of C-47s are parked nearby.

The fuselage was also new, with a maximum diameter of 3.6 m (11 ft 9^{47}/$_{64}$ in) versus the bomber's 2.9 m (9 ft 6^{3}/$_{16}$ in). Unlike Boeing, which developed the Model 377 Stratoliner from the same basic bomber, the Tupolev OKB opted for a conventional circular-section fuselage, not a 'double-bubble' structure. For the first time in Soviet aircraft design practice the fuselage was completely pressurized, with a pressure differential of 0.57 kg/cm² (8.14 psi); flush riveting was used, with sealed rivets and Thiokol rubber strips to seal the manufacturing joints. We may as well say at this stage that upon completion the fuselage was tested for leaks by pressurising it to the design pressure differential – again for the first time in the Soviet Union.

The mock-up was completed by June 1946 and approved by the mock-up review commission of the Civil Air Fleet (GVF – *Grazhdahnskiy vozdooshnyy flot*; this was then the official name of the airline doing business as Aeroflot). In October 1946 MMZ No.156 completed the 'aircraft 70' (Tu-12) prototype; following comprehensive systems checks the wings and tail unit were detached and the aircraft was trucked to Moscow's Central airfield named after Mikhail V. Frunze (Moscow-Khodynka) for reassembly and further checks. Interestingly, the aircraft wore Soviet Air Force insignia (but no serial) and 'Tu-12' nose titles. By the time 'aircraft 70' arrived at Khodynka the B-29 pattern aircraft

('358 Black') was already there, having flown in from Izmaïlovo AB a while earlier; thus any snags associated with the airliner's equipment and systems could be resolved on site by checking up against the B-29.

The ADP was officially completed on 21st October 1946. According to the project documents the Tu-12 was a four-engined airliner designed for international air services (including intercontinental ones) and long-haul domestic air services. The ability to carry a large number of passengers in comfort over long distances at high speed made the Tu-12 a first-rate heavy airliner.

Comfortable conditions for the passengers and crew were ensured by the pressurized cabin and flight deck which enabled the aircraft to cruise at up to 10,000 m (32,800 ft), staying well above the clouds and turbulence. The cabin was furnished with comfortable seats, featuring a heating/ventilation system and a galley. Three interior configurations were offered; one was a VIP aircraft for government use, with the rear cabin configured as an office and conference room for the 'main passenger' (that is, the head of an official delegation). The second version had a mixed layout with basically four-abreast first class seating in the rear cabin while the forward cabin was divided into two de luxe compartments. The aircraft could be configured with either 48 seats for day flights or 40 seats, six of which could be transformed into sleeping berths, for overnight flights. Finally, the third (baseline) version seated 72, featuring four-abreast first class seating in both cabins. It was decided that the Tu-12 would be used for 'special' (that is, VIP and other non-scheduled) flights during the initial production stage.

The Tu-12 had a crew of seven or eight comprising the captain, the first officer, the navigator, the flight engineer, the radio operator and two or three flight attendants, one of which was the 'ship's cook' (!). The flight crew sat in a flight deck featuring a two-level arrangement; the navigator sat, facing left, slightly below the others in the extreme nose which was extensively glazed, like the bomb-aimer's station on a heavy bomber. This arrangement (which became characteristic of subsequent Tupolev airliners up to and including the Tu-134 *Crusty* twin-turbofan short-haul airliner of 1963) allowed Air Force crews trained on the Tu-4 to fly the Tu-12 during special operations or if the airliner was requisitioned by the military in times of war for use as a troop/cargo transport. The flight engineer and radio operator sat behind the pilots, facing aft.

A crew rest area with a tip-up sofa and two bunks was located on the starboard side opposite the forward entry door. Apart from the latter, the aircraft's interior was accessible via a hatch in the nosewheel well leading to the forward vestibule; this would be used at locations where normal airstairs were unavailable, such as military or industry airfields. Further aft was the forward passenger cabin separated from the rear cabin by the galley located over the wing center section. In the 40-seat mixed configuration the forward cabin was divided into two luxury compartments, with the forward toilet facilities in between. The first compartment featured a pair of sofas in a club-four arrangement (with a table in between) to starboard, which could be folded down, transforming into a comfortable bed; a fold-away bunk was located above these. On the port side was a table and two revolving armchairs. The toilet facilities included a washroom to starboard, a lavatory and a wardrobe to port; the wardrobe featured a strongbox for keeping valuables.

The second luxury compartment was similar to the first one, except that a second pair of transformable club-four sofas and fold-away bunk were located on the port side. The entire forward cabin could be isolated from the crew rest area and the galley by internal doors when a delegation was carried – presumably to prevent the crew from eavesdropping! Aft of the wings was a cafe (dining area) with two tables and club-four sofas, followed by the rear cabin with first class seating for 26 (four-abreast, except the last two rows, which were three-abreast), a cabin crew rest area, a wardrobe and the rear lavatories (separate for gents and for ladies).

Three views of the Tu-12 during manufacturer's flight tests. The aircraft had a pronounced nose-up ground attitude.

Top left: The two rearmost rows of seats in the main cabin; these were three-abreast, unlike the other five rows. Note that the headrest covers are stamped 'Tu-70'.

Top: The club-four seating in one of the forward de luxe compartments.

Above left: The two revolving armchairs and table on the port side of the first de luxe compartment.

Above: The dining area, with the galley visible beyond. Note the encrusted table and the high-quality upholstery.

Left: The main cabin of the Tu-12 mock-up, looking towards the nose. Oddly, the overhead cabin luggage racks are of unequal length.

Top: The rear cabin of the actual aircraft.

Above: The first officer's workstation of the Tu-12.

Top right: The passage to the navigator's station. Note the astrodome and the emergency brake handles near the captain's seat.

Right: The flight engineer's workstation.

The flight deck and passenger cabins formed a single pressure cabin occupying almost the entire fuselage. Air for the pressurisation/ventilation system was supplied by the TK-19 engine superchargers; before entering the cabin it was cooled as required in air/air heat exchangers. Warm air was distributed in the cabins by overhead ducts, excess air exiting via automatic bleed valves under the cabin floor; this ensured that the cabins were evenly heated and ventilated. Two automatic pressure regulators maintained a constant cabin air pressure from 2,500 m (8,200 ft) right up to 9,000 m (29,530 ft); if these failed the flight engineer could still control the air pressure by means of a manual regulator and an emergency bleed valve. The maximum air delivery rate of 2,000 m³/sec (70,630 cu. ft/sec) made it possible to prevent total decompression even if a sizeable air leak developed. A self-contained heater and electric blowers powered by the M-10 emergency generator/APU served for pre-heating the cabins in winter prior to passenger embarkation.

In all other aspects the Tu-12 was identical to the B-4, utilising many of the bomber's airframe components, the same powerplant and much the same equipment. The wings, for instance, were stock B-4 subassemblies, save for the center section which was designed from scratch. Interestingly, the wings had manufacturing joints situated in line with the fuselage sides, which facilitated the airliner's assembly considerably. The entire wing trailing edge between the fuselage and the ailerons was occupied by electrically actuated slotted flaps.

Despite having a larger diameter and different contours, the fuselage of the Tu-12 made use of the same production technologies and design features as that of the B-4 – right down to the use of certain structural components, fuselage frame cross-sections etc. The tail surfaces were again borrowed from the B-4; stabiliser incidence could be adjusted on the ground.

The main landing gear units with twin 1,422 x 508 mm (56.0 x 20.0 in) wheels were identical to those of the B-4. Conversely, the twin-wheel nose gear unit with slightly smaller 914 x 333 mm (36.0 x 13.1 in) wheels was specially designed for the Tu-12, even though some B-4 components, such as the shimmy damper, were used.

The Tu-12 was powered by four ASh-73TK engines. The 22 main fuel tanks in the wings held 22,000 litres (4,840 Imp gal) of fuel; they were divided into four groups, one for each engine, with a cross-feed system. Provisions were made for four auxiliary tanks in the wing center section holding an additional 6,000 litres (1,320 Imp gal) of fuel.

Six GS-9000 nine-kilowatt engine-driven DC generators supplied primary power, with two 12A-30 DC batteries as a backup. AC power was provided by four PK-750F converters. When the engines were shut down DC power was supplied by a 5-kW GS-5000 generator driven by the APU.

The pneumatic de-icing system was taken straight from the B-4. It enabled the airliner to operate in adverse weather conditions, enhancing operational reliability and despatch regularity.

Andrey N Tupolev and his associates, including Igor' B. Babin, Aleksandr E. Sterlin, N. S. Menashev, Mikhail M. Yegorov and two further engineers on the left, Nikolay I. Petrov, Boris M. Kondorskiy and Sergey M. Yeger on the right, pose with the Tu-12. Tupolev gives instructions to the cameraman, while his wife Yuliya N. Tupoleva discusses something with Air Marshal Fyodor A. Astakhov, the then-chief of Aeroflot.

Apart from the usual complement of avionics and flight instrumentation, the Tu-12 was to feature an AP-5 electric autopilot, a high-range radio altimeter and an RV-2 low-range radio altimeter, an RUSP-45 blind landing system, an MRP-45 marker beacon receiver, two ARK-45 ADFs (main and backup), a gyro-flux gate compass, a drift sight and a co-ordinate indicator. Just in case, however, an astrodome was provided for the navigator who could plot the aircraft's course, using an astrosextant and an astrocompass if all else failed.

Two-way communication with air traffic control (ATC) centers and other aircraft was provided by a 1RSB-70 communications radio and 12RSU-10 HF and SCR-274-N command radios. An SPU-BS-36 intercom was fitted for communication within the crew (including the cabin crew). The Tupolev OKB intended to equip the Tu-12 with an indigenous Gneys-5S (Gneiss) radar, a Ton-3 (Tone) radar warning receiver and a new SCh-3 IFF transponder (SCh = *svoy/choozhoy* – friend/foe).

The real prototype did not exactly conform to the ADP specifications as far as the equipment was concerned. For instance, larger wheels were fitted to all three landing gear units – 1,450 x 520 mm (57.08 x 20.47 in) mainwheels and 950 x 350 mm (37.40 x 13.77 in) nosewheels. A production SPU-14 intercom and a Bariy IFF transponder were installed, the radar was missing etc. Later, as the flight tests progressed and the aircraft underwent repairs, some equipment items gave place to newer ones that had just been mastered by the Soviet aircraft industry – a process which paralleled the upgrading of the production Tu-4.

Officially Stage A of the manufacturer's flight tests lasted from 19th October 1946 to 16th February 1947. On 25th November the reassembled and checked aircraft was taken on charge by the OKB's flight test facility, making its maiden flight two days later with Fyodor F. Opadchiy in the captain's seat and Aleksey D. Perelyot as co-pilot; Mikhail M. Yegorov was the engineer in charge of the test program.

The first three flights from Moscow-Khodynka proceeded normally but the fourth (on 16th February 1947) nearly ended in disaster. The day's task was to check the airliner's stability and handling at maximum speed, and as the Tu-12 accelerated at 4,000 m (13,120 ft) over the suburbs of Moscow near Chkalovskaya AB the No.2 engine disintegrated. Bits and pieces of the engine flew in all directions and a fire erupted, quickly burning through the engine cowling. The crew managed to extinguish the flames, using the fire suppression system, but then trouble comes in cartloads; the propeller of the stricken engine overspeeded and, try as they would, the crew could not feather it.

A loud bang resounded through the aircraft and the runaway propeller froze violently as the No.2 engine's reduction gearbox failed and jammed. As a line from a song by Rod Stewart goes, 'Act One is over without costume change'. But the drama kept unfolding: even as Yegorov advised the captain that now they could head back to Khodynka, the three good engines started losing power inexplicably and the aircraft could no longer maintain altitude. All attempts to restore normal engine operation failed and the Tu-12 continued slowly losing altitude.

Realising he would not make it to Khodynka, Opadchiy opted for a wheels-up landing at the Medvezh'yi Ozyora ('Bear Lakes')

Basic specifications of the Tu-12 according to the ADP documents	
Length overall	35.4 m (116 ft 1¾ in)
Wing span	44.25 m (145 ft 2⅛ in)
Height on ground	9.75 m (31 ft 11¹⁵⁄₆₄ in)
Wing area	166.1 m² (1,786.0 sq ft)
Empty weight	37,140 kg (81,878 lb)
All-up weight:	
normal	50,000 kg (110,230 lb)
maximum	59,000 kg (130,070 lb)
Top speed:	
at sea level	465 km/h (289 mph)
at 10,000 m (32,800 ft)	582 km/h (361 mph)
Service ceiling	10,400 m (34,120 ft)
Climb time to 5,000 m (16,400 ft)	17.3 minutes
Maximum range:	
with normal AUW	2,500 km (1,552 miles)
with maximum AUW	5,000 km (3,105 miles)
Take-off run	960 m (3,150 ft)
Take-off field length	1,800 m (5,900 ft)

airfield located about halfway between Chkalovskaya AB and the Moscow city limits. In fact, the aircraft did not even make it to the airfield, putting down in a snow-covered field short of the runway, 1.5 km (0.93 miles) from the highway that runs east from Moscow to the town of Shcholkovo. With the engines cut and the fuel fire shut-off valves closed, the airliner was still slithering fast and a high-voltage power line was directly ahead.

Still, the prototype and the crew were miraculously saved at the last moment. Fortuitously the starboard flap hit a large snow-covered mound; the aircraft swung 90° to starboard and came to rest directly under the power line, the port wing pointing between two power line pylons. Jumping clear of the aircraft, the crew quickly ascertained there was no fire and imminent danger of an explosion and then set about assessing the damage. When the cowlings of the Nos. 1, 3 and 4 engines were opened the cause of the trouble was immediately apparent: all of the inlet pipes had become separated from the cylinders (!) and the engines were simply 'out of breath', not getting enough air.

On learning of the crash landing Andrey N. Tupolev and the top officers of GK NII VVS rushed to the scene; the area was promptly cordoned off by security police. Apart from the investigation of the accident, the OKB was now facing the task of returning the airliner to airworthy condition ASAP, since the Tu-12 was slated to participate in that year's May Day parade in Moscow. After defuelling the aircraft was lightened as much as possible by removing the engines, flaps and all equipment items. Then the Tu-12 was evacuated from under the power line and towed on its undercarriage along a specially built dirt road to the Medvezh'yi Ozyora airfield. There it was trestled, heated temporary sheds were erected around the wings where the engine nacelles and flaps had been, and an AOG (aircraft on ground) team set about making field repairs.

Two months later, on 20th April 1947, the aircraft (which in the meantime had received the new designation Tu-70 and appropriate nose titles) made its first post-repair flight. (The designation Tu-12 would later be re-used for a twin-turbojet experimental bomber

Above left: The Tu-70 passes above the cheering crowd of spectators at Moscow-Tushino during the Aviation Day flypast on 3rd August 1947.

Top: Tupolev OKB engineering staff pose with the aircraft, which is seen here parked on a perforated steel planking (PSP) hardstand, wearing new 'Tu-70' nose titles. Note the stepladder for access via the nose-wheel well.

Above: The Tu-70 starts its engines before a test flight.

Left: OKB-156 Chief Designer Andrey N. Tupolev in the forward luxury compartment of the Tu-70 in 1947.

Right: The Tu-12 after its belly landing in a field on 16th February 1947 after a multiple engine failure. The tracks in the snow show that the aircraft swung before coming to a halt; note the power line in the distance.

Below and below right: The propellers are bent by the impact and the engine nacelles show superficial damage (note the exhaust pipe that has fallen off the No. 4 engine). As the logos on the blades show, the Tu-12 had the Hamilton Standard 6526A-6 props of B-29 '365 Black'.

Above: With the engines removed and the nacelles wrapped in tarpaulins, the Tu-12 is towed on its gear by a ChTZ S-65 Stalinets tractor to Medvezh'yi Ozyora airfield for repairs. The loading hatch of the forward baggage compartment can be seen ahead of the starboard wing root.

Right and above right: The Tu-12 in the makeshift shed that was erected around it at Medvezh'yi Ozyora airfield to ensure passable working conditions for the repair crew. Note the support under the tail and the rear baggage door ahead of it.

('aircraft 77') based on the Tu-2.) Thus began Stage B of the manufacturer's flight tests which lasted until 10th October 1947. The mission was accomplished: on 1st May 1947 the Tu-70 prototype led a large formation of Il'yushin IL-12 airliners – an equally new type – in a low-level pass over Moscow's Red Square. This spectacular move is perfectly understandable, since the Tu-70 was regarded as Aeroflot's future flagship at the time. On 3rd August 1947 the aircraft participated in the traditional Aviation day flypast at Moscow-Tushino along with the first production Tu-4 bombers.

Meanwhile, both the Tupolev OKB and Arkadiy D. Shvetsov's OKB-19 were trying to determine the cause of the accident. A special accident investigation board was set up to find out why the No.2 engine had disintegrated and what had caused the mysterious damage to the other three engines. Three areas came under close scrutiny: the ruined ASh-73TK engine, the quality of the fuel and the design of the RTK-46 electronic supercharger governor system copied from the B-29's R-3350-23 engine.

Examination of the wrecked No.2 engine revealed severe scoring on all 18 pistons; this had caused several pistons to jam and the connecting rods to break, smashing the crankcase. As the engine came apart, flying debris ruptured the oil line for the propeller feathering system pump and damaged the supercharger controls.

Analysis of the fuel from the tanks of 'aircraft 70' showed things were not all well here either. Sloppy work practice at Moscow-Khodynka's fuel depot often caused different grades of aviation gasoline to get mixed inside the fuel bowsers; as a result, aircraft would often be serviced with 70-72 octane Avgas instead of the required 92-93 octane fuel.

Finally, it turned out that the RTK-46 supercharger governor had a design defect. The supercharger control circuits of all four engines were connected to a common potentiometer, which was a flawed design. If any of the supercharger control circuits shorted, the system would automatically shut all the superchargers' air bleed valves, which is exactly what happened when the No.2 engine of the Tu-70 failed. As a result, the three remaining engines were immediately overpressurized, the inlet pipes breaking away violently from the cylinder heads like a cork popping out of a champagne bottle. (Quite possibly Boeing had encountered the same problems during the B-29's flight test program when the reliability of the Wright R-3350 still left a lot to be desired. The Superfortress had several cases of uncontained engine failure and

in-flight fires which paralleled the scenario of the Tu-70's 16th February 1947 accident; one of these resulted in the loss of the first prototype XB-29. Apparently the Americans had never discovered the bug in the system – especially since the Wright Aeronautical Company quickly brought the Duplex Cyclone up to an adequate reliability level. No more investigative effort was spent on this design aspect and all B-29s left the production lines with a flawed and potentially dangerous supercharger control system.

When the findings of the accident investigation board had been formulated, the Shvetsov OKB made a promise to improve the ASh-73TK by introducing adequate manufacturing tolerances for the pistons. As an interim measure it was decided to exercise strict control of the octane number of the Avgas used by the ASh-73TK in order to ensure better working conditions for the engines. It was further decided to modify the supercharger control system by introducing individual potentiometers (one for each engine) and upgrade all existing Tu-4s forthwith. Within a short period the engine had been brought up to scratch, the Tu-4 fleet equipped with the new system and everything was back to normal. Meanwhile, on the other side of the Pacific the B-29 fleet continued operating successfully with the old system and pretty soon no one (except the manufacturers, of course) remembered or cared how many potentiometers there were.

In October 1947 the Tu-70 completed its manufacturer's flight tests which involved 35 flights totalling 36 hours. Between October 1947 and 2nd July 1948 the aircraft was in layup, sitting engineless pending the delivery of improved ASh-73TKs with an engine life extended to 100 hours. Equipping the airliner with off-the-shelf ASh-73TKs having a 50-hour service life, as fitted to the Tu-4, was out of the question: the Civil Air Fleet refused to admit the Tu-70 for state acceptance trials with these engines. On 6th August 1948 the aircraft was finally submitted for state acceptance trials; these officially began on 11th September, lasting until 14th December. The trials report said the Tu-70 met the requirements to which it was designed in accordance with the Council of Ministers directive.

A short while earlier, on 16th June 1948, the Council of Ministers had issued directive No.2051 ordering the Tu-70 into production, followed by MAP order No.424 to the same effect. In the second half of 1948 the Kazan' aircraft factory began preparing to build the aircraft under the product code '*izdeliye* 70D'. The tooling-up was basically completed by the end of the year; yet, on 16th November 1948 MAP issued order No.1587 cancelling all further work on the Tu-70.

There were two reasons for this decision. Firstly, the orderbook for the Tu-4 was so large that the three production plants building the type were not in a position to build any Tu-70s due to lack of capacity. Secondly, the Civil Air Fleet really was not interested in such a large and luxurious airliner at the time; the relatively cheap and abundant Lisunov Li-2 and IL-12 twin-engined airliners could cope with the existing passenger traffic. The need for a Soviet airliner seating 50 to 100 passengers did not arise until seven or ten years later, by which time gas turbine engines had appeared on the commercial scene, powering the Tu-104 twinjet and the An-10 and IL-18 four-turboprop airliners. The Tu-70 and the original 'first-generation' IL-18 of 1947 (incidentally, powered by the same

Performance of the Tu-70 obtained in October 1947	
Empty weight	38,290 kg (84,413 lb)
All-up weight:	
normal	51,400 kg (113,315 lb)
maximum	60,000 kg (132,275 lb)
Top speed:	
at sea level	424 km/h (263.3 mph)
at 9,000 m (29,530 ft)	568 km/h (352.8 mph)
Service ceiling	11,000 m (36,090 ft)
Climb time to 5,000 m (16,400 ft)	21.2 minutes
Take-off run	670 m (2,200 ft)
Landing run	600 m (1,970 ft)

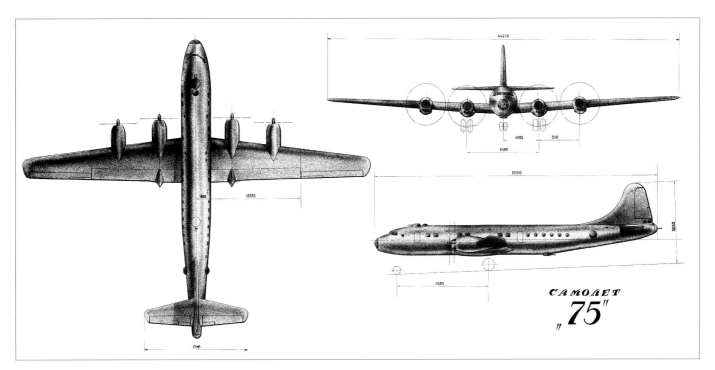

A three-view drawing of the 'aircraft 75' transport from the ADP documents.

ASh-73TK radials) were first victimised by bad timing and then rendered obsolete by new technology; yet they generated valuable experience used in designing future Soviet commercial aircraft.

Still, this was not the end of the road for the Tu-70. On 17th October 1949 it was submitted for check-up tests which continued until 20th January 1950. Then it caught the eye of Lt.-Gen. Vasiliy I. Stalin, Commander of the Moscow Military District's air force, who wanted the Tu-70 for himself as a staff transport. In preparation for the planned transfer to the Air Force the aircraft underwent a lengthy refurbishment at MMZ No.156 which was completed in the summer of 1950, but it was never delivered. The OKB managed to keep the prototype; Andrey N. Tupolev proved that the aircraft had experimental status and was powered by development engines (a rumour was even passed that the Tu-70 was fitted with Wright R-3350 engines removed from a B-29) and hence he could not put the precious life of Stalin Jr. at risk. Thus the Tu-70 was saved from the ignominy of carrying football players, horsies and dames, to all of which the generalissimo's son was more than partial.

In 1949 the Tupolev OKB considered the possibility of re-engining the Tu-70 with 4,000-hp ASh-2TK four-row radials, but the proposal was not proceeded with.

In December 1951 the Tu-70 was turned over to GK NII VVS for check-up tests with a view to determining the type's suitability as a military transport. After that the aircraft was used in various test and development programs and for passenger/cargo transport duties until finally struck off charge and scrapped in 1954.

Tu-75 military transport ('aircraft 75', Tu-16 – first use of designation)

In September 1946 the Tupolev OKB started work on a military transport derivative of the Tu-70 (Tu-12). The initiative was supported by the Soviet government: the Council of Ministers direc-

tive No.493-192 of 11th March 1947 and MAP order No.223 of 16th April tasked OKB-156 with developing a military transport based on the Tu-70 airliner. Designated Tu-16, a single prototype was to be submitted for state acceptance trials in August 1948.

The ADP was completed by 1st December 1947. Again, the Soviet approach differed from the American one: unlike the Boeing C-97 (Model 367) military transport and Boeing Model 377 Stratoliner, which were almost identical in airframe design, the new transport was not a simple adaptation of the Tu-70, even though it used many of the latter's airframe components and equipment items. The wing span and wing area were increased, and the fuselage was stretched while its maximum diameter was marginally reduced to 3.5 m (11 ft 5^{51}/$_{64}$ in). The height on the ground was reduced by a modified landing gear design (the aircraft sat lower over the ground due to the need to facilitate loading). The 'aircraft 75' was to be powered by ASh-73TKVN engines featuring direct fuel injection instead of carburettors.

The fuselage featured a ventral loading/paradropping hatch which was closed by a hinged cargo ramp forming the rear part of the cargo cabin floor. Additionally, smaller ventral hatches were provided fore and aft of the wings for dropping paratroopers. A tail gunner's station and a pair of remote-controlled cannon barbettes mounted dorsally ahead of the wings and ventrally aft of the wings were envisaged, all mounting twin Berezin B-20E cannons; the defensive armament was borrowed straight from the Tu-4.

The cargo hatch was the most interesting design feature. The aperture was closed by a downward-opening cargo ramp and a rear door segment which hinged upwards into the fuselage. The ramp could be used for troop embarkation or loading/unloading vehicles and permitted paradropping of the latter. Thus 'aircraft 75' presaged the cargo handling system used later by the Antonov OKB on the An-8 and An-12 military transports.

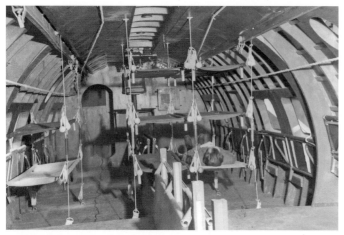

All photos on this page show the full-size mock-up of the Tu-75's cargo cabin. Top row: A view towards the open forward paradropping hatch (left) and the mock-up loaded with heavy paradroppable materiel.

Center row: Part of the cabin configured for casualty evacuation (left) and the troop seats in unfolded position. The ones along the walls were permanently installed, while the double row on the centerline was removable.

Bottom row: The cabin occupied by paratroopers seated in four rows. Two rows of troopers would jump at a time, the outer rows going first.

The drawings on the opposite page are from the Tu-75's ADP, showing the payload options. Note the Tu-4 style lateral sighting blisters at the rear. Top to bottom:

Airlift version with two OSU-76 SP guns, or STZ-NATI caterpillar tractors, or six GAZ-67B jeeps, each with an 82-mm mortar;

Airlift version with six 120-mm mortars and a 122-mm howitzer, or two 85-mm field guns and a 37-mm AA gun, or three 57-mm ZiS-2 anti-tank guns (76-mm ZiS-3 field guns) and a GAZ-63 4x4 lorry;

Paradrop version with 90 paratroopers or 64 PDMM-47 bags with small arms and other materiel.

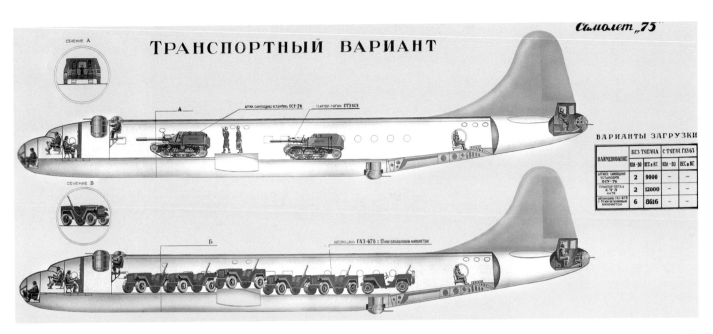

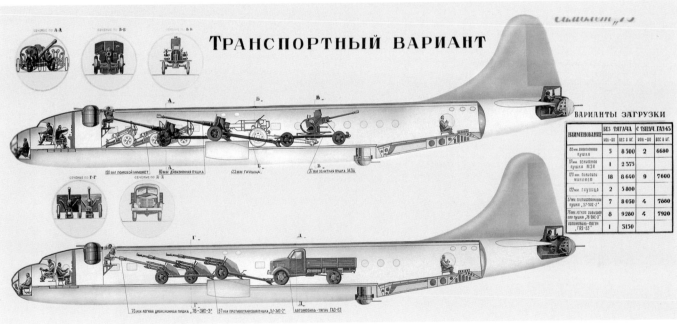

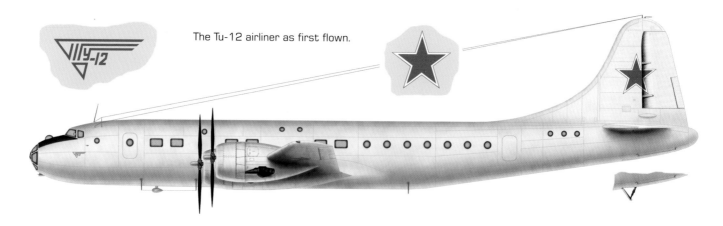

The Tu-12 airliner as first flown.

'Aircraft 75' was the Soviet Union's first successful attempt to create a fully capable military transport carrying heavy and bulky vehicles internally. The aircraft was to operate in three configurations to suit different missions – transport, troopship and medevac – and be readily convertible to any of these configurations.

Typical payload options for the basic transport configuration included two OSU-76 SP guns, or two STZ-NATI caterpillar tractors, or up to seven GAZ-67B army jeeps, or five 85-mm D-44 field guns without tractors, or two artillery pieces with tractors, or other fighting and transport vehicles and weaponry in various combina-

Two views of the sole Tu-75 transport. The rear view shows the cargo ramp and the fairing where the tail turret should have been.

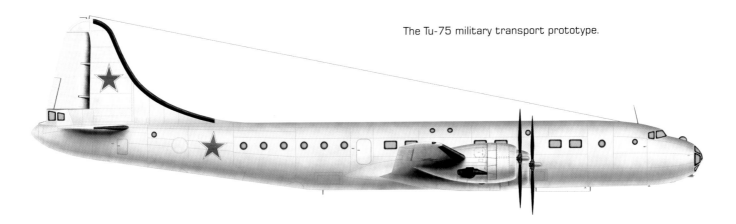

The Tu-75 military transport prototype.

tions. The maximum payload was 12 tons (26,455 lb). A cargo hoist with a of 3-ton (6,610-lb) lifting capacity moving along rails on the cargo cabin roof was provided to facilitate loading and unloading.

When configured as a troopship the aircraft could carry 120 fully armed troops or 90 paratroopers. Alternatively, small cargo items could be paradropped in 64 PDMM-47 flexible bags suspended under the cargo cabin roof. The medevac version accommodated 31 stretcher patients on three rows of stretchers (two three-tier rows and one two-tier row) and four medical attendants.

A full-size mock-up was built in December 1947 and presented to the mock-up review commission in January 1948. In June MAP issued an order postponing the beginning of the state acceptance trials until June 1949. Concurrently the transport received the provisional service designation Tu-16. Once again wearing Air Force insignia but no serial, the aircraft was completed at the Kazan' aircraft factory in November 1949, making its first flight on 21st January 1950. Manufacturer's flight tests continued until May 1950, with Maj.-Gen. Aleksandr I. Kabanov (chief of the OKB's flight test facility) and M. L. Mel'nikov as project test pilots; the aircraft was test flown with a crew of six. By then the transport had received its ultimate designation, Tu-75.

Contrary to the ADP documents, the actual aircraft had only the tail gunner's station, the upper and lower gun barbettes being omitted; no armament was ever installed. Some airframe design features also differed slightly from the project.

Despite its advanced features and fairly high performance in its day, the Tu-75 did not enter production because the Air Force opted for a 'quick fix', converting 100 (or 300?) *Bulls* into Tu-4D transports, which was cheaper. At the same time the Air Force placed an order for the development of modern military transport aircraft powered by turboprops. Additionally, the VVS demanded that new airliner designs be provided with quick-install medical and paradropping equipment and that alternative rear fuselage sections be designed for these airliners, allowing military transport versions to be built on the same production lines! These requirements proved largely impracticable and were not complied with.

Upon completion of the trials program the sole Tu-75 was used by the Tupolev OKB for several years as a transport, operating from the OKB's flight test facility in Zhukovskiy, and occasionally participated in the testing of new paradropping systems. Sadly, the aircraft was lost on 24th October 1954. The aircraft, captained by

The Tu-75's performance as recorded during manufacturer's flight tests	
Wing span	44.83 m (147 ft 1 in)
Length overall	35.61 m (116 ft 10 in)
Height on ground	9.05 m (29 ft 8⁵⁄₆ in)
Wing area	167.2 m² (1,797.8 sq ft)
Empty weight	37,810 kg (83,355 lb)
All-up weight:	
normal	56,660 kg (124,912 lb)
maximum	65,400 kg (174,180 lb)
Top speed	545 km/h (338.5 mph)
Service ceiling	9,500 m (31,180 ft)
Maximum range	4,140 km (2,571 miles)
Take-off run	1,060 m (3,480 ft)
Landing run	900 m (2,950 ft)

Kabanov, had arrived from Moscow on 23rd October to uplift Mikulin AM-3 turbojets for a Tu-16 prototype. Soon after take-off an engine failed, forcing a return to Kazan'. Once the engine had been fixed, the Tu-75 departed again for Moscow but crashed near Biryuli village shortly shortly afterwards, killing Kabanov, co-pilot M. L. Mel'nikov, flight engineer M. M. Semchoorin and radio operator N. T. Ivannikov. Several crewmembers bailed out and survived.

The Tu-75 disgorging a load of paratroopers through the front and rear paradrop doors.

Appendix 1. Tu-4 production list

The Tu-4's production run totalled 1,296 – 655 Kazan'-built aircraft, 481 Kuibyshev-built aircraft and 160 Moscow-built aircraft. Unfortunately, while the construction number systems are basically clear, few c/n to serial tie-ups are known. Therefore, to save space, production batches with no identified aircraft are listed as 'from this to that c/n'; only more or less positively identified aircraft are given. The split presentation of the c/ns is for the sake of convenience only.

	Tu-4 production list		
C/n	**Serial/registration**	**Version**	**Notes**
1. Kazan' aircraft factory No.22 named after Sergey P. Gorbunov **220501 = plant No.22, Batch 05, 01st aircraft in the batch**			
22 00 01	no serial	Tu-4	First flight 21-5-1947
22 00 02	202 Black	Tu-4	F/F 2-7-1947
22 01 01	no serial	Tu-4	C/n painted on as 101. F/F 31-7-1947; crashed near Kolomna 18-9-1947
22 01 02	no serial	Tu-4	C/n painted on as 102
22 01 03	no serial	Tu-4	F/F 6-9-1947
22 02 01	1 Black	Tu-4	F/F 13-9-1947
22 02 02	no serial;		
	later 7 Black	Tu-4	C/n painted on as 202
22 02 03	8 Black?	Tu-4	F/F 10-10-1947
22 02 04	9 Black	Tu-4	F/F 19-10-47. Converted to, see next line
		Tu-4LL	Dobrynin VD-3TK development engines (Nos. 1 & 4)
22 02 05	1000 Red	Tu-4	Should have been '10 Black'. F/F 19-10-1947
22 03 01	11 Black?	Tu-4	
22 03 02	12 Black?	Tu-4	
22 03 03	no serial	Tu-4	Serial '13 Black' not used for obvious reasons. Crashed near Kazan' 5-11-1947
22 03 04	14 Black?	Tu-4	F/F 3-7-1948
22 03 05	15 Black?	Tu-4	
22 04 01	16 Black?	Tu-4	F/F 3-7-1948
22 04 02	17 Black?	Tu-4	
22 04 03	18 Black?	Tu-4	
22 04 04	19 Black	Tu-4	
22 04 05	20 Black?	Tu-4	Serial to c/n tie-up not 100% sure but probable
22 05 01	21 Black	Tu-4	
22 05 02	22 Black?	Tu-4	
22 05 03	23 Black	Tu-4	
22 05 04	24 Black	Tu-4	
22 05 05	25 Black?	Tu-4	
22 06 01	26 Black?	Tu-4	Serial to c/n tie-up not 100% sure but probable
22 06 02	27 Black?	Tu-4	
22 06 03	28 Black?	Tu-4	
22 06 04	29 Black?	Tu-4	
22 06 05	no serial	Tu-4	Should have been '30 Black'
22 07 01	31 Black?	Tu-4	
22 07 02	32 Black	Tu-4	
22 07 03	33 Black?	Tu-4	
22 07 04	34 Black?	Tu-4	
22 07 05	35 Black?	Tu-4	
22 08 01	36 Black?	Tu-4	
22 08 02	37 Black	Tu-4	
22 08 03	38 Black?	Tu-4	
22 08 04	39 Black	Tu-4	
22 08 05	40 Black?	Tu-4	
22 09 01	41 Black	Tu-4	
22 09 02	42 Black?	Tu-4	
22 09 03	43 Black?	Tu-4	
22 09 04	44 Black?	Tu-4	
22 09 05	45 Black	Tu-4	Serial to c/n tie-up not 100% sure but probable
22 10 01	46 Black	Tu-4	Converted to 'mother ship' under Burlaki program

22 10 02	47 Black?	Tu-4	
22 10 03	48 Black?	Tu-4	
22 10 04	49 Black?	Tu-4	
22 10 05	50 Black?	Tu-4	
22 11 01	51 Black?	Tu-4	
22 11 02	52 Black?	Tu-4	
22 11 03	53 Black?	Tu-4	
22 11 04	54 Black?	Tu-4	
22 11 05	55 Black?	Tu-4	
22 12 01	56 Black?	Tu-4	
22 12 02	57 Black?	Tu-4	
22 12 03	58 Black?	Tu-4	Converted to, see below
	22 Red?	Tu-4LL	LII, Ivchenko AI-20 development engines (Nos. 1 & 4)
22 12 04	59 Black?	Tu-4	
22 12 05	60 Black?	Tu-4	
22 13 01	61 Black?	Tu-4	
22 13 02	62 Black?	Tu-4	
22 13 03	63 Black?	Tu-4	
22 13 04	64 Black?	Tu-4	
22 13 05	65 Black?	Tu-4	
22 14 01	66 Black	Tu-4	
22 14 02	67 Black?	Tu-4	
22 14 03	68 Black	Tu-4	Serial to c/n tie-up not 100% sure but probable
22 14 04	69 Black?	Tu-4	
22 14 05	70 Black?	Tu-4	
22 15 01	71 Black?	Tu-4	
22 15 02	72 Black?	Tu-4	
22 15 03	73 Black?	Tu-4	
22 15 04	74 Black?	Tu-4	
22 15 05	75 Black?	Tu-4	
22 16 01	76 Black?	Tu-4	
22 16 02	77 Black?	Tu-4	
22 16 03	78 Black?	Tu-4	
22 16 04	79 Black?	Tu-4	
22 16 05	80 Black?	Tu-4	
22 17 01	81 Black?	Tu-4	
22 17 02	82 Black?	Tu-4	
22 17 03	83 Black?	Tu-4	
22 17 04	84 Black	Tu-4	Serial to c/n tie-up not 100% sure but probable
22 17 05	85 Black?	Tu-4	
22 18 01	no serial	Tu-4	Should have been '86 Black'. IFR-capable ('wing-to-wing' system)
22 18 02	87 Black?	Tu-4	
22 18 03	88 Black?	Tu-4	
22 18 04	89 Black?	Tu-4	
22 18 05	90 Black?	Tu-4	
22 19 01	no serial	Tu-4	Should have been '91 Black'. Equipped with refuelling receptacle ('wing-to-wing' system)
22 19 02	92 Black	Tu-4	
22 19 03	93 Black	Tu-4	Serial to c/n tie-up not 100% sure but probable
22 19 04	94 Black?	Tu-4	
22 19 05	95 Black?	Tu-4	
22 20 01	96 Black?	Tu-4	
22 20 02	97 Black?	Tu-4	
22 20 03	98 Black?	Tu-4	
22 20 04	99 Black?	Tu-4	
22 20 05	100 Black?	Tu-4	
22 21 01 to 22 21 05			
22 22 01 to 22 22 05:			
22 22 02	no serial	Tu-4	IFR-capable ('wing-to-wing' system)
22 23 01 to 22 23 05			
22 24 01 to 22 24 05:			
22 24 01	no serial	Tu-4	IFR-capable (Vakhmistrov system)
22 24 02	no serial	Tu-4	Converted to tanker ('wing-to-wing' system)

22 24 05	no serial	Tu-4	Converted to tanker (Vakhmistrov system)
22 25 01 to 22 25 05			
22 26 01 to 22 26 05			
22 27 01 to 22 27 05			
22 28 01 to 22 28 05			
22 29 01 to 22 29 05			
22 30 01 to 22 30 05			
22 31 01 to 22 31 05			
22 32 01 to 22 32 05:			
22 32 04	no serial	Tu-4D?	Demilitarised/converted to ice reconnaissance aircraft
22 33 01 to 22 33 05			
22 34 01 to 22 34 05:			
22 34 02	02 Blue	Tu-4	
22 35 01 to 22 35 05			
22 36 01 to 22 36 05			
22 37 01 to 22 37 05			
22 38 01 to 22 38 05			
22 39 01 to 22 39 05			
22 40 01 to 22 40 05			
22 41 01 to 22 41 05			
22 42 01 to 22 42 05:			
22 42 03	no serial	Tu-4K	Prototype, Tupolev OKB
22 43 01 to 22 43 05			
22 44 01 to 22 44 05			
22 45 01 to 22 45 10			First confirmed Kazan'-built batch of ten aircraft
22 46 01 to 22 46 10			
22 47 01 to 22 47 10			
22 48 01 to 22 48 10			
22 49 01 to 22 49 10			
22 50 01 to 22 50 10:			
22 50 08	not known	Tu-4	Transferred to China 1953; see next line
	4134 Red		Re-engined with WJ-6A turboprops; converted to drone launcher. Preserved PLAAF Museum, Datangshan AB
22 51 01 to 22 51 10			
22 52 01 to 22 52 10:			
22 52 01	48	Tu-4	Colour of tactical code not known
22 53 01 to 22 53 10			
22 54 01 to 22 54 10:			
22 54 02		Tu-4LL	Kuznetsov TV-2 development engines (Nos. 1 & 4). Crashed 8-10-1951
22 55 01 to 22 55 10			
22 56 01 to 22 56 10			
22 57 01 to 22 57 10			
22 58 01 to 22 58 10:			
22 58 01	no serial	Tu-4	
22 59 01 to 22 59 10			
22 60 01 to 22 60 10:			
22 60 02	08 Red	Tu-4	
22 61 01 to 22 61 10			
22 62 01 to 22 62 10			
22 63 01 to 22 63 10:			
22 63 05	no serial	Tu-4K	
22 64 01 to 22 64 10:			
22 64 04	no serial	G-310	
22 65 01 to 22 65 10			
22 66 01 to 22 66 10:			
22 66 09 Б-07	09 Red	Tu-4	(that is, c/n 226609B-07)
22 67 01 to 22 67 10			
22 68 01 to 22 68 10			
22 69 01 to 22 69 10			
22 70 01 to 22 70 10			
22 71 01 to 22 71 10			
22 72 01 to 22 72 10			
22 073 01 to 22 073 10			

22 074 01 to 22 074 10			
22 075 01 to 22 075 10:			
22 075 10	29 Red	Tu-4NM	Drone launcher
22 076 01 to 22 076 10			
22 077 01 to 22 077 10			
22 078 01 to 22 078 10			
22 079 01 to 22 079 10			
22 080 01 to 22 080 10:			
22 080 09	CCCP-H1155	Tu-4	Polar Aviation, transport/ice reconnaissance aircraft
22 081 01 to 22 081 10			
22 082 01 to 22 082 10			
22 083 01 to 22 083 10			
22 084 01 to 22 084 10:			
22 084 07	CCCP-H1156	Tu-4	Polar Aviation, transport/ice reconnaissance aircraft

2. Kuibyshev aircraft factory No.18
System 1: 184213 = plant No.18, Tu-4, 2nd aircraft in Batch 13

18 4 1 01 to 18 4 3 01			
18 4 1 02 to 18 4 5 02			
18 4 1 03 to 18 4 5 03			
18 4 1 04 to 18 4 5 04			
18 4 1 05 to 18 4 5 05			
18 4 1 06 to 18 4 5 06			
18 4 1 07 to 18 4 5 07			
18 4 1 08 to 18 4 5 08			
18 4 1 09 to 18 4 5 09			
18 4 1 10 to 18 4 5 10			
18 4 1 11 to 18 4 5 11:			
18 4 4 11	58 Red	Tu-4	
18 4 1 12 to 18 4 5 12			
18 4 1 13 to 18 4 5 13			
18 4 1 14 to 18 4 5 14			
18 4 1 15 to 18 4 5 15			
18 4 1 16 to 18 4 5 16			
18 4 1 17 to 18 4 5 17			
18 4 1 18 to 18 4 5 18:			
18 4 2 18	28 Blue	Tu-4	
18 4 1 19 to 18 4 5 19			
18 4 1 20 to 18 4 5 20			
18 4 1 21 to 18 4 5 21			
18 4 1 22 to 18 4 5 22			
18 4 1 23 to 18 4 5 23			
18 4 1 24 to 18 4 5 24			
18 4 1 25 to 18 4 5 25			
18 4 1 26 to 18 4 5 26			
18 4 1 27 to 18 4 5 27			
18 4 1 28 to 18 4 5 28			
18 4 1 29 to 18 4 5 29			
18 4 1 30 to 18 4 5 30			
18 4 1 31 to 18 4 5 31			
18 4 1 32 to 18 4 5 32			
18 4 1 33 to 18 4 5 33			
18 4 1 34 to 18 4 5 34			
18 4 01 35 to 18 4 10 35			
18 4 01 36 to 18 4 10 36			
18 4 01 37 to 18 4 10 37			
18 4 01 38 to 18 4 10 38			
18 4 01 39 to 18 4 10 39:			
18 4 05 39	05 Red	Tu-4	
18 4 01 40 to 18 4 10 40			
18 4 01 41 to 18 4 10 41:			

18 4 08 41	18 Red	Tu-4D	
18 4 01 42 to 18 4 10 42			
18 4 01 43 to 18 4 10 43			
18 4 01 44 to 18 4 10 44			
18 4 01 45 to 18 4 10 45			
18 4 01 46 to 18 4 10 46			
18 4 01 47 to 18 4 10 47:			
18 4 03 47	26 Red	Tu-4D	
18 4 01 48 to 18 4 10 48:			
18 4 08 48	41 Red	Tu-4	Converted to 'mother ship' under Burlaki program; later to tug/single-point hose-and-drogue tanker
18 4 01 49 to 18 4 10 49?			Or 2 8 049 01 to 2 8 049 10? Not known if Batch 49 was numbered under the old system or the new one

System 2: 2805103 = year of production 1952, plant No. [1]8, Batch 051, 03rd aircraft in the batch

2 8 050 01 to 2 8 050 10:			
2 8 050 02	22 Blue	Tu-4	
2 8 050 03	not known	Tu-4	Converted to 'mother ship' under Burlaki program
2 8 050 05	not known	Tu-4	Converted to 'mother ship' under Burlaki program
2 8 051 01 to 2 8 051 10:			
2 8 051 03	01 Red	Tu-4	Preserved Central Russian Air Force Museum, Monino
2 8 051 10		Tu-4	Converted to 'mother ship' under Burlaki program
2 8 052 01 to 2 8 052 10:			
2 8 052 03	not known	Tu-4	Converted to 'mother ship' under Burlaki program
2 8 052 04	no serial	Tu-4	Converted to two-point hose-and-drogue tanker
2 8 052 10	not known	Tu-4	Converted to tanker
2 8 053 01 to 2 8 053 10:			
2 8 053 05	56 Red	Tu-4	
2 8 054 01 to 2 8 054 10			
2 8 055 01 to 2 8 055 10			
2 8 056 01 to 2 8 056 10:			
2 8 056 06	not known	Tu-4	IFR-capable ('wing-to-wing' system)
2 8 056 08	not known	Tu-4	IFR-capable ('wing-to-wing' system)
2 8 056 10	not known	Tu-4	IFR-capable ('wing-to-wing' system)
2 8 057 01 to 2 8 057 10:			
2 8 057 01	not known	Tu-4	Converted to tanker ('wing-to-wing' system)
2 8 057 02	not known	Tu-4	Converted to tanker ('wing-to-wing' system)
2 8 057 03	25 Blue	Tu-4	Converted to tanker ('wing-to-wing' system)
2 8 057 08	7 Black	Tu-4	
2 8 057 10	not known	Tu-4	Became, see next line
	CCCP H-1139		Polar Aviation, transport/ice reconnaissance aircraft
2 8 058 01 to 2 8 058 10:			
2 8 058 08	25 Red	Tu-4	
2 8 059 01 to 2 8 059 10:			
2 8 059 01	21 Red	Tu-4	
2 8 060 01 to 2 8 060 10:			
2 8 060 07	not known	Tu-4	Transferred to China 1953; see next line
	4074 Red		PLAAF
2 8 060 08	not known	Tu-4	Transferred to China 1953; see next line
	4005 Red		PLAAF
2 8 060 10	not known	Tu-4	Transferred to China 1953; see next line
	4104 Red		PLAAF
2 8 061 01 to 2 8 061 10:			
2 8 061 01	02 Red	Tu-4	
2 8 062 01 to 2 8 062 10:			
2 8 062 02	23 Red	Tu-4D	
2 8 062 07	not known	Tu-4	Transferred to China 1953; see next line
	4003 Red		PLAAF
2 8 062 08	not known	Tu-4	Transferred to China 1953
2 8 062 10	not known	Tu-4	Transferred to China 1953
2 8 063 01 to 2 8 063 10:			
2 8 063 01	not known	Tu-4	Transferred to China 1953; see next line
	4001 Red		PLAAF

2 8 063 03	28 Blue	Tu-4	
2 8 064 01 to 2 8 064 10			
2 8 065 01 to 2 8 065 10:			
2 8 065 01	not known	Tu-4	Transferred to China 1953; see next line
	4114 Red		Re-engined with WJ-6A turboprops; converted to KJ-1 AEW testbed. Preserved PLAAF Museum, Datangshan AB
2 8 065 08	not known	Tu-4	Transferred to China 1953; see next line
	4124 Red		PLAAF
2 8 066 01 to 2 8 066 03			

3. Moscow Machinery Plant No.23
System 1: 230106 = plant No.23, 01st aircraft of five in Batch 06

23 01 01 to 23 05 01:			
23 05 01		Tu-4LL	LII, 'aircraft 94/1' (TV-2M development engine No.3), later 'aircraft 94/2' (2TV-2F development engine)
23 01 02 to 23 05 02			
23 01 03 to 23 05 03:			
23 05 03	no serial	Tu-4	LII, 'mother ship' for DFS 346 experimental aircraft
23 01 04 to 23 05 04			
23 01 05 to 23 05 05			
23 01 06 to 23 05 06			
23 01 07 to 23 05 07:			
23 02 07	207 Black	Tu-4A	
23 01 08 to 23 05 08			
23 01 09 to 23 05 09			
23 01 10 to 23 05 10			
23 01 11 to 23 05 11			
23 01 12 to 23 05 12			
23 01 13 to 23 05 13:			
23 01 13		Tu-4LL	Myasishchev OKB, DR-1 testbed (Lyul'ka AL-5 engine); later DR-2 testbed (Mikulin AM-3 engine)
23 01 14 to 23 05 14:			
23 03 14		Tu-4LL	LII, with jet engine
23 01 15 to 23 05 15			
23 01 16 to 23 05 16:			
23 02 16		Tu-4	OKB-51, weapons testbed. Crashed 19-2-1952
23 01 17 to 23 05 17:			
23 02 17	35 Red	Tu-4	
23 01 18 to 23 05 18			
23 01 19 to 23 05 19			
23 01 20 to 23 05 20			
23 01 21 to 23 05 21:			
23 01 21	not known	Tu-4D	
23 01 22 to 23 05 22:			
23 03 22	no serial	Tu-4	Myasishchev OKB, ShR-1/ShR-2 landing gear testbed
23 01 23 to 23 05 23			
23 01 24 to 23 05 24?			Or 23 024 01 to 23 024 05? Not known if Batch 24 was numbered under the old system or the new one

System 2: 2303001 = plant No.23, Batch 030, 01st aircraft of five in the batch

23 025 01 to 23 025 05:			
23 025 01	no serial	Tu-4LL	LII, with jet engine
23 026 01 to 23 026 05:			
23 026 04	not known	Tu-4	Crashed 23-4-1956
23 027 01 23 027 05			
23 028 01 to 23 028 05:			
23 028 01	CCCP H-1138	Tu-4	Polar Aviation, transport/ice reconnaissance aircraft
23 029 01 to 23 029 05:			
23 030 01 to 23 030 05:			
23 030 01	no serial	Tu-4LL	LII, TV-12 (NK-12) development engine (No.3)
23 031 01 to 23 031 05			
23 032 01 to 23 032 05:			
23 032 01	no serial	Tu-4D	*Izdeliye* 76 prototype

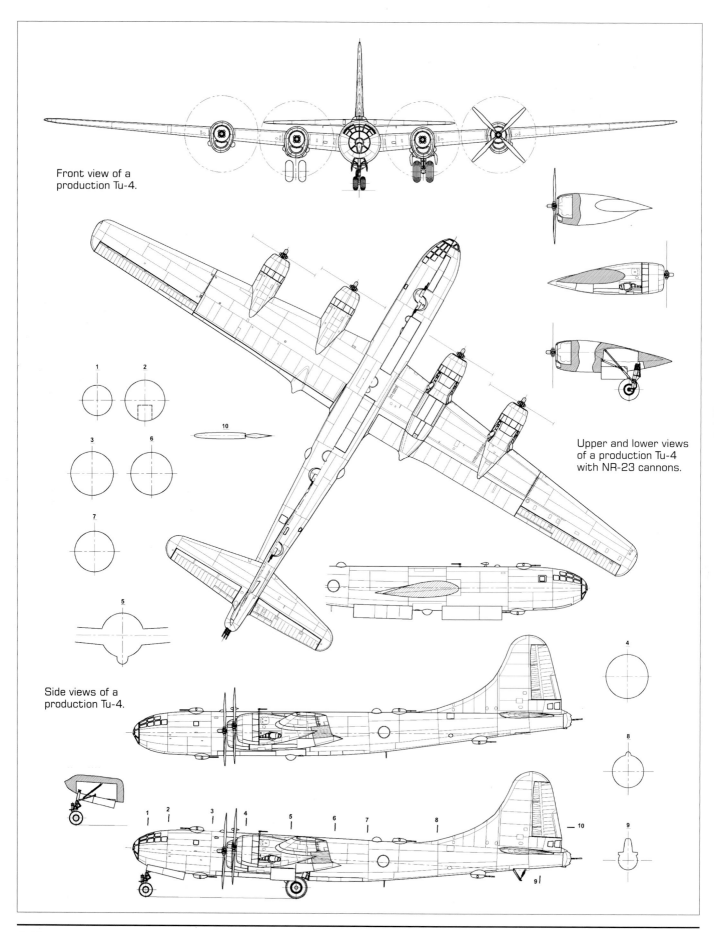

Front view of a
production Tu-4.

Upper and lower views
of a production Tu-4
with NR-23 cannons.

Side views of a
production Tu-4.

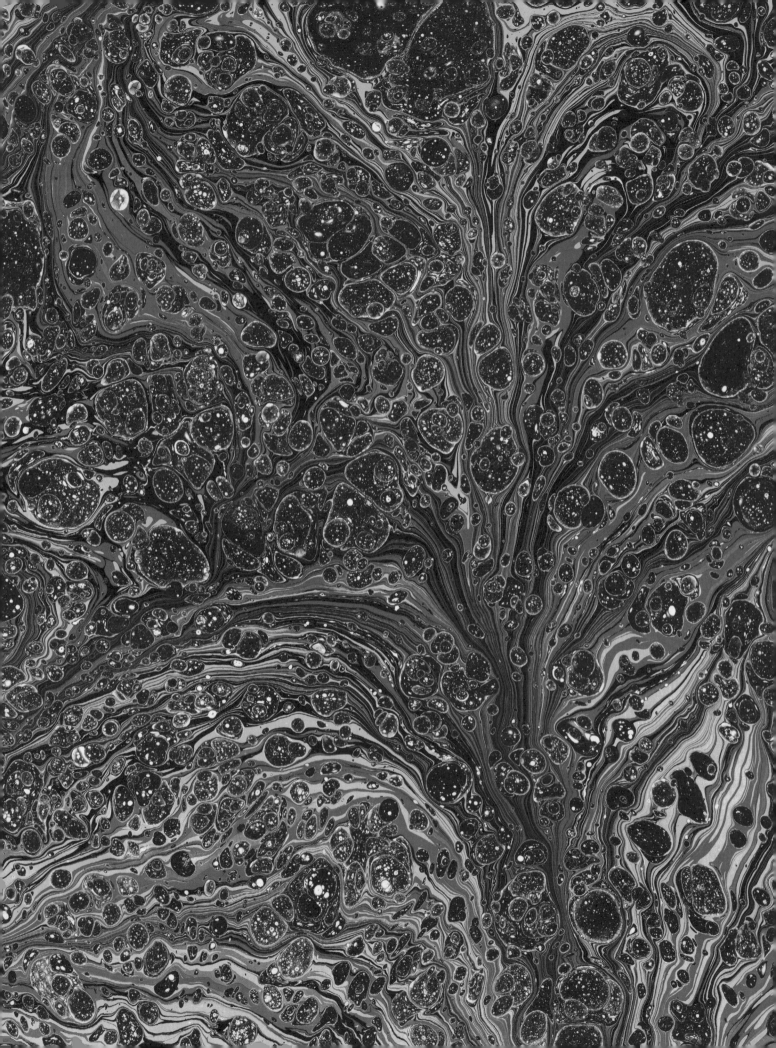